Search Theory

Steve Alpern • Robbert Fokkink
Leszek Gąsieniec • Roy Lindelauf
V.S. Subrahmanian
Editors

Search Theory

A Game Theoretic Perspective

Springer

Editors
Steve Alpern
Warwick Business School
University of Warwick
Coventry, UK

Robbert Fokkink
Technical University Delft
Institute of Applied Mathematics
Delft, Netherlands

Leszek Gąsieniec
Complexity Theory and Algorithmics Group
Department of Computer Science
University of Liverpool
Liverpool, UK

Roy Lindelauf
Netherlands Defense Academy
Breda, Netherlands

V.S. Subrahmanian
Department of Computer Science
University of Maryland
College Park, MD, USA

ISBN 978-1-4939-0067-1   ISBN 978-1-4614-6825-7 (eBook)
DOI 10.1007/978-1-4614-6825-7
Springer New York Heidelberg Dordrecht London

Preface

Ten years have passed since Alpern and Gal's monograph on Search Games and Rendezvous appeared. Over these years, the research on Search Games and Rendezvous, which started as a purely mathematical endeavor, has meandered into other disciplines. This is reflected by the changing background of the participants of the yearly workshop that Steve Alpern has organized at the London School of Economics for over a decade now. Originally these participants were mathematicians only, but slowly and steadily ever more computer scientists, biologists, and even an occasional management scientist, turned up, eventually equalling the number of mathematicians. The topics of the workshop changed accordingly. Classic mathematical problems on search games, that remain open today and can be found in this book, were supplemented by new problems that take a more algorithmic point of view, following the current trend in game theory, or that even go as far as trying to understand real life behavior of humans and other animals. The change of focus is such that we felt the need to prepare an update of Alpern and Gal's monograph, presenting a wider view of Search Games and Rendezvous, with an emphasis on open problems and future directions of research.

The preparation of a monograph is a time-consuming project, and so we decided to take a short cut, by inviting others to join in. In April 2012 we organized a workshop at the Lorentz Conference Center in Leiden, Netherlands. Around 50 researchers joined hands here, to discuss recent progress and new ideas. The result of this lies before you. We intend this book to be either used for self study, or as a guide to the literature for a graduate course, in an applied mathematics or a computer science programme. It is divided into four parts: Search Games, Geometric Games, Rendezvous, and Search Games in Biology – starting from the mathematical and ending at the applied. One gets the gist of the book's gradient, if one compares Shmuel Gal's review in the initial chapter to Jon Pitchford's open problems on search in biology at the end of the book. The first chapter of each section gives a survey and the final chapter of each section presents open problems. In selecting these problems, we have chosen those that seem to be solvable, rather than the ones that are notoriously hard. The exception to this are a few rendezvous problems in

Chap. 14 that have been around for a long time, but they make up for this notoriety by being very entertaining. There are also plenty of open problems that are mentioned in the other chapters. We expect that readers will find these problems to their liking, and solve them, so we can write another sequel to Search Games and Rendezvous within the foreseeable future.

We would like to thank all participants of the workshop on Search and Rendezvous 2012 for lively discussions and fruitful interactions. In particular we would like to thank all speakers of that conference: Rob Arculus, Mark Broom, Jérôme Casas, Shantanu Das, John Dickerson, María Jose Fernández-Sáez, Shmuel Gal, Leonhard Geupel, Thomas Gorry, Mohammad Hajiaghayi, Lora Janse, Ken Kikuta, Jun Kiniwa, Evangelos Kranakis, Tom Lidbetter, Katerina Papadaki, David Peleg, Christos Pelekis, Jon Pitchford, Lyn Thomas, Richard Weber, Noemí Zoroa. The staff of the Lorentz Center make every workshop into an enjoyable experience, and we thank them for taking all organizational troubles out of our hands. Finally, we would like to thank Michael Gubanski for preparing the witty cartoons that enliven the text.

Coventry, UK	Steve Alpern
Delft, Netherlands	Robbert Fokkink
Liverpool, UK	Leszek Gąsieniec
Breda, Netherlands	Roy Lindelauf
College Park, MD, USA	V.S. Subrahmanian

Contents

Part III Rendezvous

Part IV Search in Biology

Part I
Search Games

Chapter 1
Search Games: A Review

Shmuel Gal

Abstract This review presents an update on the area of Search Games, highlighting recent developments in the field, as well as presenting some new problems for further research. The search space is either a graph, a bounded domain, a mixture of the above, or an unbounded set. The search process is presented as a two-player zero-sum game between the searcher and the hider. The searcher moves along a continuous trajectory and the cost function is the time needed to find the hider. Our review emphasises general results concerning minimax search trajectories and optimal search strategies.

We present a review on mathematical modeling of hide and seek situations that we all know from our childhood. These situations are common in military and anti-terror activity, as well as many other areas. Isaacs [30] provided a mathematical foundation to study of such situations in the last chapter of his book, introducing the concept of a continuous search game, which was further developed by Gal [24]. That work, and the updated version by Alpern and Gal [4], has stimulated much research, with applications to Computer Science, Economics and Biology. The present chapter reviews different types of search games, putting an emphasis on recent results. As such, it is an update of a previous survey article on search games that appeared in the encyclopedia of Operations Research [26]. The present chapter also contains some new observations on searching a graph using the Traveling Salesman Tour and on searching mixed spaces which did not appear in print before, outlining possible directions of future research.

We consider a zero-sum game played between a *searcher* and a *hider* in a search space Q which is either a compact set in a Euclidean space or an unbounded connected set, e.g., the real line. The searcher usually starts moving from a specified

S. Gal (✉)
Department of Statistics, University of Haifa, 31905 Haifa, Israel
e-mail: sgal@univ.haifa.ac.il

S. Alpern et al. (eds.), *Search Theory: A Game Theoretic Perspective*,
DOI 10.1007/978-1-4614-6825-7__1, © Springer Science+Business Media New York 2013

point O called the origin but choosing the starting point arbitrarily is also considered. It is assumed that the searcher is free to choose any continuous trajectory inside Q, subject to a maximal velocity constraint which is normalized to 1. The hider can be either stationary or mobile. It will always be assumed that neither the searcher nor the hider has any knowledge about the movement of the other player until their distance apart is less than or equal to the discovery radius r, and at this moment capture occurs. The discovery radius r is assumed to be small with respect to the dimension of Q. For one dimensional sets, r is assumed zero. A search trajectory will be denoted by S and a hiding point or trajectory (depending whether the hider is immobile or mobile) by H. The cost function (payoff to the maximizing hider) $c(S,H)$, is the time spent until the hider is found (the search time). A mixed strategy s (resp. h) of the searcher (resp. hider) is a probability measure on the set of the searcher (resp. hider) pure strategies with expected payoff $c(s,h)$. The existence of a value, v, (minimax theorem) for such problems, and an optimal searcher mixed strategy are established in [24] and [3]. An optimal hiding strategy need not exist, only ε-optimal strategies always exist. The existence theorem holds both for immobile and mobile hider.

1.1 Search Games for an Immobile Hider

In this section we consider games with an immobile hider, and the search space is usually a network. In some sense, these are the simplest search games, but even their solution can be very subtle and, as we shall see below, there still remain some open problems to be solved. We assume that the search space Q is a finite connected graph in which each arc has a given length. The sum of these lengths will be denoted by μ, which is called the total length of the graph. A pure strategy for the hider is simply a (hiding) point H in Q, which can be a node or a point on one of the **arcs**. A pure strategy $S = S(t)$ for the searcher is a continuous trajectory in Q with speed 1. The payoff c corresponding to pure strategies S and H is the *search time* $c(S,H) = min\{t : S(t) = H\}$.

1.1.1 The Searcher Starts at a Fixed Point

A natural way to search a graph is to use a **Chinese postman tour** (see [21]) i.e., a closed trajectory that visits all the arcs of Q and has minimal length. Such a tour will be denoted by S^*. Obviously, if Q is Eulerian, then S^* can be chosen as any Eulerian tour with length μ. In this case, if $S^*(t)$, $0 \leq t \leq \mu$ is an Eulerian tour, then visiting it with probability 1/2 in the forward direction and probability 1/2 in the backward direction (using $S^*(\mu - t)$ as the search trajectory) is an optimal search strategy.

The search strategy used for Eulerian graphs can be extended to non-Eulerian graphs by using a **random Chinese postman tour** that encircles S^* equiprobably in each direction. It was proven by Gal [25] that a random Chinese postman tour is an optimal search strategy *if and only if* Q is weakly Eulerian (a set of Eulerian subgraphs connected in a tree-like fashion). The optimal hiding strategy is obtained by an attractive algorithm using a tree which is equivalent in some sense to the original graph.

If the graph is not weakly Eulerian then the optimal solution of the search game is very complicated. The following deceptively simple example illustrates the difficulties. Assume that Q consists of just three arcs of unit length. connecting two nodes, the origin O and another node A (Fig. 1.1).

Fig. 1.1 The three arcs graph

It is obvious that an optimal search strategy has to pick a random arc and go to A but what to do then? Surprisingly it is *not* optimal to pick another arc at random and fully traverse it (and later the remaining arc). It turns out that the optimal strategy, upon reaching A, is to randomly pick one of the unvisited arcs, go a random distance on it and return to A. Only then fully traverse the third arc and the partly visited arc. (For details, see [4] p. 33.) This strategy was suggested by Donald J. Newman in the late 1970s but the rigorous proof of its optimality is very complicated and took about 15 years to establish [40]. In general, the problem of finding the optimal search strategy is NP–hard as shown (see [43]). The search game on a graph can, in general, be formulated as an infinite-dimensional linear program. This formulation with an algorithm for obtaining its (approximate) solution is presented by Anderson and Aramendia [7].

1.1.2 The Searcher Starts at an Arbitrary Point

If we allow the searcher to start at any point in the network Q, then the natural search trajectory to use is the **Chinese postman path** $\tilde{S}$ which visits all the points of Q and has minimum length but is not necessarily closed. The searcher can use the following natural strategy:

Simple Strategy: Choose the starting point randomly, starting equiprobably at either end of $\tilde{S}$, going to the other end along $\tilde{S}$.

For example, for a tree the Chinese postman path traverses the arcs of its *diameter* once, and all other arcs twice. The simple strategy for a tree (see [18]) means picking randomly an end of the diameter and using that Chinese postman path to traverse the tree. Sometimes *Simple* is optimal. For example, if $\tilde{S}$ is an Eulerian path then *Simple* is optimal. Such an example is the three arcs graph, which is very difficult for the fixed start, but very easy for the arbitrary start search game. In addition, *Simple* is optimal for any tree, [18], and also for any tree with one or several Eulerian graphs attached, [2]. However, Alpern, Baston, and Gal [5] have shown that it is *impossible* to topologically characterize the graphs, for which *Simple* is optimal. Their result is based on the graph in Fig. 1.2.

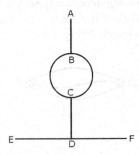

Fig. 1.2 A graph for which simple is optimal only if the base is wide

The arcs between A, B, C, D all have unit length, but the arcs DE and DF have length r. If r is large, then it is easy to see that the optimal search strategy is to randomly pick an end of the long interval and to traverse the network using the Chinese postman path, so that *Simple* is the optimal search strategy. However, if r is small then it can be shown that *Simple* is **not** optimal. Since the two networks are topologically equivalent it follows that topological characterization for the optimality of *Simple* is impossible. This is quite surprising because it contradicts the result for the fixed start search games.

1.1.3 The Hider Hides in a Node

Traditionally, the hider was allowed to hide anywhere in a graph, either in a node or somewhere in an arc. The case in which the hider can hide in the nodes only, has just recently been considered. In Chap. 3 of the book, Zoroa et al. develop a theory for such games, if the graph is an integer lattice. In general, it is natural to use **the Traveling Salesman Problem (TSP) Tour**, i.e., the trajectory that starts at O, visits all the nodes, and returns to O in minimal time. Finding the TSP tour is NP hard, in contrast to the Chinese Postman tour which can be solved in $O\left(N^3\right)$.

Definition 1. We say that the searcher uses the Random Traveling Salesman Strategy (RTSP), if he equiprobably uses either the TSP tour or the 'reverse' tour. RTSP guarantees an expected search time of at most half the tour length.

It is natural to ask if the RTSP strategy is optimal. It turns out that sometimes it is, for example if the nodes are on a circle or for a tree. However, RTSP is not always optimal, for if it would be, then it would also be optimal if the hider can hide anywhere in the network. This is not the case. The optimal searcher strategy for the three-arcs graph is much more complicated than RTSP. So if there are $n \gg 1$ equally spaced nodes along each of the arcs of the three-arcs graph, then RTSP is not optimal. One can show that RTSP need not even be optimal for an Eulerian graph, e.g., by adding one more unit arc between O and A to the three-arcs graph.

Proposition 1. *There exists no topological characterization for graphs that have the property that RTSP is optimal.*

Proof. Consider the three-arc graph G consisting of three unit arcs b_1, b_2, b_3 joining O and A with $(n \gg 1)$ nodes along each arc, that are equally spaced along b_1 and b_2, but they are not necessarily equally spaced along b_3. If the n nodes along b_3 are also equally spaced, then we already argued that RTSP is **not** optimal. However, if the n nodes along b_3 are all very close to A, then these nodes reduce more or less to one and the resulting graph is more or less a circle, in which case RTSP is optimal. Indeed, RTSP **is** optimal: go to A randomly along b_1 or b_2 − search the nodes along b_3 − return to A and then go back to O via the unvisited arc. Thus, the same topological structure can have both RTSP optimal and non-optimal.

Note. Similar examples exist with node degrees ≥ 3 rather than 2.

The **RTSP strategy with arbitrary start** is the analogous strategy to the *Simple* network search. It is based on a minimal trajectory, **not necessarily closed**, that visits all the **nodes** and use it with probability 1/2 for each direction. The RTSP leads to expected capture time equal to the half length of the minimal trajectory. It is sometimes an optimal search strategy but **for an arbitrary start there does not exist a topological characterization for the optimality of the RTSP strategy**. A proof for that can be based on a discrete version of Fig. 1.2, following the same line of argument as in [5].

1.1.4 Searching an Unknown Graph

Finding a hider in an unknown graph is equivalent to finding the exit of a maze. Assume that you find yourself in a maze. What is the best way to exit it? This has been a challenging problem since ancient times but the solution for it has been published only in 1882 [39] by an algorithm of Tremaux. In 1895 Tarry [42] described an algorithm which is now widely used in Computer Science as *depth first search* (see Even [20]). This algorithm is based on numbering the passages out of every node and using each passage by the order of this numbering, so that each passage is not traversed more than once. It guarantees 2μ search time which is the best possible for a **deterministic** strategy. However, using **random** numbering of the passages out of every node enables the searcher to exit the maze in expected time μ, as shown in [27]. This is the best possible expected time which can be guaranteed for any unknown graph.

1.1.5 Searching a Two-Dimensional Region

Searching an immobile hider in a two dimensional region with area μ is equivalent to searching an Eulerian graph. An optimal strategy for the searcher is to find a minimal length closed curve that covers the entire region. Then flip an unbiased coin and traverse either clockwise or anti-clockwise according to the toss-up result. Since the searcher discovers a strip of width $2r$ it follows that the length of a minimal covering curve for the region is $\sim \mu/2r$. By encircling this curve equiprobably in each direction the searcher can assure about $\mu/4r$ expected search time. Since the rate of discovery of new points is about $2r$ for each time unit, it follows that the area discovered by time t is at most $2rt$. Thus, if the hider uses the natural strategy of a uniform hiding distribution, then the probability of capture after time t is at least $1 - \frac{2rt}{\mu}$. Since the expected search time is the integral of that probability it follows that the hider can keep the expected search time above $\mu/4r$. Thus, $v \sim \mu/4r$ as expected. See [24] for more details.

1.1.6 Searching a Mixed Space

The case of a hider that hides in a node has been considered above. The search space then is disconnected, consisting of a finite set of points, that are connected by a set of paths. Suppose that the nodes are in the plane and that the paths are not yet given. The searcher can choose the paths. This is a type of search game that has not yet been studied and which can easily be generalized. By a *mixed space* is meant a disconnected subset of Euclidean space. An example of such a mixed space is a set of discrete points plus a region. For instance, houses and a grove. Each house can be modeled as a point or as a small rectangle. Another example is a line segment plus a set G of discrete nodes. To search the space, the searcher must connect the line segment to the nodes, and the problem is to do this in the most efficient way. The simplest example of a disconnected search space is a unit interval, with the starting point O in the middle, plus a point A above O (Fig. 1.3).

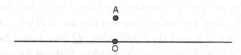

Fig. 1.3 A simple mixed space

It seems natural to connect A to O and search it as a tree. However, a better way is to construct the triangle with A as a vertex and search it as an Eulerian network, i.e., starting from O and encircling it clockwise or anti-clockwise, each with probability $1/2$. Using the triangle inequality it is easy to demonstrate the superiority of the latter strategy. This strategy, based on the triangle, is probably optimal but the problem is more complicated than it seems at first sight. We should take into consideration some other mixed strategies based on trajectories that visit either A first, or part of the interval then A and finally the remaining parts of the interval. The searching of disconnected spaces seems to be an interesting and challenging topic for further research. For example, in some cases a possible solution is the minimal value weakly Eulerian graph that spans the line segment plus the set G. It would be interesting to find some sort of characterization of these cases.

1.2 Search Games for a Mobile Hider

Search games with a mobile hider are more difficult to solve than those with an immobile hider. These games are also known as princess and monster games.

1.2.1 Isaacs' Princess and Monster Game

The **princess and monster** was a well known open problem during the 1960s and 1970s. It was introduced by Rufus Isaacs in his classical book *Differential Games* [30] as follows:

The monster P searches for the princess E, the time required being the pay-off. They are both in a totally dark room Q (of any shape) but they are each cognizant of its boundary (possibly through small light admitting perforations high in the walls). Capture means that the distance $PE \leq r$, a quantity small in comparison with the dimension of Q. The monster, supposed highly intelligent, moves at a known speed. We permit the princess full freedom of locomotion.

This game was solved by Gal [23], under the condition that the search space Q is convex. Actually a much weaker condition is needed (see [24], p. 39). The value, v, of the princess and monster game satisfies $v \sim \frac{\mu}{2r}$, where μ is the area of the room (twice the value obtained for an immobile hider). Gal's optimal strategy for the princess is to stay at random locations in Q and make the changes of location not too often but also not too rarely, (for details see [23] or [24]). His proposed optimal search strategy is based on subdividing Q into many **narrow** rectangles searching a random rectangle in a specific random fashion for some time and then moving to another random rectangle, etc. Note that this type of strategy would not be efficient if the rules of the game were changed in such a way that the hider knows the position of the searcher from time to time. A search strategy that is robust to this type of knowledge has been presented by Lalley and Robbins [35]. It is based on going in a straight line and, upon reaching the boundary, choosing a random direction going back into the region, etc. It should be noted, though, that Lalley and Robbins' strategy would be ineffective for non-convex regions but Gal's search strategy is also effective for non-convex regions and can also be adapted to more general problems, e.g., a non-constant capture radius, multi-dimensional regions, more general cost functions and search game with several searchers. Several other extensions have been presented by Garnaev (e.g. [29]).

1.2.2 Mobile Hider on a Network

The princess and monster game on an arbitrary network remains far from solved. The only case that has been settled is the game on a circle, which was originally suggested in the 1960s by Isaacs [30] as a stepping stone for the game in the region described above. This problem was independently solved by Alpern [1] and Zelikin [45]. (See also Wilson [44].) Their optimal search strategy is now a well known strategy which has been named by Alpern '**coin half tour**'. It says simply flip a symmetric coin every half tour of the circle and go clockwise or anti–clockwise. This strategy has also been used for rendezvous problems and in computer science.

The search game on the unit interval with an arbitrary start, which may look like a rather trivial problem, is surprisingly difficult. At first sight it may look like there is a simple optimal strategy: pick a random end of the interval and go, as fast as possible, to the other end. However, the hider's best response to this is to start very near to a random end, and move to the other end with unit speed. This leads to an expected search time of 0.75. Surprisingly, the searcher can do better, using a more complex mixed strategy, and the value of the game is about 0.7. This unexpected behavior, and an approximate solution of the game on the interval has been presented by Alpern et al. [6]. The search game with an mobile hider on other networks remains open. More information on these games, as well as an addendum to the computations from [6] can be found in Chap. 4.

1.3 Search in Unbounded Domains

All search games that considered so far had a bounded search region. If the region is unbounded, then the search game with an immobile hider becomes very difficult to solve, even if the probability distribution of the hider's position is known.

1.3.1 The Linear Search Problem

Computer scientists who considered the **cow path problem** in the 1990s were not aware that this problem was already being researched by mathematicians for more than 20 years. This problem was first introduced by Bellman [10] and, independently, by Beck [11] as the **linear search problem** described as follows:

> A target is located on the real line at a point H with a known probability distribution. A searcher, whose maximal velocity is 1, starts from the origin O and wishes to discover the target in minimal expected time. It is assumed that changing the direction of motion can be done instantaneously.

Much research has been carried out by Beck, e.g. [12, 13], and others, who found some general properties of the optimal solution and presented the optimal solution for several interesting specific distributions. For example, Beck has showed that if the hiding distribution is uniform, then the optimal search trajectory is to simply go to one end and then back to the other end, but if the distribution is triangular, then the optimal trajectory has, surprisingly, an infinite number of turning points.

The linear search problem for a general probability distribution is unsolved yet. However, a dynamic programming algorithm can produce a solution for any discrete distribution, [16], and also an approximate solution [4], with any desired accuracy, for any probability distribution.

The seminal paper by Beck and Newman [14] considers the linear search problem as a search game. Obviously, the hider's strategies have to be restricted, leading to assuming that the expected distance from O of the hider's location is at most λ. The constant λ is assumed to be known to the searcher but it turns out that the optimal search strategies do not depend on λ. They show that the minmax search trajectory is to **double** the step size each time. The doubling trajectory guarantees capture by time $9|H|$ and 9 is the smallest constant that can be achieved by a fixed trajectory. The optimal (mixed) search strategy uses a geometric trajectory, with the generator $\bar{g}$, where $\bar{g}$ minimizes $\frac{g+1}{\ln g}$, multiplied by a random constant $c_u = \bar{g}^u$, where u is chosen at random by the uniform distribution in the interval $[0, 2)$. The n-th step size is $c_u \bar{g}^n$. The value of the games satisfies

$$v = (1 + \bar{g})\lambda \doteq 4.6\lambda$$

so the best possible constant for mixed strategies is 4.6.

Gal [22] used a normalized cost function to find the minimax trajectory. The two approaches lead to equivalent results (see [24]). This type of normalization was new at that time but is now called the **competitive ratio** in the computer science literature, and has become a standard tool in analyzing online algorithms (see, e.g. [15]). The online solution with a turn cost, d, is given by Demaine et al. [19]. The competitive ratio remains 9 and but an additive constant of $2d$ is added to the minimax cost. It also turns out that optimal randomized strategies for searching the line in the presence of turn cost have the same competitive ratio of about 4.6.

1.3.2 On the Optimality of Geometric Trajectories

It was shown by Gal and Chazan [28] that a minimax search trajectory for search problem in unbounded domains, satisfying a unimodality condition, is always a geometric sequence (or exponential functions for continuous problems). This result has been developed in [24] as a general tool which just requires optimizing a function with one variable (the generator of the sequence) instead of minimizing the competitive ratio functional in the space of possible trajectories. This tool has been used for the linear search problem and also in searching a set of rays (star search) and searching in the plane using exponential spirals. It has also been used for minimax search on the boundary of a region consisting of two rays in the plane. That problem has been presented by Baeza-Yates [8], and solved by Gal using the general tool (see [4], pp. 154–157).

Geometric trajectories are useful in other areas to produce effective algorithms, as shown by Chrobak and others (see [17]). *Competitiveness via doubling* means designing online and offline approximation algorithms by using geometrically increasing estimates on the optimal solution to produce fragments of the algorithm's solution. This method is illustrated by discussing several applications where doubling is used. These areas, in addition to searching, include bidding, minimum latency tours, scheduling, and clustering. It seems to have a promising potential for other applications as well.

1.3.3 Searching on M Rays (Star Search)

The natural extension of the linear search problem to $M > 2$ concurrent rays, has been presented and solved by Gal [22]. It was later rediscovered in the computer science literature, [8, 32] as the 'cow path search'. Several applications in search and other areas were presented. Gal showed that the optimal trajectory is *periodic* (visiting each ray every M-th time) with *monotone* increasing step size. Thus, the general tool, discussed on the previous section, takes the following form for the star search:

For any positive sequence $X = \{x_i\}$

$$\inf_{X} \sup_{i} \frac{\sum^{i+M-1} x_j}{x_i} = \min_{g>1} \frac{\sum^{i+M-1} g^j}{g^i} = \min_{g>1} \frac{g^M}{g-1} = \frac{M^M}{(M-1)^{M-1}}$$

with the minimizing generator of the geometric sequence being $\frac{M}{M-1}$. This minimax trajectory achieves the competitive ratio $1 + 2\frac{M^M}{(M-1)^{M-1}} \sim 1 + 2eM \overset{\circ}{=} 5.4M$ for large M.

Similarly to the linear search problem, the (expected) competitive ratio can be reduced to $3.09M$ by using mixed search strategies based on geometric trajectories. In contrast to the pure strategy case, however, the optimality proof applies only to strategies that use periodic and monotone trajectories. (See [24, 33, 34] and [41].) López-Ortiz and Schuierer have solved several extensions of the star search for bounded distance, [37], and for search by several agents who move in parallel, [38].

1.4 Online Searching in the Plane

1.4.1 An Open Problem Recently Solved

An immobile hider is located at an unknown point H in the plane. The searcher chooses a trajectory, starting from O, and discovers the hider at the moment when H is covered by the area swept by the radius vector R of his trajectory. It is natural to use *periodic and monotonic* trajectories $R(\theta)$ with the angle θ of the radius vector always strictly increasing, and the trajectory does not intersect itself. Under this assumption, as we mentioned before, there exists a minimax search trajectory within the family of exponential spirals, i.e., $|R| = e^{b\theta}$, where $b > 0$ is a constant. The competitive ratio is $e^{2\pi b}\sqrt{1 + 1/b^2}$ with minimal value about 17.3, attained at $b \overset{\circ}{=} 0.16$ (see Gal [24]). Since then it has been an open problem whether the given strategy is optimal in general without the periodic and monotonic assumption. A recent work of Langetepe [36] has solved this 30 years old open problem and showed that the above described spiral search is indeed the overall optimal trajectory.

1.4.2 'Swimming in a fog' Problem

The following problem has been proposed by Bellman [9]. A person has been shipwrecked in a fog and wishes to minimize the maximum time required to reach the shore. The shape of the shore is known and some information is given about its distance. If the shore is a straight line with a known distance D then the minimax trajectory is given by Isbell [31] as follows: Imagine a clock face and a circle of radius D around O, walk towards 1 o'clock for $\frac{4}{3}D$ units then, turn on the tangent that strikes the circle at 2 o'clock. Follow the circle to 9 o' clock and continue on a tangent for length D (until reaching the tangent to the circle at 12 o'clock). At that time all the tangents of the circle with radius D have been visited. The length of this minmax trajectory is about $6.4D$. However, this trajectory is not effective unless D is known exactly.

An online framework for the general problem is presented in [24]. The goal is to minimax the ratio between the distance travelled until reaching the shore (assumed a straight line) and the unknown distance, D, to the shore. It has been suggested that the solution could be an exponential spiral but the problem seems difficult and is still open.

Acknowledgements The author would like to thank the Lorentz Center, the organizing committee of the workshop on Search and Rendezvous, and especially Robbert Fokkink and Steve Alpern for their help and support. The author is also obliged to Robbert Fokkink for his useful remarks which led to a significantly improved version.

References

1. Alpern, S. (1974). The search game with mobile hider on the circle. Pp. 181–200 in Roxin, EO, Liu, PT and Sternberg, RL, eds. Differential Games and Control Theory. Dekker, New York.
2. Alpern, S. (2008). Hide-and-seek games on a tree to which Eulerian networks are attached. Networks 52(3):109–178.
3. Alpern, S. and Gal, S. (1988). A mixed strategy minimax theorem without compactness. SIAM J. Control Optim. 26:1357–1361.
4. Alpern, S. and Gal, S. (2003). *The theory of search games and rendezvous*. Springer.
5. Alpern, S., Baston, V, and Gal, S. (2008). Network search games with immobile hider, without a designated searcher starting point. International Journal of Game Theory 37:281–302.
6. Alpern, S., Fokkink, R., Lindelauf, R. and Olsder, GJ. (2008). The Princess and Monster game on an interval. SIAM Journal on Control and Optimization. (47):1178–1190.
7. Anderson, EJ. and Aramendia, MA. (1990). The search game on a network with immobile hider. Networks. 20(7):817–844.
8. Baeza-Yates, R. A., Culberson, J. C. and Rawlins, G. J. E. (1993). Searching in the plane. Inf. Comput. 106(2):234–252.
9. Bellman, R. (1956). Minimization problem. Bull. Am. Math. Soc. 62:270.
10. Bellman, R. (1963). An optimal search problem. SIAM Rev. 5:274.
11. Beck, A. (1964). On the linear search Problem. Israel J. Mathematics. 2:221–228.
12. Beck, A. (1965). More on the linear search problem. Israel J. Mathematics. 3:61–70.
13. Beck, A. and Beck, M. (1986). The linear search problem rides again. Israel J. Mathematics. 53(3):365–372.
14. Beck, A. and Newman, DJ. (1970). Yet More on the linear search problem. Israel J. Mathematics. 8:419–429.
15. Borodin A. and El-Yaniv, R. (1998). Online Computation and Competitive Analysis. Cambridge University Press.
16. Bruce, TF. and Robertson, JB. (1988). A survey of the linear-search problem. Math. Sci. 13: 75–89.
17. Chrobak M. and Kenyon-Mathieu, C. (2006). Competitiveness via doubling. SIGACT News. 37(4):115–126.

18. Dagan, A. and Gal, S. (2008). Network search games, with arbitrary searcher starting point. Networks. 52(3):156–161.
19. Demaine, ED., Fekete, S. and Gal, S. (2006). Online searching with turn cost. Theoretical Computer Science. 361:342–355.
20. Even, S. (1979). *Graph Algorithms*. Computer Science Press, Rockville MD. Ch. 3.
21. Edmonds, J, and Johnson, EL. (1973). Matching Euler tours and the Chinese postman problem. Math. Program. 5:88–124.
22. Gal, S. (1974). Minimax solutions for linear search problems. SIAM J. Appl. Math. 27:17–30.
23. Gal, S. (1979). Search games with mobile and immobile hider. SIAM J. Control Optim. 17: 99–122.
24. Gal, S. (1980). *Search Games*. Academic Press, New York.
25. Gal, S. (2000). On the optimality of a simple strategy for searching graphs. Int. J. Game Theory. 29:533–542.
26. Gal, S. (2011). *Search Games*. Wiley Encyclopedia of Operations Research and Management Science.
27. Gal, S. and Anderson, EJ. (1990). Search in a maze. Probab. Eng. Inf. Sci. (4):311–318.
28. Gal, S and Chazan, D. (1976). On the optimality of the exponential functions for some minimax problems. SIAM J. Appl. Math. 1976 .30:324–348.
29. Garnaev, A. (1992). A Remark on the Princess and Monster Search Game. Int. J. of Game Theory. 20:269–276.
30. Isaacs, R. (1965). *Differential Games*. Wiley, New York.
31. Isbell, J. R. (1957). An optimal search pattern. Naval Res. Logist. Quart. (4):357–359.
32. Kao, M, Reif, JH, Tate, SR. (1996). Searching an unknown envitonment: an optimal randomized algorithm for the cow-path problem. Inf. Comput. 131(1):63–79.
33. Kao, MY., Ma Y., Sipser, M. and Yin, Y. (1998). Optimal construction of Hybrid algorithm. J. of Algorithms. 29:142–164.
34. Kella, O. (1993). Star search – a different show. Israel. J. Mathematics. 81:145–159.
35. Lalley, SP. and Robbins, HE. (1988). Stochastic search in a convex region. Probab. Theory Relat. Fields. 77(1):99–116.
36. Langetepe, E. (2009). On the Optimality of Spiral Search. SODA 2010.
37. López-Ortiz, A. and Schuierer, S. (2001). The ultimate strategy to search on m rays? Theoretical Computer Science. 261(2):267–295.
38. López-Ortiz, A. and Schuierer, S. (2004). Online parallel heuristics, processor scheduling, and robot searching under the competitive framework. Theoretical Computer Science. 310: 527–537.
39. Lucas, E. (1882). *Recreations Mathematique*. Paris
40. Pavlovic, L. (1993). Search game on an odd number of arcs with immobile hider. Yugosl. J. Oper. Res. 3(1):11–19.
41. Schuierer, S. (2003). A lower Bound for Randomized Searching on m Rays, Pp 264–277 in Klein, R, Six, HW and Wegner, L. eds, Computer Science in Perspective: Essays Dedicated to Thomas Ottmann, volume 2598 of Lecture Notes Comput. Sci. Springer-Verlag, Berlin.
42. Tarry, G. (1985). La problem des labyrinths. Nouvelles Annals de Mathematique. 14, 187.
43. Von Stengel, B. and Werchner, R. (1997). Complexity of searching an immobile hider in a graph. Discrete Appl. Math. 78:235–249.
44. Wilson, DJ. (1970). Isaacs' princess and monster game on the circle. JOTA. 9:265–288.
45. Zelikin, MI. (1972). On a differential game with incomplete information. Soviet Math. Dokl. 13:228–231.

Chapter 2
Search Games for an Immobile Hider

Thomas Lidbetter

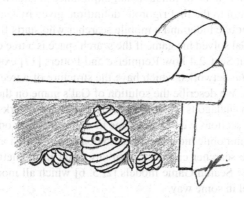

Abstract A search game for an immobile hider is a zero-sum game taking place in some search space. The hider picks a point in the space and a searcher who is unaware of the hider's location moves around attempting to find him in the least possible time. We give an overview of the theory of search games on a network with an immobile hider, starting with their conception in the Rufus Isaac's 1965 book on Differential Games, then moving on to some classic results in the field from Shmuel Gal and others. Finally we discuss some recent work on new search game models which consider, for example, what happens when the searcher does not have a fixed starting point or when the speed of the searcher depends on the direction in which he is traveling.

T. Lidbetter (✉)
Department of Mathematics, London School of Economics, London, UK
e-mail: t.r.lidbetter@lse.ac.uk

S. Alpern et al. (eds.), *Search Theory: A Game Theoretic Perspective*,
DOI 10.1007/978-1-4614-6825-7_2, © Springer Science+Business Media New York 2013

2.1 Introduction

Since the conception of *search games* almost 50 years ago, the field has expanded
and developed in many different directions, as seen in Chap. 1. In this chapter we
focus in on one particular theme: that of search games on a network with a mobile
searcher and an immobile hider. Games of this type may be described as 'hide-
and-seek' games. The classic results in this field can be found in Alpern and Gal's
monograph [4] and Gal's recent survey [13]. Here we do not aim to give an exhaustive list of all work in the field, but we follow on from Sects. 1.1.1 and 1.1.2 in
Chap. 1, taking a more detailed look at some classic results and linking them to new
work on search games with an immobile hider.

We begin in Sect. 2.2 by discussing how Isaacs [14] first introduced search games
of this type, and how he described strategies for both the hider and the searcher
which would continue to be of fundamental importance in later work in the field. In
Sect. 2.3 we then turn to the first rigorous definition, given by Gal [10], of a search
game with an immobile hider and a mobile searcher who starts from a given point.
We indicate how Gal solved his game if the search space is a tree or if it is Eulerian.

We then show in Sect. 2.4 how Reijnierse and Potters [17] extended Gal's analysis to *weakly cyclic* networks, which have the structure of a tree with some nodes
replaced by cycles. We describe the solution of Gal's game on these networks, and
how Gal proved an analagous result for weakly Eulerian networks.

In the final two sections we discuss some more recent work on search games on
networks with an immobile hider. Section 2.5 deals with a version of Gal's original
game in which the searcher can start from any point in the network. Section 2.6
describes three new Search Game models [2, 5, 6] which all modify or generalize
Gal's classic model in some way.

2.2 The Birth of Search Games

Search games were first introduced by Rufus Isaacs in his 1965 book Differential
Games [14], as indicated in Chap. 1. The book was originally motivated by combat
problems, and indeed, many of the problems discussed in the book have a military
focus to them. Earlier chapters in the book are concerned with so called *Pursuit
Games*, in which a *Pursuer* (or *Pursuers*) aim to capture an *Evader* whose location
is known to him at all times during the game. Search games are introduced later
in the book in a chapter called 'Toward a Theory with Incomplete Information'.
The model presented differs from Pursuit Games in that Pursuers now aim to capture
an Evader about whose position the Pursuers now do not have complete information.
The terminology changes: the Evader becomes the hider and the Pursuers become
the searchers. This terminology has stuck and is now widely used in the search
games literature.

Isaacs begins by defining what he calls the *simple search game*. This could be
regarded as the simplest and most general possible search game, and is described in
informal terms. In an arbitrary region $\mathscr{R}$, which may be a subset of Euclidean space

of any dimension, a hider picks a hiding point (that is a point in $\mathscr{R}$). The searcher then picks some sort of unit speed trajectory in the region. The payoff (or search time) is the time taken until the searcher's trajectory meets the hider. There is an assumption that the searcher is able to find a tour of the region that is not wasteful, so that it does not 'double back' on itself. The solution of the game Isaacs gives is simple: the searcher picks one such tour S, then follows it with probability one half and follows the reverse tour with probability one half. Supposing $\mathscr{R}$ has measure μ, if S finds a point in $\mathscr{R}$ at time t, the reverse of S will find the same point at time $\mu - t$. Hence the expected time T to find any given point is given by

$$T = 1/2t + 1/2(\mu - t) = \mu/2$$

The value of the game is therefore at most $\mu/2$. The hider can ensure the payoff is no more than $\mu/2$ by hiding uniformly in $\mathscr{R}$, so that the probability he hides in any subset of $\mathscr{R}$ is proportional to its measure. By using this strategy, the hider ensures that the probability the searcher finds him before time t is no more than t/μ for $0 \le t \le \mu$, so the probability the search time is t or more is at least $1 - t/\mu$. Hence the expected time T satisfies

$$T = \int_0^\infty Pr(\text{search time is} \ge t)dt$$
$$\ge \int_0^\mu (1 - t/\mu)dt$$
$$= \mu/2.$$

The value of the game is therefore at least $\mu/2$, and combining the bounds we have

Theorem 1 (Isaacs). *The value of the simple search game is $\mu/2$.*

These strategies given by Isaacs are important and direct a lot of the later research on search games.

2.3 Search Games on Networks

A more precise formulation of Isaac's game is given by Gal [10] and [11]. Gal focuses on the game played on a network Q, which is any connected finite set of arcs of measure μ with a distinguished starting point O, called the root. The hider picks a point H in Q and the searcher picks a unit speed path S starting from O. The payoff (or search time) is the time taken for the path to reach H. This game is mentioned in Sect. 1.1.1 of Chap. 1.

In [10], Gal uses Isaacs hider strategy to give a lower bound for the value V of the game: by hiding uniformly in the network the hider can ensure that the search time always at least $\mu/2$. We call this strategy u. However, the assumption made by Isaacs that the searcher can find a non-wasteful trajectory is not made, so the searcher strategy given in [14] is not always available and the value of the game may be greater than $\mu/2$. The searcher is also restricted to picking a path which starts from O, so it may not be possible for him to implement the 'reverse' of a path. For instance, if Q is a single unit length arc with the root O at one end and a point A at the other, the value of this game is clearly $1 > \mu/2$. The hider simply uses the pure strategy of hiding at A and the searcher picks the path from O to A.

However, adapting the searcher strategy given in [14], Gal gives an upper bound for the value. The searcher may not be able to find a non-wasteful, reversible path in Q, but he will always have some minimal time tour S of Q starting and ending at O of length $\bar{\mu} \geq \mu$. He can then use the mixed strategy where he picks S with probability 1/2 and the reverse of S with probability 1/2, ensuring that he finds every point in Q in expected time no more than $\bar{\mu}/2$. The searcher's minimal tour S is later called a Chinese Postman Tour (CPT) in [12], and the randomized strategy given here is called the Random Chinese Postman Tour (RCPT). The RCPT gives an upper bound for the value V, and combining this with the lower bound we have

$$\mu/2 \leq V \leq \bar{\mu}/2 \tag{2.1}$$

Gal examines when these two bounds are tight. Suppose Q is Eulerian, so that it has a continuous closed path that visits each point of Q exactly once. Then the searcher's CPT is one such Eulerian path starting at O. Since the length $\bar{\mu}$ of this tour is μ, the bounds in (2.1) are tight and we have $V = \mu/2 = \bar{\mu}/2$. The uniform strategy u is optimal for the hider. It is easy to see that Eulerian networks are the only networks for which $\bar{\mu} = \mu$.

We can also consider the game played on a tree, that is a network without any cycles. In a sense, a tree is the opposite of an Eulerian network since the CPT of a tree has the maximum possible length, $\bar{\mu} = 2\mu$, as all arcs must be traversed in both directions. The inequalities (2.1) therefore become $\mu/2 \leq V \leq \mu$. Clearly the uniform hider strategy u is not optimal for the hider, since every point H of Q is dominated in strategies by a leaf node (a node of degree 1). Hence an optimal hider strategy must be some distribution on the leaf nodes. In [10] Gal defines a hider distribution later called the Equal Branch Density (EBD) distribution in [12], and shows that it is optimal for the hider, guaranteeing him an expected search time of no less than $\mu = \bar{\mu}/2$, which is the value of the game. The RCPT is optimal for the searcher.

The EBD distribution can be defined in terms of a concept called *search density*, which extends to general search spaces Q that may not be networks. For a connected subset A of Q and a hider hidden on Q according to a fixed distribution, the search density $\rho(A)$ is defined as the time taken for the searcher to tour A divided by the probability the hider is in A. Consider a tree Q and a node x of Q that has degree at least 3. We call x a branch node. The arcs touching x consist of one arc on the path from x to O and some other arcs. For each of these other arcs a, we define a branch at x which consists of a together with all arcs whose unique path to O intersects with a. The EBD distribution is the unique hider distribution on the leaf nodes of Q that ensures that at every branch node of Q, all branches have equal search density.

We illustrate the EBD distribution with an example. In Fig. 2.1 nodes are labelled by letters and arc lengths indicated by numbers. To calculate the EBD distribution on this network, first note that there are two branches at O, which must have equal search density. This can be achieved by assigning hider probability $3/9 = 1/3$ to the branch consisting of the arc OC, and probability $2/3$ to the other branch. The branch node D has two branches, and to ensure these have equal search density, the hider probability assigned to the arcs AD and BD must be proportional to 2 and 3, respectively. Hence the probabilities the hider is at nodes A and B are $2/5 \cdot 1/3 = 2/15$ and $3/5 \cdot 1/3 = 3/15$ respectively. The probability the hider is at C is $1/3$.

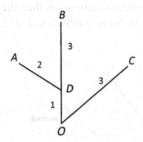

Fig. 2.1 A tree network

In [10], Gal shows that if the hider uses the EBD distribution, this ensures that any depth-first search of Q, and in particular any CPT finds the hider in expected time exactly $\mu = \bar{\mu}/2$, which must therefore be the value of the game. In the case of the network in Fig. 2.1, the value of the game is $\mu = 9$.

Hence we have

Theorem 2 (Gal). *If Q is an Eulerian network or a tree then the value of the search game with an immobile hider played on Q is $\bar{\mu}/2$.*

As discussed in Sect. 1.1.1 of Chap. 1, the RCPT is not optimal for all networks, in particular the 3-arc network depicted in Fig. 1.1, though this was not shown for another 15 years [15].

2.4 Weakly Cyclic and Weakly Eulerian Networks

Solutions of the game described in the previous section are not limited to trees and Eulerian networks. In [17] Reijnierse and Potters solve the game for *weakly cyclic* networks, showing that the RCPT is optimal for the searcher, so that the value is $\bar{\mu}/2$. A weakly cyclic network can be thought of as a tree network for which some of the nodes have been replaced with cycles. Alternatively, a weakly cyclic network can be defined more precisely as a network for which there are at most two disjoint paths between any two nodes. Weakly cyclic networks cannot contain any subnetwork that is topologically homeomorphic to the three arc network depicted in Fig. 1.1. A weakly cyclic network is depicted on the left hand side of Fig. 1.2; the cycles are indicated by the dotted lines.

Reijnierse and Potters give an algorithm to calculate the optimal hider distribution, in which the hider hides with some probability on leaf nodes and with some probability hides uniformly on the cycles. Alpern and Gal [4] later give an alternative version of the algorithm, in which every cycle in the network is replaced with a leaf arc of half the length of the cycle, and the EBD distribution is calculated on the new network. The network depicted on the right hand side of Fig. 1.2 is the modification of the weakly cycle network on the left. The hider probability that should be assigned to a cycle in the original network is then the probability assigned to the end of the associated leaf arc in the new network (Fig. 2.2).

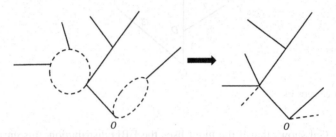

Fig. 2.2 A weakly cyclic network and its modifications

Reijnierse [16] later showed that the equivalent result holds if we replace 'weakly cyclic' with 'weakly Eulerian'. A network is weakly Eulerian if it can be obtained from a tree by replacing some nodes with Eulerian networks. Gal [12] found a simple proof of this result, showing not only that the value V of the game is $\bar{\mu}/2$ for weakly Eulerian networks, but, as mentioned in Chap. 1, these are the only networks for which this is the value, and the RCPT is optimal. In summary,

Theorem 3 (Gal). *The value of the search game with an immobile hider played on a network Q is $\bar{\mu}/2$ if and only if Q is weakly Eulerian.*

Notice that the class of weakly Eulerian networks includes both trees and Eulerian networks, so Theorem 3 generalizes Theorem 2.

2.5 Search Games Without a Fixed Searcher Starting Point

In a recent paper [9] Dagan and Gal define a new Search Game model on a network Q in which the assumption that the searcher has a fixed starting point O is dropped, and the searcher can begin his search from any point on Q. This model has already been discussed in Sect. 1.1.2 in Chap. 1, where it was noted that provided the searcher has some Eulerian path (one which visits every point of the network exactly once), Isaac's result holds and the value of the game is $\mu/2$. The searcher can simply choose the Eulerian path with probability 1/2 and its reverse with probability 1/2; the hider can hide uniformly on the network. The networks that have an Eulerian path include the 3 arc network in Fig. 1.1, whose solution in Gal's classic model was so elusive. For the arbitrary start model, the value of this game is $\mu/2 = 3/2$. Just as we define Chinese Postman Tours, we can define a *Chinese Postman Path* of a network Q as a minimal time path that visits all the points of Q. We can then define $\tilde{\mu}$ as the length of a Chinese Postman Path, and we obtain a result analgous to (2.1) for the value V of the Search Game played on networks with an arbitrary starting point:

$$\mu/2 \leq V \leq \tilde{\mu}/2. \tag{2.2}$$

The arbitrary start model was further studied in [3] in which the authors call a network *simply searchable* if the upper bound on V in (2.2) is tight. They give sufficient conditions for a network to be simply searchable, and in particular they show that trees are simply searchable and that the hider should use the EBD distribution, with respect to a root located at the *center* of the tree: that is the point c whose greatest distance from any other point in the tree is minimal. For example, in Fig. 2.1 the center c is located halfway between nodes O and D. If we add a node at c, then at this point there are two branches of lengths 7/2 and 11/2, which the hider chooses with probabilities proportional to 7 and 11, respectively. Hence the hider chooses the node C with probability 7/18 and nodes A and B with total probability 11/18.

2.6 Other Search Game Models

Recently some alternative models of search games on networks have been proposed. In the models we have discussed so far the searcher's strategy space is a set of unit speed paths. However we might consider associated games in which the searcher has a different strategy set.

In [2] Alpern defines a new model called *find-and-fetch* in which he considers a searcher who not only wishes to find a hider but also wishes to return to the root O. This models common problems such as search-and-rescue and foraging problems in which an animal must find food and then return to its lair. As in Gal's model, the searcher follows a unit speed path from O, but then upon reaching the hider takes the shortest path back to O at speed ρ. The payoff is the total time to find the hider and return to O. In the case of a bird being weighed down by food he is taking back to his nest we might have $\rho < 1$, whilst $\rho > 1$ might be more appropriate for the case of someone searching for a contact lens, in which the return speed would be quicker.

Alpern finds that if Q is a tree, the optimal strategy for the hider is still the EBD distribution in this game. However, the RCPT is no longer optimal for the searcher. Instead, he randomizes between all possible depth-first searches using a type of strategy called a branching strategy. Upon reaching a node for the first time the searcher chooses which outward branch to take according to a certain probability. Alpern proves the following.

Theorem 4 (Alpern). *The value V of the find-and-fetch game on a tree is*

$$V = \mu + D/\rho, \tag{2.3}$$

where $D = D(Q)$ is the mean distance from O to the leaf nodes of Q, weighted according to the EBD distribution.

To illustrate how D is calculated, consider the network in Fig. 2.1. The probabilities that the hider is at nodes A, B and B are 2/15, 3/15 and 10/15, respectively, and the distances of these nodes from O are 3, 4 and 3. Hence $D = 2/15(3) + 3/15(4) + 5/15(3) = 11/5$. For $\rho = 1/2$, the value V of the find and fetch game played on the tree in Fig. 2.1 is $V = 9 + (11/5)/(1/2) = 10.1$. Note that as ρ tends to infinity so that the searcher can return instantaneously to the root after finding the hider, the value V in (2.3) tends to μ, Gal's classic result (Theorem 2).

A different model of search is given in [6], in which the authors suppose that the searcher can use an *expanding search*. This is defined as a sequence of unit speed paths on a network Q, starting at O, each of which starts from a point already reached by the searcher. Another way to think of expanding search is as a family of connected subsets of Q starting with O and expanding at unit speed. To differentiate expanding search from the type of search used in Gal's model, we call the latter pathwise search. Expanding search provides a model of mining, in which the time taken to recommence mining from a location already reached in small compared to the time taken up by the mining itself. As before, the hider simply picks a point on Q and the searcher picks an expanding search. The search time is the time taken for the searcher to reach the hider.

Again, if Q is a tree it turns out that the EBD distribution is optimal for the hider, and the searcher's optimal strategy is a branching strategy. The authors show that

Theorem 5 (Alpern and Lidbetter). *The value V of the expanding search game on a tree is*

$$V = 1/2(\mu + D), \tag{2.4}$$

The variable D is defined as above. In the case of the network in Fig. 2.1 where $D = 11/5$ and $\mu = 9$, the value is $V = 1/2(9 + 11/5) = 5.6$.

In [6] the expanding search game is also solved for 2-arc-connected networks. These are networks that cannot be disconnected by the removal of fewer than two arcs. The authors show that on these networks it is optimal for the hider to hide uniformly, and for the searcher to make an equiprobable choice of a reversible expanding search and its reverse. A reversible expanding search is simply one whose reverse is also an expanding search, analogous to an Eulerian circuit in Gal's model. The authors show that such a search always exists on a 2-arc-connected network, and the randomized choice of this search and its reverse ensures that the searcher finds the hider in expected time no greater than $\mu/2$, which is the value of the game. For example, the 3-arc network depicted in Fig. 1.1 in Chap. 1 is 2-arc-connected, and hence has value $\mu/2 = 3/2$.

For trees, the find and fetch model and the expanding search model can both be encapsulated in a single overarching model. In [1] Alpern examines the Search Game on a network with asymmetric travel times, meaning that the speed it takes for the searcher to traverse an arc depends on the direction in which he travels. An equivalent formulation is given in [5] in which the searcher moves with a speed that depends on his direction of travel. We therefore call this the *variable speed* model. The game is then defined as usual: the searcher picks a path in the network starting at O, the hider picks a point on the network and the payoff is the time taken for the searcher to reach the hider. The model clearly encompasses Gal's model of networks with symmetric travel times if the travel times of each arc are set to be the same in either direction.

For a tree Q we can give every point x on Q a height $\delta(x)$, equal to the time taken to travel from O to x (along the shortest path) minus the time taken to travel from x to O. This definition is motivated by the assumption that it is quicker to travel uphill than downhill. In [1] Alpern shows that the EBD is once again optimal for this game played on a tree, and he gives recursive formulae for the optimal branching strategy for the searcher. In [5], the authors derive a closed form expression for the optimal searcher strategy as well as a formula for the value V of the game:

Theorem 6 (Alpern and Lidbetter). *The value V of the variable speed search game is*

$$V = 1/2(\tau + \Delta), \tag{2.5}$$

where τ is defined as the time taken for the searcher to tour the network, and $\Delta = \Delta(Q)$ is defined as the mean height of the leaves, weighted with respect to the EBD distribution.

If the network is symmetric, then all leaf nodes have height 0, and $\tau = 2\mu$, so (2.5) reduces to Gal's classic result, $V = \mu = \bar{\mu}/2$ given in Theorem 2. In fact, in the case that the network Q is a tree, the variable speed network model also encompasses both the find and fetch model and the expanding search model, as we now explain.

We first consider the find and fetch game, in which the searcher must return to O along the shortest path at speed ρ after finding the hider. It is optimal for the hider to choose a leaf node x, and for any such choice of x at shortest distance $d(x, O)$ from O, the searcher must travel for additional time $d(x, O)/\rho$ after finding the hider. We therefore form a new network Q' from Q by adding an asymmetric arc from x to a new leaf node x^+ with forward travel time (from x to x^+) of $d(x, O)/\rho$ and backward travel time $-d(x, O)/\rho$. The variable speed game played on Q' is then equivalent to the find and fetch game played on Q: traveling to x^+ in Q' is equivalent to traveling to x in the original network and then back to O at speed ρ, and if the hider is not at x the extra arc from x to x^+ makes no contribution to the search time. Hence the two models are equivalent.

The total tour time τ of Q' is equal to twice the length 2μ of Q, and in the Q' the leaf node x^+ has height $2d(x, O)/\rho$, so $\Delta = \Delta(Q')$ is the mean value of $2d(x, O)/\rho$, weighted with respect to the EBD distribution, which is equal to $2D(Q)/\rho$. Hence by (2.5), the value is

$$V = 1/2(2\mu + 2D/\rho)$$
$$= \mu + D/\rho,$$

as given in (2.3).

We now return to the expanding search model played on a tree Q. Suppose we form a new network Q'' by replacing each arc of Q of length λ with an asymmetric arc with forward travel time (away from O) of λ and backward travel time 0. Then a depth-first pathwise search on Q'' is equivalent to an expanding search on Q. It can be shown that it is optimal to use a depth-first search in the expanding search game, so that the two models are equivalent. The total tour time of the new network is the length μ of the original network, and the height of a leaf node in the new network is the distance from that node to O in the old network, so $\Delta(Q'') = D(Q)$. Hence, by (2), the value is

$$V = 1/2(\mu + D),$$

as given in (2.4).

2.7 Conclusion

We have seen how an idea in [14] sparked a field of research which has produced many elegant results, and continues to develop and expand. We have focused here on search games on a network with an immobile hider, but search games are not limited to this paradigm. Much has been achieved in the field of search games with a mobile hider (also originally motivated by Isaacs [14]), as well as many other variations on the classic models. The connected field of search games in unbounded domains, initiated independently by Bellman [8] and Beck [7], has also been extensively studied. Many unanswered questions in search games remain and new problems arise, capturing the imaginations of those who have taken the childhood game of hide-and-seek to its mathematical extreme.

References

1. Alpern, S. (2010). Search games on trees with asymmetric travel times. SIAM J. Control Optim. 48, no. 8, 5547–5563.
2. Alpern, S. (2011). Find-and-fetch search on a tree. Operations Research 59 (2011): 1258–1268.
3. Alpern, A., Baston, V. and Gal, S. (2008). Network Search Games with an immobile hider without a designated searcher starting point, Int. J. Game Theory 37, 281–302.
4. Alpern S. and Gal S. (2003). The Theory of Search Games and Rendezvous (Kluwer International Series in Operation Research and Management Sciences, Kluwer, Boston).

5. Alpern, S. and Lidbetter, T. (2013) Searching a Variable Speed Network. Mathematics of Operations research (in press).
6. Alpern, S. and Lidbetter, T. (2013). Mining Coal or Finding Terrorists: the Expanding Search Paradigm. Operations Research (in press).
7. Beck, A. (1964). On the linear search Problem. Naval Res. Logist. 2, 221–228
8. Bellman, R. (1963). An optimal search problem. SIAM Rev. 5, 274.
9. Dagan, A. and Gal, S. (2008). Networks Search Games with Arbitrary searcher starting point, Networks 52, 156–161.
10. Gal, S. (1979). Search games with mobile and immobile hider. SIAM J. Control Optim. 17, 99–122.
11. Gal, S. (1980). Search Games. Academic Press, New York.
12. Gal, S. (2000). On the optimality of a simple strategy for searching graphs. Int. J. Game Theory 29, 533–542.
13. Gal, S. (2011). Search games. Wiley Encyclopedia of Operations Research and Management Sci. (James J. Cochran, ed.), Wiley.
14. Isaacs, R. (1965). Differential Games. Wiley, New York.
15. Pavlovic, L. (1993). Search game on an odd number of arcs with immobile hider. Yugosl. J. Oper. Res. 3, no. 1, 11–19.
16. Reijnierse, J. H. (1995). Games, graphs and algorithms. Ph.D Thesis, University of Nijmegen, The Netherlands.
17. Reijnierse, J. H. and Potters, J. A. M. (1993). Search games with immobile hider. Int. J. Game Theory 21, 385–394.

Chapter 3
Tools to Manage Search Games on Lattices

Noemí Zoroa, María-José Fernández-Sáez, and Procopio Zoroa

Abstract Search games on a network or a graph have been widely studied in the literature. In some of these situations, the search space can be represented by the lattice

$$L = \{1, 2, \ldots, n\} \times \{1, 2, \ldots, m\}$$

and the strategies for the players are subsets of this lattice. We develop a method to simplify the resolution of games of this kind when they satisfy some general conditions. Some games are solved, these are interesting in themselves, and their resolution illustrates the usefulness of the obtained results.

3.1 Introduction

In this chapter we suggest a general approach to solve games on a lattice. Search games where the search space can be thought as a network or graph have been widely studied in the literature, [2, 4–6]. In some of these situations, the search space can be represented by the lattice

$$L = \{1, 2, \ldots, n\} \times \{1, 2, \ldots, m\}.$$

It is clear that the lattice L can be applied to discretize a game in which the search space is a rectangular region. The lattice can also be applied to search games that are played over time. Consider, for example, a game like the patrolling game in [3] in which the search space is the linear set $\{1, 2, \ldots, m\}$ and players move over n consecutive periods of time. The lattice L represents the space-time network, by depicting the nodes of the linear set vertically and time horizontally. If the set is

N. Zoroa (✉) • M.-J. Fernández-Sáez • P. Zoroa
Faculty of Mathematics, Department of Statistics and Operational Research,
University of Murcia, Campus of Espinardo, 30071, Murcia, Spain
e-mail: nzoroa@um.es; majose@um.es; procopio@um.es

S. Alpern et al. (eds.), *Search Theory: A Game Theoretic Perspective*,
DOI 10.1007/978-1-4614-6825-7__3, © Springer Science+Business Media New York 2013

not linear but cyclic, then we may take $\{1, 2, \ldots, m\}$, considering point 1 as next to point m. Therefore, if the situation is carried out in a finite set of points placed on a perimeter, the strategies for the players can also be represented by subsets of L. Figure 3.1 shows how the moving of a point on a linear set over time can be represented on the lattice. Games where the search is developed on a star graph, as

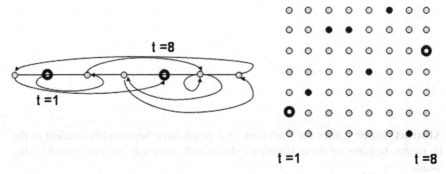

Fig. 3.1 Representation, on the lattice $L = \{1, 2, \ldots, 8\} \times \{1, 2, \ldots, 7\}$, of the moving of a point on the linear set $\{1, 2, \ldots, 7\}$ over eight consecutive periods of time

considered by Gal in Chap. 1, Sect. 1.3 of the book, can also be modeled on L by imposing appropriate constraints on the strategies for the players. The nice book by Ruckle, *Geometric games and their applications*, which is discussed in Chap. 6 of the book, also includes a number of games on the lattice L. Most of the problems on lattice games that can be found in that book remain open. More games on the lattice L can be found in [2, 3, 8, 9, 11].

Here we are going to deal with two-person search games on the lattice L; they are win-loose games, and, therefore, zero-sum games. One interesting problem is to obtain results or to develop methods to attack the resolution of games with common characteristics. In the following section we present a method to simplify the resolution of games on a finite set that satisfy some general properties. Section 3.3 shows how this method can be applied when the finite set is a lattice L. Two games of this kind are solved in Sect. 3.4. These games are interesting in themselves, but their resolution is also useful to illustrate the proposed method.

Let us consider the following situation. A hacker gets information from 20 computers, C_1, C_2, $\ldots$, C_{20}, of an enterprise. Each day he picks information from each of the computers, but for the information obtained for one computer to prove trustworthy, he has to verify it at least k times. The company performs one inspection each day of all the computers. If a leak is detected, a protection system is set in motion which invalidates the information that can be obtained during that day. The hacker must select, every day, the computers and the hours at which he will make the incursions. He has to do this bearing in mind, first, that the importance of the data depends on the responsibility of the person who uses the computer and, secondly, that he cannot make more than s contacts during a day. This problem can be modelled as a two-person zero-sum game on the lattice

$L = \{1, 2, \ldots, 20\} \times \{1, 2, \ldots, 24\}$ where column i corresponds to computer C_i and row j to the one-hour period beginning at j o'clock. The pure strategies for the hacker are the subsets of L of cardinality equal to s; the pure strategies for the company are the subsets of L with just one point in each column; if the hacker is not detected, he receives a quantity c_i for every computer C_i where he has made at least k incursions, and zero otherwise. This payoff can be formalized as follows, when the hacker uses his strategy A, and the company its strategy B, $M(A,B)$ is defined by

$$M(A,B) = \begin{cases} \displaystyle\sum_{j \in H} c_j & \text{if } A \cap B = \varnothing, \\ 0 & \text{if } A \cap B \neq \varnothing, \end{cases}$$

where H is the set of the columns of A with k elements at least. The study of this problem can be simplified with the method developed in this chapter and it is studied in [11].

A two-person zero-sum game will be expressed by $G = (X, Y, M)$ where X, Y are the sets of pure strategies for players I and II, respectively, and

$$M : X \times Y \to \mathbb{R} \tag{3.1}$$

is the payoff function which represents the winnings of player I and the losses of player II. Player I chooses a strategy $A \in X$, player II chooses a strategy $B \in Y$ and these choices determine the payoff $M(A,B)$ to player I and $-M(A,B)$ to player II.

Throughout this chapter X and Y are finite sets, therefore a probability distribution on X, that is to say, a mixed strategy for player I, can be written as a function

$$x : X \to \mathbb{R}$$

such that $x(C) \geq 0$ for all $C \in X$ and $\sum_{C \in X} x(C) = 1$. Similarly, a mixed strategy for player II will be given by a function

$$y : Y \to \mathbb{R}$$

such that $y(C) \geq 0$ for all $C \in Y$ and $\sum_{C \in Y} y(C) = 1$. When the players use their mixed strategies x and y, the payoff $M(x,y)$ is the expected value of $M(A,B)$.

3.2 Transformations on the Strategy Space

In this section we develop a method to facilitate the study and resolution of games where the strategies for the players are subsets of a finite set and these and the payoff function satisfy certain conditions. This method is based on a well known invariance property [1, 8]: we show how, given a game G, we can build a game $\overline{G}$, which is easier to study than G and has the same value as G. This game has fewer strategies than the original and the optimal strategies for the players in the game G are easily obtained from the optimal strategies in $\overline{G}$. This method can be applied to finite games in general, but we focus on the lattice and solve two general search games of this kind. Games on different sets, where the method has been used, appear in [3, 8, 10, 11].

The method is formalized as follows. Let $G = (X, Y, M)$ be a two person zero sum game on a finite set L, that is X and Y are subsets of the power set of L, and let

$$T_i : L \longrightarrow L \qquad i = 1, 2, 3 \ldots, n$$

be a set of transformations defined on L. Notice that a transformation T induces on the power set of L, which we denote by T as well. Now suppose that the following conditions are satisfied for every i, j:

1. $T_i T_j = T_j T_i$.
2. T_i is a bijection on L.
3. $T_i X = X$, $T_i Y = Y$.
4. $M(T_i A, T_i B) = M(A, B)$, $A \in X$, $B \in Y$.
5. For every $A \subset L$ there exists an integer r_{iA} such that

$$T_i^{r_{iA}} A = A,$$

6. For every set of integers $s_1, s_2, \ldots, s_n$, which do not vanish simultaneously and such that $0 \leq s_i < r_{iA}$, the inequality

$$T_1^{s_1} T_2^{s_2} \ldots T_n^{s_n} A \neq A$$

holds.

Given a subset $C \subset L$, let $\overline{C}$ be the class defined by

$$\overline{C} = \left\{ D = T_1^{i_1} T_2^{i_2} \ldots T_n^{i_n} C, \ \text{for all } i_1, i_2, \ldots, i_n \text{ integers} \right\}.$$

Then we have partitions

$$\overline{X} = \left\{ \overline{A} : A \in X \right\},$$
$$\overline{Y} = \left\{ \overline{B} : B \in Y \right\},$$

of X and Y, respectively.

As usual, the cardinality of a set C will be written as $|C|$. We can define the function

$$\overline{M} : \overline{X} \times \overline{Y} \longrightarrow \mathbb{R}$$

by

$$\overline{M}\left(\overline{A}, \overline{B}\right) = M\left(x_{\overline{A}}, B\right) = \frac{1}{|A|} \sum_{C \in \overline{A}} M\left(C, B\right), \tag{3.2}$$

where $x_{\overline{A}}$ means the probability distribution on X uniformly concentrated on $\overline{A}$. That is to say,

$$x_{\overline{A}}(C) = \frac{1}{|A|}, \quad \text{if} \quad C \in \overline{A},$$

$$x_{\overline{A}}(C) = 0, \quad \text{if} \quad C \in X - \overline{A}.$$

The last member of (3.2) does not depend on $B \in \overline{B}$, in other words, it takes the same value for any $B' \in \overline{B}$. In fact, bearing in mind properties 3 and 4, we have that $M(x_{\overline{A}}, B) = M(x_{\overline{A}}, T_1 B) = M\left(x_{\overline{A}}, T_1^{s_1} B\right) = M\left(x_{\overline{A}}, T_1^{s_1} T_2^{s_2} \dots T_n^{s_n} B\right)$, from which we can easily obtain the independence. The expression (3.2) can also be computed by $M(A, y_{\overline{B}})$, where $y_{\overline{B}}$ is the distribution on Y uniformly concentrated on $\overline{B} \subset Y$. The game $\overline{G} = (\overline{X}, \overline{Y}, \overline{M})$, constructed above, will be called the associated (averaged) game of $G = (X, Y, M)$.

Theorem 1. *Given the game $G = (X, Y, M)$, let $\overline{x}$, $\overline{y}$ be the optimal mixed strategies for players I and II respectively in the associated game $\overline{G} = (\overline{X}, \overline{Y}, \overline{M})$. Then the mixed strategies of game G defined by*

$$x(A) = \frac{\overline{x}(\overline{A})}{|\overline{A}|}, \quad A \in X,$$

$$y(B) = \frac{\overline{y}(\overline{B})}{|\overline{B}|}, \quad B \in Y$$

are optimal in the game G and both games have the same value.

Proof. By definition (3.2), we easily obtain that

$$\sum_{\overline{A} \in \overline{X}} \overline{x}(\overline{A}) \overline{M}(\overline{A}, \overline{B}) = \sum_{\overline{A} \in \overline{X}} \overline{x}(\overline{A}) M(x_{\overline{A}}, B)$$

$$= \sum_{\overline{A} \in \overline{X}} \overline{x}(\overline{A}) \sum_{C \in \overline{A}} \frac{1}{|A|} M(C, B), \quad \overline{B} \in \overline{Y}.$$

And therefore we have

$$\overline{M}(\overline{x}, \overline{B}) = M(x, B).$$

A similar reasoning leads to

$$\overline{M}(\overline{A}, \overline{y}) = M(A, y).$$

Hence the inequalities

$$\overline{M}(\overline{A}, \overline{y}) \leq \overline{M}(\overline{x}, \overline{y}) \leq \overline{M}(\overline{x}, \overline{B}), \quad \overline{A} \in \overline{X}, \ \overline{B} \in \overline{Y}$$

become

$$M(A, y) \leq \overline{M}(\overline{x}, \overline{y}) \leq M(x, B), \quad A \in X, \ B \in Y$$

which completes the proof. $\square$

3.3 Transformations on the Lattice

We apply the symmetry principle of the previous section to the lattice L. We will refer to the subset $L_i = \{i\} \times \{1, 2, \ldots, m\}$ as the column i of L. Let $\mathscr{F}_{n,m} = \mathscr{F}$ represent the family of all subsets of $L = \{1, 2, \ldots, n\} \times \{1, 2, \ldots, m\}$ with just one point in each column, that is, the set of all functions from $\{1, 2, \ldots, n\}$ to $\{1, 2, \ldots, m\}$. In the games we are interested in, the pure strategies for one of the players are elements of $\mathscr{F}_{n,m}$ satisfying different constraints, depending on the game. An $A \in \mathscr{F}_{n,m}$ may be identified by the subset of L $\{(i, A(i)) : i = 1, 2, \ldots, n\}$ and it can also be represented simply by the vector $(A(1), A(2), \ldots, A(n))$. Thus, A can be interpreted as

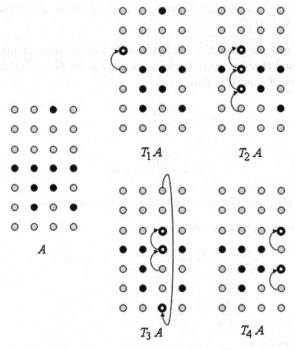

Fig. 3.2 Effect of transformations T_i over a subset $A \subset L$

a *walk* along a linear set of m points at moments $1, 2, \ldots, n$ and also as a path from the first to the last column of L which does not double back on itself. The transformations

$$T_s : L \longrightarrow L, \quad s = 1, 2, \ldots, n$$

defined by

$$T_s(i, j) = \begin{cases} (i, j), & \text{if } i \neq s, \\ (s, j+1), & \text{if } i = s, \quad j < m \\ (s, 1), & \text{if } i = s, \quad j = m \end{cases} \tag{3.3}$$

have properties 1, 2, 5 and 6. Figure 3.2 shows the effect of transformations T_1, T_2, T_3 and T_4 on the subset $A = \{(1,4),(2,2),(2,3),(2,4),(3,3),(3,4),(3,7),$ $(4,2),(4,4)\}$ of the lattice $\{1,2,3,4\} \times \{1,2,\ldots,7\}$. It is also easy to see that

$$T_s\mathscr{F} = \mathscr{F}, \quad s = 1,2,\ldots,n.$$

Therefore we can state the following theorems.

Theorem 2. *Let (X,Y,M) be a game on a lattice satisfying*

$$\begin{aligned}
X &= \mathscr{F} \\
T_sY &= Y, \\
M(T_sA, \, T_sB) &= M(A,B) \\
(A &\in X, \, B \in Y, \, s = 1,2,\ldots n).
\end{aligned} \tag{3.4}$$

An optimal strategy for player I is the uniform distribution on X.

$$x(A) = x_{\mathscr{F}}(A) = \frac{1}{|\mathscr{F}|} = \frac{1}{m^n}, \; A \in X. \tag{3.5}$$

Let $B_0 \in Y$ such that

$$\min_{B \in Y} M(x_{\mathscr{F}}, \, B) = \min \frac{1}{m^n} \sum_{A \in \mathscr{F}} M(A,B)$$

$$= \frac{1}{m^n} \sum_{A \in \mathscr{F}} M(A,B_0) = M(x_{\mathscr{F}},B_0). \tag{3.6}$$

Thus an optimal strategy for player II is the distribution on Y uniformly concentrated in $\overline{B}_0$:

$$y_{\overline{B}_0}(B) = \begin{cases} \frac{1}{|\overline{B}_0|}, & \text{if } B \in B_0 \\ 0 & \text{if } B \notin B_0 \end{cases} \tag{3.7}$$

and the value of the game is $M(x_{\mathscr{F}},B_0)$ given by (3.6).

Proof. Since properties 1–6 of transformations T_s work here, we can apply Theorem 1. In the associated game $\overline{G} = (\overline{X}, \, \overline{Y}, \, \overline{M})$ the set $\overline{X} = \{\mathscr{F}\}$ contains the only element $\mathscr{F}$ which will be the optimal strategy for player I. The optimal strategy for player II will be a pure strategy $\overline{B}_0$ such that

$$\min_{\overline{B}} \overline{M}(\mathscr{F},\overline{B}) = \overline{M}(\mathscr{F},\overline{B}_0).$$

The two strategies $x_{\mathscr{F}}, y_{\overline{B}_0}$ are obtained from the two optimal strategies $\mathscr{F}, \overline{B}_0$, of the associated game $\overline{G}$, completing the proof. $\square$

Theorem 3. *Let* (X, Y, M) *be a game on a lattice satisfying*

$$
\begin{aligned}
Y &= \mathscr{F} \\
T_s X &= X, \\
M\,(T_s A,\ T_s B) &= M\,(A, B) \\
(A &\in X,\ B \in Y,\ s = 1, 2, \ldots n).
\end{aligned}
\tag{3.8}
$$

An optimal strategy for player II is the uniform distribution on $Y = \mathscr{F}$,

$$
y\,(B) = y_{\mathscr{F}}\,(B) = \frac{1}{|\mathscr{F}|} = \frac{1}{m^n},\ \ B \in \mathscr{F}.
\tag{3.9}
$$

If we call $A_0 \in X$ *a strategy which fulfills*

$$
\max_{A \in X} M\,(A,\ y_{\mathscr{F}}) = M\,(A_0,\ y_{\mathscr{F}}),
\tag{3.10}
$$

then an optimal strategy for player I is the uniformly concentrated distribution on $\overline{A}_0$:

$$
x_{\overline{A}_0}\,(A) = \begin{cases} \frac{1}{|\overline{A}_0|}, & \text{if } A \in \overline{A}_0 \\ 0 & \text{if } A \notin \overline{A}_0 \end{cases}
\tag{3.11}
$$

and the value of the game is $M\,(A_0,\ y_{\mathscr{F}})$ *given by (3.10).*

Proof. Similar to the proof of the Theorem 2. □

These theorems give a general method for solving those games satisfying (3.4) or (3.8). In each case it will be sufficient to determine either $B_0 \in Y$ satisfying (3.6) or $A_0 \in X$ satisfying (3.10).

Example 1. In a conflict situation an intruder (the hider) has to carry out a sabotage on the perimeter of a protected zone. He has to perform the action over n consecutive days, and has to position himself each day at one of m strategic points placed on this border in order to set a device on it. These points are represented by 1, 2, 3, $\ldots$, m, considering point 1 as next to point m. The first day the hider can take his place at any of the m points, on successive days he can either stay, move one step to the right or move one step to the left. This constraint on the movements of the hider can be considered as a limit on his maximum speed, and expresses that his movements are difficult, e.g. because he has to use safe ways to go from one point to another. Furthermore, the perimeter is protected by a patroller (the searcher) who every day selects one of the m strategic points to inspect. This selection has to be done satisfying different constraints depending on the situation. If it is assumed $n = 6$ and $m = 9$, Fig. 3.3 shows a representation of strategy $\{(1,7),(2,7),(3,8),(4,9),(5,9),(6,1)\}$ of the intruder on the lattice $L = \{1,2,\ldots,6,\} \times \{1,2,\ldots,9\}$ and on the cyclic set $\{1,2,\ldots,9\}$.

This problem, modeled as a two-person zero-sum search game developed on the lattice L is studied in [9] . It is called ambush game over time on a cyclic set and the set of strategies for the hider X is equal to

$$\mathscr{F}_{n,m}^1 = \{A \in \mathscr{F}_{n,m} : A(i+1) - A(i) \in \{0, 1, -1, m-1, 1-m\},$$

$$i = 1, \ldots, n-1\}, \quad (3.12)$$

the set of strategies for the searcher is $Y = \mathscr{F}_{n,m}$ and the payoff function

$$M(A,B) = \begin{cases} 1 & \text{if } A \cap B = \varnothing \\ 0 & \text{if } A \cap B \neq \varnothing \end{cases}.$$

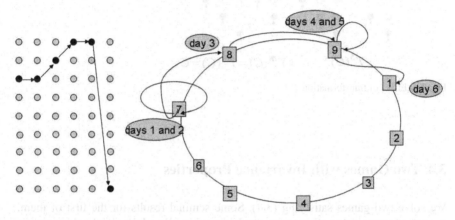

Fig. 3.3 Representation of strategy $\{(1,7),(2,7),(3,8),(4,9),(5,9),(6,1)\}$ on the lattice $L = \{1,2,\ldots,6,\} \times \{1,2,\ldots,9\}$ and on the cyclic set $\{1,2,\ldots,9\}$

Note that the results obtained in Theorems 2 and 3 cannot be applied to this game because conditions $T_s X = X, s = 1, 2, \ldots, n$ are not satisfied. Now we are going to define a new transformation that can be applied to make the handling of games where one of the sets of strategies for the players is the set defined by (3.12) or a subset of it easier. Let

$$T : L \longrightarrow L, \quad s = 1, 2, \ldots, n$$

be the transformation defined by

$$T(i,j) = \begin{cases} (i, j+1) & \text{if } j < n, \\ (i, 1) & \text{if } j = n. \end{cases} \quad (3.13)$$

Figure 3.4 shows the effect of transformation T over a subset of $L = \{1,2,\ldots,6,\} \times \{1,2,3,4\}$. If $G = (X,Y,M)$ is a game satisfying $X \subset \mathscr{F}_{n,m}^1, TY = Y$ (or $Y \subset \mathscr{F}_{n,m}^1, TX = X$) and $M(TA, TB) = M(A,B)$ then Theorem 1 can be applied and it is easy to prove that there exist optimal mixed strategies for the players such that if a pure strategy C has positive probability, all the pure strategies obtained from C by successive applications of transformation T, TC, T^2C, T^3C, …, $T^{m-1}C$, have the same probability. A similar result is obtained in [1].

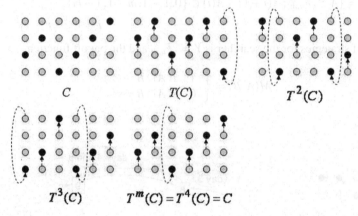

Fig. 3.4 Effect of transformation T

3.4 Two Games with Invariance Properties

We solve two games satisfying (3.4). Some seminal results for the first of them, the one-element intersection game, can be found in [8] and the weighted inspection game, which is the second game, has been studied in [11].

3.4.1 One-Element Intersection Game

The one-element intersection game (OEIG), is a game (X, Y, M) on a lattice L satisfying the following conditions:

$$X = \mathscr{F}$$

$$Y = \{B : B \subset L, \quad |B| = s\}$$

$$M(A, B) = \begin{cases} 1, & \text{if} \quad |A \cap B| = 1 \\ 0, & \text{otherwise.} \end{cases}$$

The intepretation of this game is that Player I wins if he finds just one point of B. When he finds this first point, however, an alarm system is activated, so that if he finds a new point, he is detected and cannot carry out his mission. This game has a trivial solution when $s \geq 2m$. In fact, player II can choose B filling up two columns, which is an optimal strategy for him and the value of the game is zero. The game satisfies hypothesis of Theorem 2. Therefore, an optimal strategy for player I is the uniform distribution on $X = \mathscr{F}$. To obtain an optimal strategy for player II we must determine B_0 in such a way that

$$
\begin{aligned}
M(x_{\mathscr{F}}, B_0) &= \min_{B \in Y} M(x_{\mathscr{F}}, B) = \min_{B \in Y} \frac{1}{m^n} \sum_{A \in \mathscr{F}} M(A, B) \\
&= \min_{s_1, s_2, \ldots, s_n} \frac{1}{m^n} f(s_1, s_2, \ldots, s_n) = \frac{1}{m^n} \min_{s_1, s_2, \ldots, s_n} \sum_{i=1}^{n} s_i \prod_{j \neq i} (m - s_j),
\end{aligned}
$$

where $s_i = |B \cap L_i|$ and $\sum_{i=1}^{n} s_i = s$. And the value of the game is given by $M(x_{\mathscr{F}}, B_0)$.

To solve the game when $s < 2m$, we have to solve the following problem. Given integers s and m such that $s < 2m$, find the minimum of the function

$$
f(s_1, s_2, \ldots, s_n) = \sum_{i=1}^{n} s_i \prod_{j \neq i} (m - s_j), \tag{3.14}
$$

for $s_1, s_2, \ldots, s_n$ integers satisfying

$$
0 \leq s_i \leq m \qquad \sum_{i=1}^{n} s_i = s. \tag{3.15}
$$

In the following $[x]$ denote the integer part of x. To obtain the solution for the OEIG when $n = 2$ we will need the next result.

Lemma 1. *If $n = 2$, the minimum of the function $f = f(s_1, s_2)$ defined by (3.14) is obtained with the values*

$$
\bar{s}_1 = s - [s/2],
$$

$$
\bar{s}_2 = [s/2].
$$

Proof. Given integers s_1, s_2 satisfying (3.15), the symmetry of the function (3.14) allows us to assume that

$$
s_1 \geq s_2
$$

We will prove that either $s_1 = \bar{s}_1$ and $s_2 = \bar{s}_2$ or $f(s_1, s_2)$ is not the minimum of f. If $s_1 - s_2 = 0$ or $s_1 - s_2 = 1$, then s_i should be equal to $\bar{s}_i$ for $i = 1, 2$. If $s_1 - s_2 \geq 2$, we can build

$$
s_1' = s_1 - 1, \qquad s_2' = s_2 + 1
$$

and compute the difference

$$
f(s_1, s_2) - f(s_1', s_2') = (s_1 - s_2 - 1) > 0,
$$

therefore $f(s_1, s_2)$ is not a minimum and the proof is complete. $\square$

We can now obtain the solution for the OEIG when $n = 2$. We have to compute (3.6), which in this case is equal to

$$\frac{1}{m^2} \min \left(s_1(m - s_2) + s_2(m - s_1) \right)$$

$$= \frac{1}{m^2} \min f(s_1, s_2)$$

where $s_i = |B \cap L_i|$. From Lemma 1 it follows that this minimum is given by

$$\frac{1}{m^2} f(\bar{s}_1, \bar{s}_2) = \frac{1}{m^2} \left(ms - 2 [s/2] (s - [s/2]) \right)$$

which is the value of the game. A set B_0 which determines the optimal strategy for player II is defined by

$$B_0 = \{(i, j) : 1 \leq j \leq s - [s/2], \quad \text{for} \quad i = 1, \quad 1 \leq j \leq [s/2], \quad \text{for} \quad i = 2 \}.$$

The solution for the OEIG when $n \geq 3$ is established by a number of lemmas and theorems.

Lemma 2. *Let s and m be integers such that $0 < s < 2m$. For each set of integers $s_1, s_2, \ldots, s_n$ satisfying*

$$0 \leq s_i \leq m \qquad \sum_{i=1}^{n} s_i = s,$$

$$s_1 \geq s_2 \geq \ldots \geq s_n, \tag{3.16}$$

the inequality

$$\frac{s_2}{m - s_2} + \ldots + \frac{s_n}{m - s_n} \leq \frac{s - [s/n]}{m - [s/2]},$$

holds.

Proof. From (3.16) it follows that

$$s_2 \leq [s/2] \quad \text{and therefore} \quad s_i \leq [s/2] \quad \text{for all } i = 2, \ldots, n$$

then

$$\frac{1}{m - s_i} \leq \frac{1}{m - [s/2]} \quad \text{for all } i = 2, \ldots, n.$$

Multiplying both members by s_i and adding

$$\sum_{2}^{n} \frac{s_i}{m - s_i} \leq \frac{s - s_1}{m - [s/2]},$$

assumptions $\sum_{i=1}^{n} s_i = s$ and $s_1 \geq s_2 \geq \ldots \geq s_n$ imply that $s_1 \geq [s/n]$, then

$$\sum_2^n \frac{1}{m-s_i} \leq \frac{s-[s/n]}{m-[s/2]}.$$

□

Lemma 3. *Let* s, n, m, $s_1, s_2, \ldots, s_n$ *be integers such that* $0 < s < 2m$, $n \geq 2$,

$$0 \leq s_i, \qquad \sum_{i=1}^n s_i = s,$$

$$s_1 \geq s_2 \geq \ldots \geq s_n.$$

and f *the function defined by (3.14). If for indexes* h *and* k, *one of them equal to* 1, *the inequalities* $s_h > 0$ *and* $s_k < m$ *are satisfied and we build the new set of integers given by*

$$s'_h = s_h - 1,$$

$$s'_k = s_k + 1,$$

$$s'_i = s_i \quad i \neq h, i \neq k,$$

then the difference

$$f(s_1, s_2, \ldots, s_n) - f(s'_1, s'_2, \ldots, s'_n)$$

is equal to

$$(s_h - s_k - 1)P(2 - \sum_{i \neq h,k} \frac{s_i}{m-s_i}),$$

where

$$P = \prod_{j \neq h,k} (m - s_j).$$

Proof. By computing the difference

$$f(s_1, s_2, \ldots, s_n) - f(s'_1, s'_2, \ldots, s'_n)$$

$$= \sum_i (s_i \prod_{j \neq i}(m - s_j) - s'_i \prod_{j \neq i}(m - s'_j)), \qquad (3.17)$$

we pay first attention only on those terms involving $i = h$ and $i = k$, so that we find that

$$s_h(m - s_k)P + s_k(m - s_h)P - s'_h(m - s'_k)P - s'_k(m - s'_h)P$$

$$= 2(s_h - s_k - 1)P, \qquad \text{where} \quad P = \prod_{j \neq h,k} (m - s_j).$$

Bearing in mind that one of the indexes h or k is equal to 1, we can assure that $m - s_i > 0$ for each $i \neq h, i \neq k$. Then, for the remaining terms in (3.17) we can write

$$\sum_{i\neq h,k} s_i \frac{(m-s_h)(m-s_k)}{m-s_i}P - \sum_{i\neq h,k} s_i \frac{(m-s_h')(m-s_k')}{m-s_i}P$$

$$= -\sum_{i\neq h,k} \frac{s_i}{m-s_i}(s_h - s_k - 1)P.$$

Then (3.17) can be rewritten as

$$(s_h - s_k - 1)P(2 - \sum_{i\neq h,k} \frac{s_i}{m-s_i})$$

and the proof is finished.   □

To state the following lemma we will write

$$\bar{s}_i = p+1, \quad \text{for } i \leq q,$$

$$\bar{s}_i = p \quad \text{for } i > q.$$

(3.18)

$$\text{where} \quad p = [s/n] \quad \text{and} \quad q = s - pn,$$

(3.19)

Lemma 4. *Let s and m be integers, $0 < s < 2m$. The minimum of function f defined by (3.14) for integers $s_1, s_2, \ldots, s_n$ satisfying (3.15) and*

$$\sum_{i=2}^{n-1} \frac{s_i}{m-s_i} \leq 2,$$

(3.20)

is achieved with the integers $s_i = \bar{s}_i$ defined by (3.18) and it is equal to

$$(ms - np(p+1))(m-p-1)^{q-1}(m-p)^{n-q-1}.$$

(3.21)

with p and q defined by (3.19).

Proof. Let $\mathscr{S}$ be the set defined by

$$\mathscr{S} = \left\{ \{s_1, s_2, \ldots, s_n\} : s_i \quad \text{integers}, 0 \leq s_i \leq m, \sum_{i=1}^{n} s_i = s, \sum_{i=2}^{n-1} \frac{s_i}{m-s_i} \leq 2 \right\}$$

We want to obtain the minimum

$$\min_{(s_1, s_2, \ldots, s_n) \in \mathscr{S}} f(s_1, s_2, \ldots, s_n).$$

Given a set of integers $\{s_1, s_2, \ldots, s_n\} \in \mathscr{S}$, the symmetry of the function (3.14) allows us to assume that

$$s_1 \geq s_2 \geq \ldots \geq s_n.$$

If $s_1 - s_n = 0$ or $s_1 - s_n = 1$, then s_i should be equal to $\bar{s}_i$ (for each index i). If $s_1 - s_n \geq 2$, we can build the new set of integers $\{s_1', s_2', \ldots, s_n'\} \in \mathscr{S}$, by

$$s_1' = s_1 - 1$$

$$s_i' = s_i, \quad 2 \leq i \leq n-1$$

$$s_n' = s_n + 1$$

Now, from Lemma 3 with $h = 1$ and $k = n$ it follows

$$f(s_1, s_2, \ldots, s_n) - f(s_1', s_2', \ldots, s_n')$$

$$= (s_1 - s_n - 1)P(2 - \sum_{i=2}^{n-1} \frac{s_i}{m - s_i}) \geq 0.$$

If this difference is greater than 0 then $f(s_1, s_2, \ldots, s_n)$ is not a minimum. If the difference is equal to 0 then $f(s_1, s_2, \ldots, s_n) = f(s_1', s_2', \ldots, s_n')$ and we can remove the set of integers $\{s_1, s_2, \ldots, s_n\}$ from set $\mathscr{S}$ to find the minimum of f and the proof of the lemma is complete. □

To state the following theorem we will write

$$\begin{align} \hat{s}_1 &= a = s - b, \\ \hat{s}_2 &= b = [s/2], \\ \hat{s}_i &= 0, \quad i > 2 \end{align} \tag{3.22}$$

Theorem 4. *Let s, n, m, be integers $0 < s < 2m$, $n \geq 3$. Then the minimum of function $f(s_1, s_2, \ldots, s_n)$ defined by (3.14) for integers $s_1, s_2, \ldots, s_n$, satisfying (3.15) is either*

$$K = (ms - np(p+1))(m - p - 1)^{q-1}(m - p)^{n-q-1} \tag{3.23}$$

where $p = [s/n]$ and $q = s - pn$, achieved with the values $s_i = \bar{s}_i$ defined by (3.18), or

$$K' = (ms - 2ab)m^{n-2}, \tag{3.24}$$

where $a = s - b$, and $b = [s/2]$, achieved with the values $s_i = \hat{s}_i$ defined by (3.22).

Proof. Given a set of integers $s_1, s_2, \ldots, s_n$ satisfying (3.15), the symmetry of the function (3.14) allows us to assume that

$$s_1 \geq s_2 \geq \ldots \geq s_n.$$

If inequality $\sum_{i=2}^{n-1} \frac{s_i}{m - s_i} \leq 2$ then it follows from Lemma 4 that

$$f(s_1, s_2, \ldots, s_n) \geq f(\bar{s}_1, \bar{s}_2, \ldots, \bar{s}_n)$$

$$= (ms - np(p+1))(m - p - 1)^{q-1}(m - p)^{n-q-1}$$

with $\bar{s}_i$ defined by (3.18) and p and q defined by (3.19). Now let us suppose that

$$\sum_{i=2}^{n-1} \frac{s_i}{m-s_i} > 2$$

and let l be the last index for which $s_l > 0$, that is $s_i > 0$ for $i = 1, 2, \ldots, l$ and $s_i = 0$ for $i = l+1, l+2, \ldots, n$. Let us suppose $l > 2$. We distinguish two cases, $s_1 < m$ and $s_1 = m$. If $s_1 < m$, then we can define the new set of integers

$$s_1' = s_1 + 1, \quad s_l' = s_l - 1 \quad \text{and} \quad s_i' = s_i \quad i \neq 1, \text{ and } i \neq l,$$

and applying Lemma 3 with $h = l$ and $k = 1$ it follows that

$$f(s_1, s_2, \ldots, s_n) - f(s_1', s_2', \ldots, s_n') = (s_l - s_1 - 1)P(2 - \sum_{i \neq 1, l} \frac{s_i}{m - s_i})$$

$$\geq (s_l - s_1 - 1)P(2 - \sum_{i \neq 1}^{n-1} \frac{s_i}{m - s_i}).$$

Bearing in mind $s_1 \geq s_l$, $P > 0$ and $\sum_{i=2}^{n-1} \frac{s_i}{m-s_i} > 2$ it follows that $f(s_1, s_2, \ldots, s_n) - f(s_1', s_2', \ldots, s_n') > 0$. Therefore $f(s_1, s_2, \ldots, s_n)$ is not a minimum.

If $s_1 = m$, bearing in mind that $s < 2m$ we can assert $s_2 < m$, then we can define the new set of integers

$$s_2' = s_2 + 1, \quad s_l' = s_l - 1 \quad \text{and} \quad s_i' = s_i \quad i \neq 2, \text{ and } i \neq l,$$

In this case, to compute the difference

$$f(s_1, s_2, \ldots, s_n) - f(s_1', s_2', \ldots, s_n')$$

we can not apply Lemma 3 because neither index 2 nor index l is equal to 1, but it is not difficult to compute it directly,

$$f(s_1, s_2, \ldots, s_n) - f(s_1', s_2', \ldots, s_n') = \sum_i (s_i \prod_{j \neq i} (m - s_j) - s_i' \prod_{j \neq i} (m - s_j'))$$

$$= s_1 \prod_{j \neq 1} (m - s_j) - s_1' \prod_{j \neq 1} (m - s_j') = m(\prod_{j=2}^{l-1} (m - s_j))(m - s_l - m + s_l')$$

$$= m \prod_{j=2}^{l-1} (m - s_j) > 0$$

therefore $f(s_1, s_2, \ldots, s_n)$ is not a minimum. Now we can assert that, if $f(s_1, s_2, \ldots, s_n)$ is the minimum of function f, and $s_n = 0$ then $s_3 = s_4 = \ldots = s_n = 0$. Moreover, given a set of integers $s_1, s_2, \ldots, s_n$ satisfying (3.15) and $s_3 = s_4 = \ldots = s_n = 0$, if $s_1 - s_2 \geq 2$ we can define the new set of integers

$$s_1' = s_1 - 1, \qquad s_2' = s_2 + 1$$
$$s_i' = s_i = 0 \quad \text{for} \quad i = 3, 4, \dots, n$$

and from Lemma 3, with $h = 1$ and $k = 2$ it is obtained that $f(s_1, s_2, \dots, n) - f(s_1', s_2', \dots, s_n') = (s_1 - s_2 - 1) > 0$. Therefore $f(s_1, s_2, \dots, s_n)$ is not a minimum. From the above considerations we obtain that the minimum of f is either

$$f(\bar{s}_1, \bar{s}_2, \dots, \bar{s}_n) = (ms - np(p+1))(m - p - 1)^{q-1}(m - p)^{n-q-1},$$

with the values $\bar{s}_1, \bar{s}_2, \dots, \bar{s}_n$ defined by (3.18), or $f(\hat{s}_1, \hat{s}_2, \dots, \hat{s}_n) = (ms - 2ab)m^{n-2}$, where the values $\hat{s}_1, \hat{s}_2, \dots, \hat{s}_n$ are defined by (3.22). The second part of the theorem follows straightforwardly from Lemma 4, and the proof is complete. $\square$

Corollary 1. *Let s, n, m, be positive integers, $n \geq 3$ and suppose that*

$$[s/2] \leq \frac{2m}{3} \tag{3.25}$$

Then the minimum of function f defined by (3.14) for integers $s_1, s_2, \dots, s_n$, satisfying (3.15) is K defined by (3.23) achieved with the values $s_i = \bar{s}_i$ defined by (3.18).

Proof. From Theorem 4 it follows that the minimum of function f is achieved with the values $\bar{s}_1, \bar{s}_2, \dots, \bar{s}_n$ defined by (3.18), or with the values $\hat{s}_1, \hat{s}_2, \dots, \hat{s}_n$ defined by (3.22). But if (3.25) is satisfied it follows that

$$\sum_{i=2}^{n-1} \frac{\hat{s}_i}{m - \hat{s}_i} = \frac{b}{m - b} = \frac{[s/2]}{m - [s/2]} \leq 2$$

is satisfied. Then, from Lemma 4, it follows that $f(\hat{s}_1, \hat{s}_2, \dots, \hat{s}_n)$ is not a minimum, therefore the minimum is K defined by (3.23) achieved with the values $s_i = \bar{s}_i$ defined by (3.18) and the proof is complete. $\square$

Now we can solve the OEIG. We will write

$$\{(i,j) : 1 \leq j \leq p+1, \quad \text{for} \quad i \leq q, \quad 1 \leq j \leq p, \quad \text{for} \quad i > q\}, \tag{3.26}$$

$$H = \frac{1}{m}(s - \frac{np(p+1)}{m})(1 - \frac{p+1}{m})^{q-1}(1 - \frac{p}{m})^{n-q-1} \tag{3.27}$$

where $p = [s/n]$ and $q = s - np$,

$$\{(i,j) : 1 \leq j \leq a, \quad \text{for} \quad i = 1, \quad 1 \leq j \leq b, \quad \text{for} \quad i = 2\} \tag{3.28}$$

$$H' = \frac{1}{m^2}(ms - 2ab) \tag{3.29}$$

where $b = [s/2]$ and $a = s - b$.

Theorem 5. *In the OEIG with $n \geq 3$ and $0 < s < 2m$, if*

$$[s/2] \leq \frac{2m}{3}$$

is satisfied, then an optimal strategy for player II is the uniform distribution on the set $\overline{B}_0$, where B_0 is the set defined by (3.26) and the value of the game is equal to H defined by (3.27). If, on the contrary,

$$[s/2] > \frac{2m}{3}, \tag{3.30}$$

then the value of the game is

$$v = \min \{H, H'\}$$

with H defined by (3.27) and H' defined by (3.29). If $v = H$ then an optimal strategy for player II is the uniform distribution on the set $\overline{B}_0$, where B_0 is the set defined by (3.26), if $v = H'$ then an optimal strategy for player II is the uniform distribution on the set $\overline{B}_0$, where B_0 is the set defined by (3.28).

Proof. As we know, the OEIG satisfies hypothesis of Theorem 2, therefore, an optimal strategy for player I is the uniform distribution on $X = \mathscr{F}$ and the value of the game is given by

$$v = \frac{1}{m^n} \min \sum_{A \in \mathscr{F}} M(A, B)$$

which in this case is written as

$$\frac{1}{m^n} \min f(s_1, s_2, \ldots, s_n) = \frac{1}{m^n} \min \sum_{i=1}^{n} s_i \prod_{j \neq i} (m - s_j).$$

If $[s/2] \leq \frac{2m}{3}$ is satisfied, then from Corollary 1 it follows that the minimum of f is equal to K defined by (3.23), therefore

$$v = H$$

and an optimal strategy for player II is the uniform distribution on the set $\overline{B}_0$, where B_0 is the set defined by (3.26).

If, on the contrary, (3.30) is satisfied, then, Theorem 4 proves that the minimum for f is the minimum of the values K, defined by (3.23), and K' defined by (3.24). If this minimum is K, then the value of the OEIG is equal to H defined by (3.27) and an optimal strategy for player II is the uniform distribution on the set $\overline{B}_0$, where B_0 is the set defined by (3.26). If this minimum is K', then the value of the OEIG is equal to H' defined by (3.29) and an optimal strategy for player II is the uniform distribution on the set $\overline{B}_0$, where B_0 is the set defined by (3.28). This proves the theorem. $\square$

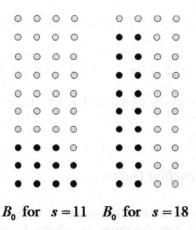

$$B_0 \ \text{ for } \ s = 11 \qquad B_0 \ \text{ for } \ s = 18$$

Fig. 3.5 Sets B_0 for $n = 4$, $m = 10$ and the two values of s, 11 and 18

Example 2. Let us consider the OEIG with $n = 4$ and $m = 10$. An optimal strategy for player I is the uniform distribution on the set of his pure strategies, $\mathscr{F}$. If $s = 18$, then (3.30) is satisfied, therefore the value of the game is equal to $v = \min\{H, H'\}$ where $H = \dfrac{K}{m^n}$ and $H' = \dfrac{K'}{m^n}$, K and K' defined by (3.23) and (3.24) respectively. In this case

$$K = 3,000, \quad K' = 1,800,$$

therefore $K' < K$, $H' < H$. Then the value of the game is

$$v = H' = \frac{1,800}{10^4} = 0.18$$

and an optimal strategy for player II is the uniformly concentrated distribution on $\bar{B}_0$, where B_0 is

$$B_0 = \{(i,j) : 1 \le j \le 9, \quad \text{for} \quad i = 1, 2\}.$$

A representation of this set can be seen in Fig. 3.5. If $s = 11$, then (3.30) is also satisfied. In this case

$$K = 1,274, \quad K' = 5,000,$$

therefore $K < K'$, $H < H'$. Then the value of the game is

$$v = H = \frac{1,274}{10^4} = 0.1274.$$

and an optimal strategy for player II is the uniformly concentrated distribution on $\overline{B}_0$, where B_0 is

$$B_0 = \{(i,j) : 1 \leq j \leq 3, \quad \text{for} \quad i = 1,2,3, \quad 1 \leq j \leq 2, \quad \text{for} \quad i = 4\}.$$

A representation of this set can be seen in Fig. 3.5. Figure 3.6 shows some of the elements of $\overline{B}_0$.

3.4.2 Weighted Inspection Game

An inspection game is a mathematical model of a situation where an inspector verifies that another part, the inspectee, adheres to certain rules. Typically, the inspector's resources are limited, so the verification can only be partial. The weighted inspection game (WIG) is a game (X,Y,M) on the lattice L satisfying

$$X = \mathscr{F}$$

$$Y = \{B : |B| = s\}$$

$$M(A,B) = \sum_{i=1}^{n} c_i \, |B \cap L_i| \, |A \cap B \cap L_i|$$

where c_i are constants such that

$$0 < c_1 \leq c_2 \leq \ldots \leq c_n. \tag{3.31}$$

Therefore, in this game the inspector, player I, makes a single inspection on each column of L (each column can represent a different day, zone, product, etc.), player II, the inspectee, chooses a subset of L of cardinality equal to s to hide one object at each one of its points. If player I finds one of the objects hidden by player II in column L_i, then he receives a quantity c_i for every object hidden by player II in this column. This game has been studied in [11], where the case $c_1 = c_2 = \ldots = c_n = c$ is completely solved.

Example 3. A farmer has 800 cows distributed in 8 cowsheds, $C_1, C_2, \ldots, C_8$, 100 in every cowshed. He has decided to administer an illegal substance to 80 of his cows. The public health agency will test one cow from each cowshed every month. If the illegal substance is detected, then the sanitary inspector will test all the cows of that cowshed and the farmer has to pay a fine of c_i times the number of positive results. The coefficient c_i depends on the village where the cowshed is allocated.

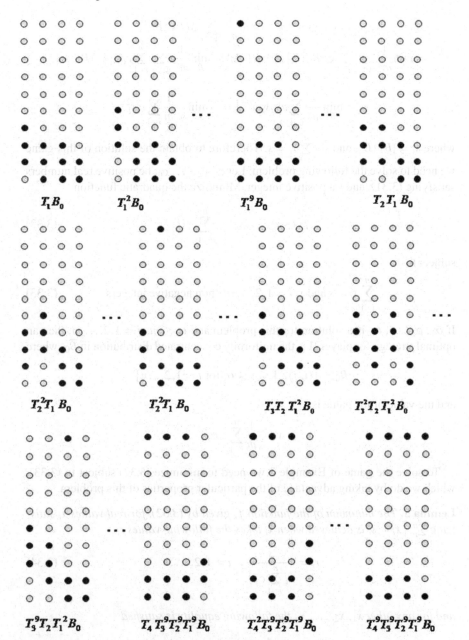

Fig. 3.6 Some elements of $\overline{B}_0$ for $n = 4$, $m = 10$ and $s = 11$

To solve this WIG we can apply Theorem 2. Hence an optimal strategy for player I is the uniform distribution on $\mathscr{F}$. To obtain an optimal strategy for player II we must determine B_0 in such a way that

$$M\left(x_{\mathscr{F}}, B_0\right) = \min_{B \in Y} M\left(x_{\mathscr{F}}, B\right) = \min_{B \in Y} \frac{1}{m^n} \sum_{A \in \mathscr{F}} M\left(A, B\right)$$

$$= \min_{B} \frac{1}{m^n} \sum_{A \in \mathscr{F}} \sum_{i=1}^{n} c_i \left|B \cap L_i\right| \left|A \cap B \cap L_i\right| = \min_{B} \frac{1}{m^n} \sum_{A \in \mathscr{F}} \sum_{i=1}^{n} c_i s_i \left|A \cap B \cap L_i\right|$$

$$= \min_{B} \frac{1}{m^n} \sum_{i=1}^{n} c_i s_i s_i m^{n-1} = \min_{s_1, s_2, \ldots, s_n} \frac{1}{m} \sum_{i=1}^{n} c_i s_i^2$$

where $s_i = \left|B \cap L_i\right|$, and so $\sum_{i=1}^{n} s_i = s$. Therefore to obtain the solution of this game we need to solve the following problem: Let $c_1, c_2, \ldots, c_n$, be positive real numbers satisfying (3.31), and s a positive integer. Minimize the quadratic function

$$f(x_1, x_2, \ldots, x_n) = \sum_{i=1}^{n} c_i x_i^2. \tag{3.32}$$

subject to

$$\sum_{i=1}^{n} x_i = s \text{ and } x_i \ i = 1, 2, \ldots, n \quad \text{non negative integers} \tag{3.33}$$

If $\alpha_1, \alpha_2, \ldots, \alpha_n$ is a solution for this problem and $\alpha_i \leq m$, $i = 1, 2, \ldots, n$, then an optimal strategy for player II is the uniformly concentrated distribution in $\bar{B}_0$, where

$$B_0 = \{(i, j) : 1 \leq j \leq \alpha_i \text{ for } i = 1, 2, \ldots, n\}$$

and the value of the game is

$$v = \frac{1}{m} \sum_{i=1}^{n} c_i \alpha_i^2.$$

To solve the game of Example 3 we need to minimize (3.32) subject to (3.33), which we do by taking advantage of the particular properties of this problem.

Lemma 5. *The minimum of the function f, given by (3.32), for real values x_i satisfying $\sum_{i=1}^{n} x_i = s$, is achieved when x_i takes the following values r_i*

$$r_i = \frac{s}{c_i \sum_{j=1}^{n} c_j^{-1}}, \quad i = 1, 2, \ldots, n \tag{3.34}$$

and for any other $x_1, x_2, \ldots, x_n$ the following equality is satisfied

$$\sum_{i=1}^{n} c_i x_i^2 = \sum_{i=1}^{n} c_i r_i^2 + \sum_{i=1}^{n} c_i (x_i - r_i)^2. \tag{3.35}$$

Proof. It is sufficient to prove (3.35) because from this it follows that the minimum condition is satisfied. Write

$$k = \frac{s}{\sum\limits_{j=1}^{n} c_j^{-1}}$$

then

$$c_i r_i = k, \ i = 1, 2, \ldots, n. \tag{3.36}$$

and from (3.36) and the relation $\sum\limits_{i=1}^{n} x_i = \sum\limits_{i=1}^{n} r_i = s$ we have

$$\sum_{i=1}^{n} c_i x_i^2 = \sum_{i=1}^{n} c_i (x_i - r_i)^2 + \sum_{i=1}^{n} c_i r_i^2$$

and the proof is complete. $\square$

If we denote $\bar{x} = (x_1, x_2, \ldots, x_n)$ and therefore $\bar{r} = (r_1, r_2, \ldots, r_n)$, equality (3.35) can be written briefly as

$$f(\bar{x}) = f(\bar{r}) + f(\bar{x} - \bar{r}). \tag{3.37}$$

From Lemma 5 we obtain the following considerations.

Remark 1. For any non-negative $\bar{x} = (x_1, x_2, \ldots, x_n)$ and $\bar{y} = (y_1, y_2, \ldots, y_n)$ such that $\sum\limits_{i=1}^{n} x_i = s$ and $\sum\limits_{i=1}^{n} y_i = s$, (integers or real), it follows that

(a) $\sum\limits_{i=1}^{n} c_i x_i^2 \leq \sum\limits_{i=1}^{n} c_i y_i^2 \Leftrightarrow \sum\limits_{i=1}^{n} c_i (x_i - r_i)^2 \leq \sum\limits_{i=1}^{n} c_i (y_i - r_i)^2,$

(b) $f(\bar{x}) - f(\bar{y}) = f(\bar{x} - \bar{r}) - f(\bar{y} - \bar{r}).$

Therefore, finding the point $\bar{\alpha} = (\alpha_1, \alpha_2, \ldots, \alpha_n)$ with α_i integers, which minimizes $f(\bar{s}) = f(s_1, s_2, \ldots, s_n)$ is equivalent to finding the point $\bar{\alpha}$ which minimizes $f(\bar{s} - \bar{r}) = f(s_1 - r_1, s_2 - r_2, \ldots, s_n - r_n)$ where $\bar{s} = (s_1, s_2, \ldots, s_n)$, s_i are integers, $\sum\limits_{i=1}^{n} s_i = s$, and $\bar{r} = (r_1, r_2, \ldots, r_n)$, the r_i given by (3.34).

Once the real numbers given by (3.34) have been obtained we determine the integers $a_1, a_2, \ldots, a_n$, such that

$$a_i - 1/2 \leq r_i < a_i + 1/2$$

that is to say

$$a_i = \begin{cases} [r_i] & \text{if } [r_i] \leq r_i \leq [r_i] + 1/2 \\ [r_i] + 1 & \text{if } [r_i] + 1/2 < r_i < [r_i] + 1 \end{cases} \tag{3.38}$$

where $[x]$ denotes the integer part of x. Now let us write

$$b_i = r_i - a_i \tag{3.39}$$

and

$$d = s - \sum_{i=1}^{n} a_i, \tag{3.40}$$

then

$$r_i = a_i + b_i, \quad a_i \text{ and } d \text{ integers},$$

$$|b_i| \le 1/2 \quad \text{and} \quad d = \sum_{i=1}^{n} b_i$$

will be satisfied.

Theorem 6. *If d, given by (3.40) is equal to 0, then the minimum of (3.32) subject to (3.33) is obtained with the values $\alpha_i = a_i$ for $i = 1,2,\ldots,n$. Moreover, if every one of the values b_i given by (3.39) satisfies $|b_i| < 1/2$ then there is no other point $\bar{s}$ in which the minimum is attained.*

Proof. In fact, for any other point $\bar{s} = (s_1, s_2, \ldots, s_n)$, s_i integers $s_i \ge 0$, for $i = 1,2,\ldots,n$, and $\sum_{i=1}^{n} s_i = s$, we can write

$$f(\bar{s}) - f(\bar{a}) = \sum_{i=1}^{n} c_i (s_i - r_i)^2 - \sum_{i=1}^{n} c_i (a_i - r_i)^2$$

$$= \sum_{i=1}^{n} c_i (s_i - a_i - b_i)^2 - \sum_{i=1}^{n} c_i b_i^2,$$

but at least one of the integers $s_i - a_i$ is different from zero, and for every one of the terms in which $s_i - a_i \ne 0$,

$$|s_i - a_i - b_i| \ge |s_i - a_i| - |b_i| \ge 1/2,$$

is satisfied, hence

$$(s_i - a_i - b_i)^2 \ge (1/2)^2 \ge b_i^2$$

and therefore

$$\sum_{i=1}^{n} c_i (s_i - a_i - b_i)^2 \ge \sum_{i=1}^{n} c_i b_i^2$$

which proves that $f(\bar{a})$ is the minimum. $\square$

Theorem 7. *If d defined by (3.40), is different from 0 let A and B be the sets*

$$A = \left\{ \bar{s} : s_i = a_i + \delta_i, \ i = 1,2,\ldots,n, \ \delta_i \text{ integers}, \right.$$

$$\left. \delta_i = 0 \text{ or } \operatorname{sgn} \delta_i = \operatorname{sgn} d, \ \sum_{i=1}^{n} \delta_i = d \right\},$$

where a_i, $i = 1,2,\ldots,n$, are defined by (3.38), and

$$B = \left\{ \bar{s} : s_i \text{ are integers}, s_i \geq 0, \sum_{i=1}^{n} s_i = s \right\}$$

Then $A \subset B$, and the following equality

$$\min_{\bar{s} \in B} f(\bar{s}) = \min_{\bar{s} \in A} f(\bar{s}) \tag{3.41}$$

is satisfied. That is to say, the minimum of (3.32) subject to (3.33) is obtained with an $\bar{s} \in A$.

Proof. The relation $A \subset B$ is clear. To proof equality (3.41) first let us suppose $d > 0$. Let $\bar{s} \in B$ be given, then $\sum_{i=1}^{n} s_i = s$ and $s_i = a_i + \delta_i$ is satisfied, where $\delta_i, i = 1, 2, \ldots, n$, are integers. If every δ_i different from zero has the same sign (that is $\delta_i \geq 0$ for $i = 1, 2, \ldots, n$) then

$$\sum_{i=1}^{n} s_i = \sum_{i=1}^{n} a_i + \sum_{i=1}^{n} \delta_i = s - d + \sum_{i=1}^{n} \delta_i$$

$$\Rightarrow \sum_{i=1}^{n} \delta_i = d$$

and so $\bar{s} \in A$. If, on the contrary, not all the δ_i have the same sign, let us call

$$H_1 = \{i : \delta_i > 0\}, \quad H_2 = \{i : \delta_i < 0\}.$$

We can write

$$\sum_{i=1}^{n} s_i = \sum_{i=1}^{n} a_i + \sum_{i \in H_1} \delta_i + \sum_{i \in H_2} \delta_i = s - d + \sum_{i \in H_1} \delta_i + \sum_{i \in H_2} \delta_i = s.$$

Then $\sum_{i \in H_1} \delta_i = d - \sum_{i \in H_2} \delta_i > d$, and we can select $K \subset H_1$ and δ_i' for $i \in K$ such that δ_i' are integers, $1 \leq \delta_i' \leq \delta_i$ and $\sum_{i=1}^{n} \delta_i' = d$. Now we can build $\bar{s}' \in A$ as follows

$$s_i' = a_i + \delta_i' \quad \text{for } i \in K \subset H_1,$$

$$s_i' = a_i \quad \text{for } i \notin K.$$

We have

$$f(\bar{s} - \bar{r}) = \sum_{i=1}^{n} c_i (s_i - r_i)^2 = \sum_{i \in K} c_i (s_i - r_i)^2 + \sum_{i \notin K} c_i (s_i - r_i)^2$$

$$\geq \sum_{i \in K} c_i (s_i - r_i)^2 + \sum_{i \notin K} c_i (a_i - r_i)^2$$

$$= \sum_{i \in K} c_i(a_i + \delta_i - r_i)^2 + \sum_{i \notin K} c_i(a_i - r_i)^2$$

and bearing in mind $1 \le \delta_i' \le \delta_i$ and $-1/2 \le r_i \le 1/2$ it follows that $1/2 \le \delta_i' - r_i \le \delta_i - r_i$, therefore

$$f(\bar{s} - \bar{r}) \ge \sum_{i \in K} c_i(a_i + \delta_i - r_i)^2 + \sum_{i \notin K} c_i(a_i - r_i)^2$$

$$\ge \sum_{i \in K} c_i(a_i + \delta_i' - r_i)^2 + \sum_{i \notin K} c_i(a_i - r_i)^2 = f(\bar{s}' - \bar{r}).$$

Then, the minimum cannot be achieved in $\bar{s}$. The same conclusion can be drawn for $d < 0$, which finishes the proof. $\square$

Remark 2. Now, it is not difficult to develop a program for the cases where $|d| = 0, 1, 2$, or 3. Note that $|d| \le [n/2]$, so this program will solve the problem whenever $n \le 7$. Moreover, when $n > 7$, it is more probable that $|d|$ takes values less than 3 (if coefficients c_i differ) because it is easy for the values b_i to have different signs. The program will, therefore, solve the problem in a wide variety of cases.

Example 4. Let us consider the optimization problem of minimizing function (3.32) subject to (3.33) with the following data

$$n = 10;$$

$$c_1 = 1, c_2 = 1, c_3 = 3, c_4 = 5, c_5 = 7,$$

$$c_6 = 7, c_7 = 8, c_8 = 11, c_9 = 15, c_{10} = 16.$$

With these data and changing the value of s, the following values for d are obtained. We will write the pairs (s, d):

$$
\begin{array}{llll}
(10, 2) & (20, 1) & (30, 1) & (40, -2) \\
(50, 1) & (60, 0) & (70, 2) & (80, 0) \\
(90, 0) & (100, -2) & (110, 0) & (120, 0) \\
(130, -1) & (140, 0) & (150, 1) & (160, 0)
\end{array}
$$

As can be seen $|d| < 3$ is satisfied in all the cases. Observe also that the values 0, 1, -1, 2 and -2 appear with high irregularity.

Remark 3. To apply the solution of the optimization problem to solve the *WIG* it is necessary that $\alpha_i \le m$ for $i = 1, 2, \ldots, n$. It is clear that this condition will be satisfied if $m \ge a_i + d$ for $i = 1, 2, \ldots, n$, where a_i and d are given by (3.38) and (3.40) respectively.

Now we can solve some further examples.

Example 5. Let us consider the situation described in Example 3 with the following data: $s = 80$; a fine that must be paid for every positive result in cowshed C_i given by $c_1 = 2$, $c_2 = 2.5$, $c_3 = 3$, $c_4 = 2.3$, $c_5 = 1.5$, $c_6 = 1.8$, $c_7 = 2.5$, $c_8 = 2$. With these parameters, we obtain the values r_i, a_i, b_i and α_i summarized in the following table:

r_i	a_i	b_i	α_i
10.55	11	−0.45	11
8.44	8	0.44	8
7.04	7	0.04	7
9.18	9	0.18	9
14.07	14	0.07	14
11.73	12	−0.27	12
8.44	8	0.44	8
10.55	11	−0.45	11
Total 80.00	80	$d = 0$	80

An optimal strategy for player I is the uniform distribution on $\mathscr{F}$, an optimal strategy for player II is the uniformly concentrated distribution in $\overline{B}_0$, where B_0 is a set with 14 elements in column 5, 12 elements in column 6, 9 elements in column 4, 11 elements in columns 1 and 8, 8 elements in columns 2 and 7 and 7 elements in column 3, as for example

$$B_0 = \{(1,1), (1,2), \ldots, (1,11), (2,1), (2,2), \ldots, (2,8), (3,1), (3,2), \ldots, (3,7),$$
$$(4,1), (4,2), \ldots (4,9), (5,1), (5,2), \ldots, (5,14), (6,1), (6,2), \ldots, (6,12),$$
$$(7,1), (7,2), \ldots, (7,8), (8,1), (8,2), \ldots, (8,11)\}$$

and the value of the game

$$v = \frac{1690.5}{100}.$$

Example 6. Let us consider the same situation with the following data: $s = 20$; a fine that must be paid for every positive result in cowshed C_i given by: $c_1 = 10$, $c_2 = 10$, $c_3 = 15$, $c_4 = 15$, $c_5 = 15$, $c_6 = 15$, $c_7 = 20$, $c_8 = 30$. With these parameters, we obtain the values r_i, a_i, b_i and α_i summarized in the following table:

r_i	a_i	b_i	α_i
3.63	4	−0.36	4
3.63	4	−0.36	4
2.42	2	0.42	3
2.42	2	0.42	2
2.42	2	0.42	2
2.42	2	0.42	2
1.82	2	−0.18	2
1.21	1	0.21	1
Total 20.00	19	$d = 1$	20

An optimal strategy for player I is the uniform distribution on $\mathscr{F}$, an optimal strategy for player II is the uniformly concentrated distribution in $\overline{B}_0$, where B_0 can be for example

$$B_0 = \{(1,1),(1,2),(1,3),(1,4),(2,1),(2,2),(2,3),(2,4),(3,1),(3,2),(3,3),$$
$$(4,1),(4,2),(5,1),(5,2),(6,1),(6,2)(7,1),(7,2),(8,1)\}$$

and the value of the game

$$v = \frac{745}{100}.$$

Example 7. A factory distributes its product to eight different areas. The factory sends 1,000 units of the product to each one of the eight distribution centers, $C_1, C_2,$ ..., C_8, once a week. In these centers the product is submitted to a quality control by the respective inspection services. The inspection service of each one of the centers C_i chooses one unit of the product out of the 1,000 in the series to inspect. If the unit selected does not meet the requirements, then the service inspects all the series, and the company has to pay a fine of c_i times the number of faulty units in this series.

The management of the factory knows that they produce s faulty units every week and wants to distribute these among the different areas, thus minimizing the fine. The number of faulty products, $s = 85$. The fine that must be paid to center C_i for every faulty product found is $c_1 = 10$, $c_2 = 10$, $c_3 = 15$, $c_4 = 15$, $c_5 = 15$, $c_6 = 15$, $c_7 = 20$, $c_8 = 30$. With these parameters we obtain the values r_i, a_i, b_i and α_i summarized in the following table:

	r_i	a_i	b_i	α_i
	15.45	15	0.45	16
	15.45	15	0.45	16
	10.30	10	0.30	10
	10.30	10	0.30	10
	10.30	10	0.30	10
	10.30	10	0.30	10
	7.72	8	−0.27	8
	5.15	5	0.15	5
Total	85.00	83	$d = 2$	85

An optimal strategy for player I is the same that in the previous case, an optimal strategy for player II is the uniformly concentrated distribution in $\overline{B}_0$, where B_0 can be

$$B_0 = \{(1,1),(1,2),\ldots,(1,16),(2,1),(2,2),\ldots,(2,16),(3,1),(3,2),\ldots,(3,10),$$
$$(4,1),(4,2),\ldots(4,10),(5,1),(5,2),\ldots,(5,10),(6,1),(6,2),\ldots,(6,10),$$
$$(7,1),(7,2),\ldots,(7,8),(8,1),(8,2),\ldots,(8,5)\}$$

and the value of the game is

$$v = \frac{13150}{1000}.$$

3.5 Conclusions and Open Problems

In this chapter we develop a method to facilitate the resolution of games on a finite set, and particularly games on the lattice L, satisfying certain invariance properties. Two games are studied following the proposed method, the OEIG and the WIG. Some seminal results for the first can be seen in [8], where a conjecture about the value of the game is made. Here the OEIG is solved in closed form, showing that the conjecture made in [8] was true. To solve the WIG we have to solve an interesting problem of minima and once this problem is solved, the solution for the WIG is straightforward. We develop a method to solve the problem of minima which can be easily implemented in a program. This program gives the solution when $n \leq 7$ and in a wide variety of cases for $n > 7$, but, unfortunately, we have not been able to solve this game in closed form. The complete solution for the WIG with $c_1 = c_2 = \ldots = c_n = c$ is obtained in [11].

When we deal with games on the lattice we find many other open problems. Let as consider the example in the introduction where a hacker tries to get information from the computers of a big enterprise, the game that model this situation is studied in [11] and, there, it is solved when some constraints on their parameters are satisfied, but the complete solution of the game is not obtained. The patrolling games on line graphs over the time studied in [3] are solved in some particular cases, but the general solutions have not been obtained. Other examples are the lattice games that have been studied in [7] and in [9]. None of them has been completely solved. We discuss these lattice games in Chap. 7 of the book.

References

1. Alpern, S., Asic M.: The search value of a network. Networks 15(2), 229–238 (1985)
2. Alpern, S. and Gal, S.: The theory of search games and rendezvous. Kluwer Academic Publishers, Boston, Dordretch, London (2003)
3. Alpern, S., Morton, A. and Papadaki, K.: Patrolling games. Oper. Res. 59, 1246–1257 (2011)
4. Gal, S.: Search games with mobile and immobile hider. SIAM J. Control Optim. 17 99–122 (1979)
5. Gal, S.: Search games. Academic Press, New York (1980)

6. Gal, S.: On the optimality of a simple strategy for searching graphs. Int. J. Game Theory 29 533–542 (2000)
7. Ruckle, W.: Geometric games and their applications. Pitman Advanced Publishing Program (1983)
8. Zoroa, N. and Zoroa, P.: Some games of search on a lattice. Naval Res. Logistics 40, 525–541 (1993)
9. Zoroa, N., Fernández-Sáez, M.J., Zoroa, P.: Patrolling a perimeter. Eur. J. Oper. Res. 222, 571–582 (2012)
10. Zoroa, N., Zoroa, P., Fernández-Sáez, M.J.: Raid games across a set with cyclic order. Eur. J. Oper. Res. 145, 684–692 (2003)
11. Zoroa, N., Zoroa, P., Fernández-Sáez, M.J.: Weighted search games. Eur. J. Oper. Res. 195, 394–411 (2009)

Chapter 4
Network Coloring and Colored Coin Games

Christos Pelekis and Moritz Schauer

Abstract Kearns et al. introduced the Graph Coloring Problem to model dynamic conflict resolution in social networks. Players, represented by the nodes of a graph, consecutively update their color from a fixed set of colors with the prospect of finally choosing a color that differs from all neighbors choices. The players only react on local information (the colors of their neighbors) and do not communicate. The reader might think of radio stations searching for transmission frequencies which are not subject to interference from other stations. While Kearns et al. (see [10]) empirically examined how human players deal with such a situation, Chaudury et al. performed a theoretical study and showed that, under a simple, greedy and selfish strategy, the players find a proper coloring of the graph within time $O\left(\log\left(\frac{n}{\delta}\right)\right)$ with probability $\geq 1 - \delta$, where n is the number of nodes in the network and δ is arbitrarily small. In other words, the graph is properly colored within τ steps and $\tau < c\log\left(\frac{n}{\delta}\right)$ with high probability for some constant c. Previous estimates on the constant c are very large. In this chapter we substantially improve the analysis and upper time bound for the proper coloring, by combining ideas from search games and probability theory.

4.1 Notation and Definitions

The network coloring game is a stochastic process evolving on a graph, $G = (V, E)$, on n vertices and maximum degree Δ. Each vertex is thought of as a player that has k available colors. Each player has the same set of colors. As in [2] we assume that $k \geq \Delta + 2$. The game is played in rounds and in each round all players simultaneously and individually choose a color. They can only observe the colors chosen by their

C. Pelekis (✉) • M. Schauer
Institute of Applied Mathematics, Delft University, P.O. Box 5031,
2600 GA, Delft, The Netherlands
e-mail: C.Pelekis@tudelft.nl; m.r.schauer@tudelft.nl

S. Alpern et al. (eds.), *Search Theory: A Game Theoretic Perspective*,
DOI 10.1007/978-1-4614-6825-7_4, © Springer Science+Business Media New York 2013

neighbors. We say that a player is 'happy' if she chooses a color that is different from the colors of her neighbors, otherwise she is 'unhappy'. The game reaches a proper coloring when all players are happy.

We assume that once a player is happy, she chooses the same color in the next round. Knowing this, players will never choose a color that has been used by a neighbor in the previous round. Therefore, once a player is happy, she continues to be happy in all consecutive rounds by sticking to her color, i.e., happiness can only increase. Note that happy players are essentially removed from the game.

Suppose each player adopts the following strategy: if the player is happy she sticks to her color, if she is unhappy she changes her color and chooses equiprobably between the remaining colors that are not used by her neighbors. We call this the *simple* strategy. In [2] it is shown that under this strategy the expected number of unhappy players decays exponentially in each round. Note that the condition $k \geq \Delta + 2$ guarantees that for every unhappy player, there are always at least two colors that are not chosen by the neighbors.

For an individual player, $v \in V$, denote by τ_v the first round in which she is happy. The first round in which all players are happy, τ, is the maximum over all τ_v. In particular, the main result of Chaudhuri et al. says that

$$P\left[\tau \leq O\left(\log\left(\frac{n}{\delta}\right)\right)\right] \geq 1 - \delta, \tag{4.1}$$

for arbitrary small δ. It is remarkable that this estimate does not depend on the maximum degree of the network. The proof of this theorem depends on the following key lemma [2, p. 526]

Lemma 1 (Key Lemma). *There exists a constant c such that*

$$P[\tau_v \leq t + 2 | \tau_v > t] \geq c, \tag{4.2}$$

for every v.

It turns out that the constant c according to the estimates of Chaudhuri et al. is equal to $\frac{1}{1.050e^9}$. Notice that this estimate does not depend on Δ. Also notice that the estimate is over two rounds instead of one, which is because of a two-step approach to obtain the constant c.

The probability that an unhappy player v gets happy after the next round depends on two factors: the number of colors that v can choose from and the number of unhappy neighbors. Roughly, the proof of the Key Lemma is in two steps. The first step concerns the event that v, who is unhappy after round t, gets many available colors in round $t + 1$. The second step concerns the event that such a v that has many available colors gets happy in round $t + 2$. In both steps the probabilities are estimated by using Markov's inequality. Now, using the Key Lemma, the main result in [2] is proven by applying the so-called Bayes sequential formula and an union bound. We also use a two step approach, but we avoid the use of Markov's inequality (mean estimate), which is rather crude. Instead we use ideas from search games and bound the probabilities using median estimates, which give much sharper bounds and allow us to replace the constant c in the Key Lemma by $\frac{1}{2^9}$.

Our approach differs from that of [2] in the following way. First we consider a new search game that turns out to be of use to estimate the constant c in the Key Lemma. To the best of our knowledge, this is the first attempt to use search games in the context of graph coloring. We find that the optimal strategy of the searchers involves tossing colored coins. This leads to a combinatorial probability problem, which we solve in the final section. We use the solution of this problem to obtain a better estimate in the first step in the proof of the Key Lemma. Then we apply the arithmetic-geometric mean inequality to obtain a better estimate of the second step in the proof of the Key Lemma. Finally, we apply results on maximally dependent random variables to show that the global time to equilibrium, τ, is stochastically dominated by an exponential random variable.

There exists a vast literature on graph coloring algorithms. Some related work is in [13], where a graph coloring is provided in $O(\log n)$ rounds via a distributed algorithm which uses $\Delta + 1$ colors, or more, but requires that the neighbors have information on the status of a vertex. Attempts to properly color a graph via strategic games can be found in [5, 14]. Another line of research in which games are used to model conflict situations that are similar to those that are modeled by network coloring is in [1]. For a general discussion on the network coloring game see [3].

4.2 A Related Search Game

In order to estimate the first time player v is happy, τ_v, we define the following Search Game. A player, H, the hider, chooses an element from $\Omega = \{1, 2, \ldots, d\}$ with $d \geq 2$. So the strategy space of H is the set Ω. The opponent of H consists of a team of $m \geq d$ searchers (agents) that each choose a subset Ω_j containing *at least two* colors from Ω. We denote these searchers by $S_j, 1 \leq j \leq m$. Subsequently, each searcher draws a color ω_j uniformly randomly from his own Ω_j. The searchers may communicate their choice of Ω_j. If H has chosen a color that is different from all ω_j he wins, otherwise he looses. This is a finite, one round zero-sum game that has a value, which is the probability that H wins under optimal play on both sides.

Lemma 2. *The optimal strategy for H is to choose his color uniformly randomly.*

Proof. This is a standard invariance argument (see [6], page 24). The game is invariant under the group, $\mathscr{S}_d$, of permutations. To see this, let $\pi(\ell, \Omega_1, \Omega_2, \ldots, \Omega_m)$ be the payoff to H (i.e. his winning probability) provided that H has chosen ℓ and S_j has chosen $\Omega_j, j = 1, 2, \ldots, m$. Then, for any $\sigma \in \mathscr{S}_d$ we have that

$$\pi(\ell, \Omega_1, \Omega_2, \ldots, \Omega_m) = \pi(\sigma(\ell), \sigma(\Omega_1), \sigma(\Omega_2), \ldots, \sigma(\Omega_m)).$$

As the game is invariant under the group $\mathscr{S}_d$, there exist invariant optimal strategies for the players. Since for any two $\ell_1, \ell_2 \in \{1, 2, \ldots, d\}$ there exists a permutation σ that maps ℓ_1 to ℓ_2, a mixed strategy for H is invariant if it assigns the same probability to all elements of Ω. $\square$

The value of the game equals the expected proportion of the number of colors chosen by the searchers.

Lemma 3. *There exists an optimal pure strategy in which all searchers use a doubleton.*

Proof. Any searcher, S_j, picks a color uniformly randomly from his own Ω_j, i.e. with probability $\frac{1}{|\Omega_j|}$. It is equivalent to first pick a doubleton from Ω_j uniformly randomly and then equiprobably choose one of the two colors from that doubleton. This means that every pure strategy of S_j is equivalent to a mixed strategy on (some) doubletons. Now we prove that it is optimal for each searcher to choose one doubleton. Since the game is finite, there exists an optimal mixed strategy for the searchers which can be described by a probability distribution on doubletons (pure strategies). Fix some searcher, say S_1, and suppose that he chooses a collection of doubletons, $D_1, D_2, \ldots, D_k$ with probabilities $p_1, p_2, \ldots, p_k$ that add up to 1. Let, P, denote the winning probability of the searchers. Then $P = \sum p_i P_i$, where P_i denotes the probability that the searchers win, given that S_1 chooses D_i and the other searchers do not change their strategy. Choose an i_0 for which $P_{i_0} = \max_i P_i$. Then $P_{i_0} \geq \sum p_i P_i$. This means that there is a doubleton such that if it is chosen by S_1, the expected payoff does not decrease, provided that the rest of the searchers do not change their strategy. $\qquad\square$

Theorem 1. *If $2m = ad + b$, for integers a and $0 < b < d$, then the value of the game equals $\frac{2d-b}{2^{a+1}d}$.*

Proof. Clearly, it is optimal for the searchers to use coins that contain every color at least once. Let Z be the set of colors chosen by the searchers after flipping their coins, let $X_{d,m} = |Z|$. That is, $X_{d,m}$ is the number of different colors after a toss. The value of the game is equal to the expected proportion of the complement of Z, $\frac{E[|Z^c|]}{d} = 1 - \frac{E[X_{d,m}]}{d}$. Fix some strategy, s, of the searchers, let G_s be the set of colors corresponding to this strategy and let C_i be the event that color i is chosen by the searchers after they toss their coins. Note that $|G_s| = d$. Then $E[X_{d,m}] = \sum_{i \in G_s} P[C_i] = \sum_{i \in G_s} (1 - (\frac{1}{2})^{c(i)})$, where $c(i)$ is the number of times that color i appears on a coin. The searchers seek to minimize the sum $\sum_{i \in G_s} (\frac{1}{2})^{c(i)}$ under the constraint $\sum_i c(i) = 2m$. Note that whenever $l - j \geq 2$ then $(\frac{1}{2})^l + (\frac{1}{2})^j \geq (\frac{1}{2})^{l-1} + (\frac{1}{2})^{j+1}$. Iteration of this inequality shows that the minimum is achieved by choosing G_s such that all $c(i), i \in G_s$, are as equal as possible, i.e. b of them equal to $a + 1$ and the remaining $d - b$ equal to a. Then we get $\sum_{i \in G_s} (\frac{1}{2})^{c(i)} = \frac{b}{2^{a+1}} + \frac{d-b}{2^a} = \frac{2d-b}{2^{a+1}}$. $\qquad\square$

4.3 Maximizing the Median

Picking an element from a doubleton is just flipping a coin and so the searchers are using d colors to create m coins that do not use the same color on both sides. Note that for each array of coins used by the searchers, one can draw a *graph* whose vertices correspond to the colors and whose edges correspond to the coins.

More explicitly, for each color put a vertex in the graph and join two vertices if and only if they are sides of the same coin. Note that the graph is loop-less and that it might have parallel edges, because the same coin may occur more than one time. In addition, note that the graph may not be connected and that there is a one-to-one correspondence between array of coins and graphs and so one can choose not to distinguish between vertices and colors as well as between coins and edges. We call this graph the *dependency* graph of the set of coins.

Notice that in case $m = d$, the searchers strategy $\{1,2\}\{2,3\}\ldots\{d-1,d\}\{d,1\}$ corresponds to the cycle-graph on d vertices. The proof of Theorem 1 actually says that if the searchers want to maximize the mean of X_d, the number of different colors after a toss, then they have to choose coins in such a way that the corresponding graph is a cycle or a union of cycles. But what if the searchers want to maximize the *median* of X_d? By median of a random variable, X, we mean any number μ satisfying $P[X \geq \mu] \geq 1/2$ and $P[X \leq \mu] \geq 1/2$. Notice that this μ might not be unique. It turns out that the following theorem is true.

Theorem 2. *The median of X_d is $\leq \frac{3d+2}{4}$.*

The proof of this Theorem is involved and builds on ideas from combinatorial probability. We prove this theorem in the final section. Having this result, we are then able to improve on the constant of the Key Lemma. This is the content of the following section.

4.4 Back to Network Coloring

4.4.1 Probability of Individual Happiness

The lemma below improves on Lemma 4, from [2].

Lemma 4. *Consider a single player, i.e., a vertex v in the network game at a given round, t, and suppose that v is unhappy. Let Y be the set of available colors to v in the next round, $t + 1$, and let f be the number of happy neighbors of v in the next round, $t + 1$. Then*

$$P\left[|Y| \geq \frac{k-f-2}{4}\right] \geq \frac{1}{2}.$$

Proof. Let h be the number of happy neighbors of v at the start of round $t+1$. Let θ be the degree of v. Then only $\theta - h$ unhappy neighbors are active in the game. Let I be the set of colors that are not used by the happy neighbors. Then I contains $k - h$ elements. In the worst case there are $\Delta - h \leq k - h$ unhappy neighbors all choosing a color from I. That is, the neighbors are searchers in a game as the one of the previous section. We may even add more searchers and suppose that the number of unhappy neighbors in $k - h$. If Z is the set of colors chosen by the searchers, then we have that with probability $\geq \frac{1}{2}$, the cardinality of Z is less than $\frac{3(k-h)+2}{4}$, by Theorem 2. That is, with probability more than $\frac{1}{2}$ we have that $|Y| \geq \frac{k-h-2}{4} \geq \frac{k-f-2}{4}$, since the number of happy players can only increase. □

Recall that τ_v is the number of rounds needed for player v to become happy in the Network Coloring Game.

Lemma 5. *For every player, v, in the Network Coloring Game we have that*

$$P[\tau_v \leq t+2 | \tau_v > t] \geq \frac{1}{2^9}.$$

Proof. Suppose that v is unhappy after round t has been played. Let Y be the set of available colors to v after round $t+1$ and f the number of happy neighbors after this round. So v is choosing a color with probability $\frac{1}{|Y|}$. Suppose that U is the set of unhappy neighbors of v after round $t+1$. Thus $|U| \leq k-f-2$. For each $u \in U$, let $p_u(i)$ be the probability with which player u chooses color i. Define also Y_u to be the set of available colors to each $u \in U$. From the previous lemma we know that with probability more than $\frac{1}{2}$ the cardinality of Y is more than $\frac{k-f-2}{4}$. The probability that a fixed color $i \in Y$ is not chosen by the neighbors is

$$\prod_{\{u \in U : i \in Y_u\}} (1 - p_u(i)).$$

Thus the probability P_v that v is happy in the next round equals

$$P_v = \frac{1}{|Y|} \sum_{i \in Y} \prod_{u \in U : i \in Y_u} (1 - p_u(i)) \geq \left(\prod_{i \in Y} \prod_{u \in U : i \in Y_u} (1 - p_u(i)) \right)^{\frac{1}{|Y|}},$$

by the arithmetic-geometric mean inequality. For each player in $u \in U$ that has i as a choice we have that $1 - p_u(i)$ equals to $1 - \frac{1}{\ell}$, for some $\ell \geq 2$. If i is not a choice of $u \in U$, then $p_u(i) = 0$. Thus $1 - p_u(i) = 1 - \frac{1}{|Y_u|} \geq \frac{1}{2}$ for every i and so

$$\left(\prod_{i \in Y} \prod_{u \in U : i \in Y_u} (1 - p_u(i)) \right)^{\frac{1}{|Y|}} \geq \left(\prod_{u \in U} \prod_{i \in Y_u} (1 - p_u(i)) \right)^{\frac{1}{|Y|}}$$

$$\geq \left(\prod_{u \in U} \left(1 - \frac{1}{|Y_u|} \right)^{|Y_u|} \right)^{\frac{1}{|Y|}}$$

$$\geq \frac{1}{4^{|U|/|Y|}},$$

since $|Y_u| \geq 2$. Now on the event $|Y| \geq \frac{k-f-2}{4}$, and since $|U| \leq k-f-2$, we find $\frac{1}{4^{|U|/|Y|}} \geq \frac{1}{4^4} = \frac{1}{2^8}$. The result follows by noticing that $P[\tau_v \leq t+2 | \tau_v > t]$ is at least

$$P\left[\tau_v \leq t+2 | \tau_v > t, |Y| \geq \frac{k-f-2}{4} \right] \cdot P\left[|Y| \geq \frac{k-f-2}{4} | \tau_v > t \right].$$

□

Our lower bound of $\frac{1}{2^9}$ improves on the lower bound of $\frac{1}{1.050e^9}$ that is derived in [2]. In the next subsection we use this lower bound to estimate the expected time to global happiness.

4.4.2 Time to Global Happiness

So far, we have obtained a bound on the time τ_v of an individual player. Now we want to obtain a bound on the global time to happiness $\tau = \max_v \tau_v$. Unfortunately, we know nothing about the dependence structure between the τ_v, so the estimate on $\max_v \tau_v$ has to be a worst case estimate. It turns out that this worst case estimate is covered by the case of maximally dependent random variables. This is a notion that comes up in the study of stochastic order relations.

Recall that a random variable, X, is said to be stochastically smaller than another random variable, Y, if $P[X > t] \leq P[Y > t]$, for all t. Denote this as $X \leq_{st} Y$. It is known (see [15], Theorem 1.A.1) that $X \leq_{st} Y$ if and only if there exist two random variables $\hat{X}, \hat{Y}$ such that $\hat{X} \sim X$, $\hat{Y} \sim Y$ and $\hat{X} \leq \hat{Y}$ with probability 1. This will apply in our case because we will show that τ_v is stochastically smaller than S_v, where $S_v \sim 2 \cdot Exp(\lambda)$ and $\lambda := -\log(1 - \frac{1}{2^9})$. In that case $\max_v \hat{\tau}_v \leq \max_v \hat{S}_v$ with probability 1 and $\tau \sim \max_v \hat{\tau}_v$.

To see that $\tau_v \leq_{st} S_v$, note that the estimate of the previous subsection shows that $P[\tau_v > t + 2 | \tau_v > t] \leq 1 - \frac{1}{2^9}$. Notice also that, for every player v,

$$P[\tau_v > 1 | \tau_v > 0] = 1 - (1 - \frac{1}{k})^{\deg(v)} \leq 1 - (1 - \frac{1}{k})^{k-1} \leq 1 - \frac{1}{e} \leq 1 - \frac{1}{2^9}.$$

Hence, if t is odd,

$$P[\tau_v > t] = P[\tau_v > 1 | \tau_v > 0] \cdot P[\tau_v > 3 | \tau_v > 1] \cdots P[\tau_v > t | \tau_v > t - 2]$$
$$\leq \left(1 - \frac{1}{2^9}\right)^{t/2}$$
$$= P[Exp(\lambda) > \frac{t}{2}]$$
$$= P[2 \cdot Exp(\lambda) > t],$$

and similarly if t is even.

Thus $\tau_v \leq_{st} S_v$ and thus $\max_v \hat{\tau}_v \leq \max_v \hat{S}_v$ with probability 1. Define $M_n := \max_v \hat{S}_v = 2\max_v X_v$, where $X_v \sim Exp(\lambda)$. Since $\tau \sim \max_v \hat{\tau}_v$ and $\max_v \hat{\tau}_v \leq M_n$ with probability 1, we have that $E[\tau] \leq E[M_n]$.

Thus, in order to estimate $E[\tau]$, it is enough to estimate the maximum possible value of $E[M_n] = 2E[\mu_n]$, where μ_n is the maximum of n (dependent) $Exp(\lambda)$ random variables. Such ensemble maxima occur often in practical problems and have been well studied both in the independent and the dependent case (see [4, 11, 12]).

We estimate $E[\mu_n]$ using ideas from [11]. Let F be the distribution function of $X_v, v \in V$. For any real number t, we have that

$$\mu_n \le t + \sum_v (X_v - t)^+,$$

which gives that $E[\mu_n] \le h(t) := t + n \int_t^\infty [1 - F(x)]\, dx$, for any $t \in R$. Differentiating $h(\cdot)$ one finds that its minimum is at $t_n := F^{-1}(1 - \frac{1}{n})$ and so $E[\mu_n] \le t_n + n \int_{t_n}^\infty [1 - F(x)]\, dx$. Since $1 - F(x) = e^{-\lambda x}$ it follows that $E[\mu_n] \le \frac{1}{\lambda}(1 + \log n)$. Hence

$$E[\tau] \le 2 \cdot E[\mu_n] \le \frac{2}{\lambda}(1 + \log n).$$

We summarize the preceding results into a Theorem which is an improvement of the Main Theorem from [2].

Theorem 3. *Let G be a graph on n vertices and maximum degree Δ. If the number of available colors is at least $\Delta + 2$ and if all players adopt the simple strategy, then for any starting assignment of colors, the network coloring game reaches a proper coloring at time τ that is stochastically smaller than a random variable T, such that $E[T] \le \frac{2}{\lambda}(1 + \log n)$, where $\frac{2}{\lambda} \approx 1,023$.*

4.5 Proof of Theorem 2

This section is devoted to the proof of Theorem 2. We want to show that the median of X_d is $\le \frac{3d+2}{4}$, where X_d is the number of different colors after a toss of d coins that are colored using d colors. Before proving this theorem we need some notation and remarks.

Suppose that we have d coins that are colored with d colors. Let G be the dependency graph corresponding to this set of coins. We are going to orient G as follows. Toss all the coins and orient each edge towards the vertex (color) that came up in the toss. Thus a toss of the coins gives rise to an orientation on the edges of G. As a consequence, $X_d = j$ corresponds to the fact that j vertices have positive in-degree, which means that $d - j$ vertices must have in-degree 0. Also note that none of the vertices of zero in-degree can be adjacent.

We denote the in-degree of a vertex v by $\deg^-(v)$ and by Z_d the number of vertices of zero in-degree. Thus $X_d = d - Z_d$.

It turns out that the median of X_d can be estimated through the median of E_d, the number of even in-degree vertices, whose distribution is easier to determine. We will need the following two graph-theoretic results.

Lemma 6. *Suppose that G is a (possibly disconnected) graph on d vertices and m edges. Fix some orientation on the edges and let $O_{d,m}, E_{d,m}$ be the number of odd and even in-degree vertices respectively. Then the parity of $E_{d,m}$ equals the parity of $m - d$.*

Proof. The in-degree sum formula states that

$$\sum_{v \in G} \deg^-(v) = m.$$

From this we have that the parity of $O_{d,m}$ equals the parity of m. Note that $d - E_{d,m} = O_{d,m}$. Hence the parity of m equals the parity of $d - E_{d,m}$ and the lemma follows. $\square$

For any real number r, we denote $r^+ = \max\{r, 0\}$.

Lemma 7. *For every oriented graph on d vertices and m edges,*

$$-Z_d + \sum_v (\deg^-(v) - 1)^+ = m - d.$$

Proof. We use again the in-degree sum formula, $\sum_v \deg^-(v) = m$. Thus $\sum_v (\deg^- (v) - 1) = m - d$ and so $-Z_d + \sum_v (\deg^-(v) - 1)^+ = m - d$, since the sum contributes a -1 for every vertex of in-degree zero. $\square$

We denote by $\mathrm{Med}(Y)$ the median of the random variable Y.

Lemma 8. *If $\mathrm{Med}(E_d) \geq \frac{d-2}{2}$, for any graph on d vertices and d edges, then Theorem 2 holds true.*

Proof. Let $Y_d := E_d - Z_d$, then Lemma 7 gives that $Z_d = \sum_v (\deg^-(v) - 1)^+$, since $m = d$. Note that

$$\sum_v (\deg^-(v) - 1)^+ \geq \sum_{\{v : \deg^-(v) \geq 2\}} (\deg^-(v) - 1)^+ \geq \sum_{\{v : \deg^-(v) \geq 2\}} 1 \geq Y_d.$$

Since $Y_d + Z_d = E_d$, it follows that $Z_d \geq \frac{1}{2} E_d$. Now $X_d + Z_d = d$ so that $X_d = d - Z_d \leq d - \frac{1}{2} E_d$ and $\mathrm{Med}(X_d) \leq d - \frac{d-2}{4} = \frac{3d+2}{4}$. $\square$

So it remains to prove that $\mathrm{Med}(E_d) \geq \frac{d-2}{2}$. To prove this, we first compute the distribution of the number of even in-degree vertices in the case of a *connected* graph on d vertices and $m \geq d - 1$ edges. We then extend this computation to the general case by considering the connected components of the graph.

We denote by $\mathrm{Bin}(s, p)$ a Binomially distributed random variable of parameters s and p. In case $p = \frac{1}{2}$ we just write $\mathrm{Bin}(s)$. The parity of the in-degree of each particular vertex is related to the parity of the Binomial distribution for which the following is well known.

Lemma 9. *Suppose that $X_s := \mathrm{Bin}(s) \bmod 2$. Then X_s is a $\mathrm{Bin}(1)$ random variable regardless of s.*

Proof. The proof is by induction on s. When $s = 1$ the conclusion is true. Suppose that it is true for all integers up to $s - 1$ and consider X_s. Observe that $X_s \sim X_{s-1} + \mathrm{Bin}(1)$, mod 2. The induction hypothesis gives that $X_{s-1} + \mathrm{Bin}(1)$ equals $\mathrm{Bin}(1) + \mathrm{Bin}(1) \bmod 2$, for two independent $\mathrm{Bin}(1)$ random variables which finishes the proof of the lemma. $\square$

The next lemma is also well known. We include a proof for the sake of completeness.

Lemma 10. *A median of a Bin(s) random variable is its mean.*

Proof. Let $X \sim \text{Bin}(s)$. Then X and $s - X$ are identically distributed. Thus, for any t,

$$P[X \geq t] = P[s - X \geq t] = P[X \leq s - t].$$

Now apply this with $t = \frac{s}{2}$ to get the result.   □

Lemma 11. *Fix some vertex v of the graph. Let C be any set of edges (coins) that does not contain some edge incident to v. Then the parity of $\deg^-(v)$ is independent of the orientation of the edges in C.*

Proof. Suppose the coins corresponding to C have been flipped. Let C^- be the number of edges in C which are oriented towards v after the toss. By the previous lemma, C^- is even or odd with probability $\frac{1}{2}$. Since there is at least one edge incident to v that does not belong to C, we have that

$$\mathbb{P}[\deg^-(v) \text{ is even}|C^-] = \frac{1}{2}\mathbb{P}[\deg^-(v) \text{ is even}|C^- \text{ is odd}]$$

$$+ \frac{1}{2}\mathbb{P}[\deg^-(v) \text{ is even}|C^- \text{ is even}] = \frac{1}{2}.$$

So this conditional probability does not depend on C^-. Similarly for the odd outcomes.   □

We will also need a special enumeration on the vertices and edges of a tree which, combined with the previous lemma, allows us to compute the distribution of the number of even in-degree vertices.

Lemma 12. *For any tree, T, on d vertices, there exists an enumeration, $v_1, v_2, \ldots, v_d$, of the vertices and an enumeration, $e_1, e_2, \ldots, e_{d-1}$, of the edges such that the only edge incident to vertex $v_i, i = 1, 2, \ldots, d - 1$, among the set of edges $\{e_i, e_{i+1}, \ldots, e_{d-1}\}$ is e_i.*

Proof. Fix a tree, T, on $d > 1$ vertices and choose any of its vertices. Call this vertex v_d. If v_d is a leaf, then consider the vertex set L of leaves in T except v_d and enumerate them $v_1, v_2, \ldots, v_\ell$. If v_d is not a leaf, then consider all leaves of T and enumerate them in the same manner. Note that L is not empty even if v_d is a leaf since any tree with at least two vertices has at least two leaves. Enumerate each edge incident to v_j by $e_j, j = 1, 2, \ldots, \ell$. Now consider the tree $T' := T \setminus \{v_1, v_2, \ldots, v_\ell\}$ and repeat this process on the leaves of T' again sparing v_d if it is a leaf of T'. We continue enumerating the leaves and edges of the subtrees until we end up with the graph consisting of vertex v_d only. It is evident that the enumeration satisfies the required condition.   □

We are now ready to compute the distribution of the number of even in-degree vertices for any connected graph.

Theorem 4. *Suppose that G is a connected graph on d vertices and $m \geq d - 1$ edges. Let $E_{d,m}$ be the number of even in-degree vertices after a random orientation on the edges. Then $E_{d,m}$ has the probability distribution of a Bin(d) random variable conditional on the event that the outcome of Bin(d) has the parity of $m - d$. To be more precise,*

$$\mathbb{P}[E_{d,m} = k] = \binom{d}{k} \frac{1}{2^{d-1}},$$

where k runs over the odd integers up to d, if $m - d$ is odd and over the even integers if $m - d$ is even.

Proof. Fix some spanning tree, T, of G and toss the coins corresponding to the edges that do not belong to T. Enumerate the vertices of the tree $v_1, v_2, \ldots, v_d$ and the edges $e_1, e_2, \ldots, e_{d-1}$ as in Lemma 12. Now toss the coins $e_1, e_2, \ldots, e_{d-1}$ in that order. The enumeration on the vertices and edges gives that once the coin e_j is flipped, then the parity of vertex v_j is determined. Lemma 11 gives that once the parity of some vertex v_j is determined, the parity of the next vertex v_{j+1} is independent of the parity of $v_1, v_2, \ldots, v_{j-1}$. Only the parity of v_d is deterministic given the parities of the previous vertices. Thus, if we set $\delta_i := \deg^-(v_i) \bmod 2$, for $i = 1, 2, \ldots, d$, we have that each δ_i is distributed as a Bin(1) random variable which, by independence, means that $\sum_{i=1}^{d-1} \delta_i \sim \text{Bin}(d-1)$. Let $O_{d,m}$ be the number of odd in-degree vertices. Then $O_{d,m} = \delta_1 + \cdots + \delta_{d-1} + \delta_d \sim X + \delta_d$, where $X \sim \text{Bin}(d-1)$ and δ_d depends on the outcome of X. From the relation $O_{d,m} + E_{d,m} = d$ and the fact that X is symmetric, i.e. $X \sim d - 1 - X$, we get that $E_{d,m} = d - X - \delta_d \sim X + 1 - \delta_d$. Suppose that $m - d$ is even. In case $m - d$ is odd, the argument is similar. Then $E_{d,m}$ is also even, by Lemma 6, and thus $1 - \delta_d$ equals 0, if X is even and equals 1, if X is odd. Hence, we have that $E_{d,m} = k$, for some even k, if and only if either $X = k$ or $X = k - 1$. This means that

$$\mathbb{P}[E_{d,m} = k] = P[\text{Bin}(d-1) = k] + P[\text{Bin}(d-1) = k - 1] = \binom{d}{k} \frac{1}{2^{d-1}}.$$

$\square$

For any positive integer, s, we write $W \sim \text{Bin}(s, \text{even})$ (resp. $\text{Bin}(s, \text{odd})$) whenever the random variable W is distributed as a Bin(s) random variable conditioned to be even (resp. odd). We will also write $\text{Bin}(s, \odot)$ whenever we don't want to specify the exact parity and refer to it as a *half-Binomial*.

Note that the proof of the last Theorem says that if we are interested in an outcome of, say, $\text{Bin}(s, \text{even})$ (resp. $\text{Bin}(s, \text{odd})$), we can toss $s - 1$ fair 0/1 coins and if the result is even, add a 0 (resp. a 1), if it is odd add 1 (resp. a 0). Call such a toss an *even-sum* (resp. *odd-sum*) toss of s coins.

We now consider the general case of a disconnected graph, G. Suppose that it consists of connected components, $G_1, G_2, \ldots, G_t$ each having d_i vertices and m_i

edges such that $\sum d_i = d$ and $\sum m_i = m$. Recall that we assume $d = m$. Let $E_i, 1 \leq i \leq t$ be the number of vertices of even in-degree in each graph after a toss. The E_i's are independent random variables and the total number of even in-degree vertices is given by $E = E_1 + \cdots + E_t$. Now, the distribution of each E_i is given by the previous theorem and thus E is the sum of independent $\mathrm{Bin}(d_i, \odot)$ random variables. Note that if these were pure Binomials instead of half-Binomials, then we would be done. In that case E would also be Binomial whose median is known. The problem is that we have a sum of independent half-Binomials and it is not immediately clear how to analyze a sum like $\mathrm{Bin}(7, \mathrm{odd}) + \mathrm{Bin}(6, \mathrm{even})$. We analyze such sums by breaking down each term of the sum, $\mathrm{Bin}(s, \odot)$, into a sum of $\mathrm{Bin}(2, \odot)$ and $\mathrm{Bin}(3, \odot)$. More specifically, $\mathrm{Bin}(s, \odot)$ will be a convex combination (mixture) of such sums. Recall that a mixture of random variables Z_i is defined as a random selection of one of the Z_i according to a probability distribution on the index set of i's. It is clear that if all these Z_i have a median that is $\geq \mu$, then also the mixture has a median $\geq \mu$.

Lemma 13. *For any $s \geq 2$, let $s = s_1 + s_2 + \cdots + s_l$ be a partition of s into $s_i = 2$ or $s_i = 3$, with at most one part equal to 3 in case s is odd. Then $\mathrm{Bin}(s, \odot)$ is a mixture of sums $\mathrm{Bin}(s_1, \odot) + \cdots + \mathrm{Bin}(s_l, \odot)$, where the parities of all these half-Binomials, $\mathrm{Bin}(s_i, \odot)$, add up to the given parity of $\mathrm{Bin}(s, \odot)$.*

Proof. Suppose we want to decompose a $\mathrm{Bin}(s, \mathrm{even})$ random variable. The other case is similar. We get an outcome of such a half-Binomial by tossing $s - 1$ independent coins and add a deterministic one to fix the parity, i.e., by tossing s even-sum $0/1$ fair coins. This is equivalent to partition s into $s_1, \ldots, s_l$, where all s_i are equal to 2, except possibly one that is equal to 3, and then toss l even-sum $0/1$ fair coins, assign the parity of the j-th coin, $j = 1, \ldots, l$, to s_j and then this parity to $\mathrm{Bin}(s_j, \odot)$. To be more precise, suppose that $Y_j \in \{\mathrm{even}, \mathrm{odd}\}$ is the parity of the j-th coin. Then for each $j = 1, \ldots, l$, toss s_j Y_j-sum coins to get an outcome from $\mathrm{Bin}(s_j, Y_j)$. Then the parity of $\sum_{j=1}^{l} Y_j$ is even and thus the independent sum $\sum_{j=1}^{l} \mathrm{Bin}(s_j, Y_j)$ has as even number of terms of the form $\mathrm{Bin}(s_j, \mathrm{odd})$ which means that it is an outcome from $\mathrm{Bin}(s, \mathrm{even})$.

To see that this is equivalent, notice that the probability of each particular outcome equals $\frac{1}{2^{l-1}} \cdot \frac{1}{2^{s_1-1}} \cdot \frac{1}{2^{s_2-1}} \cdots \frac{1}{2^{s_l-1}} = \frac{1}{2^{s-1}}$, which is exactly the probability of each particular outcome from $\mathrm{Bin}(s, \mathrm{even})$. So it remains to prove that the number of outcomes for which $\mathrm{Bin}(s, \mathrm{even}) = k$, for some even k, equals the number of outcomes for which $\sum_{i=1}^{l} \mathrm{Bin}(s_i, Y_i) = k$, given the parities $Y = (Y_1, \ldots, Y_l)$. But this is immediate. Every outcome of $\mathrm{Bin}(s, \mathrm{even})$, that is, every toss of s even-sum $0/1$ fair coins with k 1's gives rise to a vector of parities $Y = (Y_1, \ldots, Y_l)$ such that the parity of $\sum_{j=1}^{l} Y_j$ is even, $\sum_{i=1}^{l} \mathrm{Bin}(s_i, Y_i) = k$ and vice versa. □

If we apply the last Lemma to each $E_i \sim \mathrm{Bin}(d_i, \odot), i = 1, \ldots, t$ we get the following.

Corollary 1. *E is a mixture of sums of independent half-Binomials $\mathrm{Bin}(2, \odot)$ and $\mathrm{Bin}(3, \odot)$.*

The reason to partition each d_i into sums of 2's and at most one 3 is the following.

Lemma 14. $Bin(2, \odot)$ *and* $Bin(3, \odot)$ *can be interpreted as Binomials of biased coins. More precisely, they are distributed like the sum of a Binomial and a scalar.*

Proof. It is easy to check that $Bin(3, \text{odd}) \sim 1 + 2 \cdot Bin(1, \frac{1}{4})$ and $Bin(2, \text{odd}) \sim Bin(1, 1)$, as well as $Bin(3, \text{even}) \sim 2 \cdot Bin(1, \frac{3}{4})$ and $Bin(2, \text{even}) \sim 2 \cdot Bin(1, \frac{1}{2})$. $\square$

Corollary 2. *E has the distribution of a mixture of a sum of a scalar and a sum of independent Binomials.*

Having this corollary, we can then apply a well known result of Hoeffding [8].

Theorem 5 (Hoeffding). *If* $X_{p_1}, X_{p_2}, \ldots, X_{p_\ell}$ *are independent Bernoulli trials with parameters* $p_1, p_2, \ldots, p_\ell$ *respectively, then*

$$P[b \le \sum_{i=1}^{\ell} X_{p_i} \le c] \ge P[b \le Bin(\ell, \bar{p}) \le c], \text{ when } 0 \le b \le \ell \bar{p} \le c \le \ell,$$

where $\bar{p} = \frac{1}{\ell} \sum_{i=1}^{\ell} p_i$.

Recall that we are interested in a lower bound on the median of the independent sum $E \sim \sum_{i=1}^{t} E_i \sim \sum_{i=1}^{t} Bin(d_i, \odot)$. We know that E is a mixture of independent sums of $Bin(2, \odot)$ and $Bin(3, \odot)$, which are (rescaled) biased coins. We finish the proof of Theorem 2 by proving that every particular independent sum of this mixture has a median that is $\ge \frac{d-2}{2}$. Suppose that we have an independent sum, Ξ, consisting of $r, z, a, w \in \{0, 1, 2, \ldots\}$ terms from $Bin(3, \text{odd}), Bin(3, \text{even}), Bin(2, \text{even})$ and $Bin(2, \text{odd})$ respectively. Notice that $3r + 3z + 2a + 2w = d$.

Lemma 15. *A median of* Ξ *is* $\ge \frac{d-2}{2}$.

Proof. Suppose first that $z \ge r$. In that case we show that $\text{Med}(\Xi) \ge \frac{d-1}{2}$. Denote by Ψ the independent sum $Bin(r, \frac{1}{4}) + Bin(a, \frac{1}{2}) + Bin(z, \frac{3}{4})$. Then $\Psi = j$ if and only if $\Xi = r + 2j + w$. Thus a median of Ξ can be estimated through a median of Ψ and so a median of Ξ is $\ge \frac{d-1}{2}$ if and only if a median of Ψ is $\ge \frac{r+2a+3z-1}{4}$. We apply Hoeffding's result with $\bar{p} = \frac{1}{r+a+z}\left(\frac{r+2a+3z}{4}\right), \ell = r + a + z$ and $c = r + a + z, b = \frac{r+2a+3z-1}{4}$. This gives that

$$P[\Psi \ge \frac{r+2a+3z-1}{4}] \ge P[Bin(r+a+z, \bar{p}) \ge \frac{r+2a+3z-1}{4}]$$

$$\ge P[Bin(r+a+z, \bar{p}) \ge \frac{r+2a+3z}{4}].$$

Hence the lemma will follow once we prove that $P[Bin(r+a+z, \bar{p}) \ge \frac{r+2a+3z}{4}] \ge \frac{1}{2}$. Note that the mean of $Bin(r+a+z, \bar{p})$ equals $\frac{r+2a+3z}{4}$. Now, if $z \ge r$ then $\bar{p} \ge \frac{1}{2}$ and thus $Bin(r+a+z, \bar{p})$ is stochastically larger than $Bin(r+a+z, \frac{1}{2})$. This means that a median of $Bin(r+a+z, \bar{p})$ is bigger than or equal to a median of $Bin(r+a+z, \frac{1}{2})$.

But a median of $\text{Bin}(r+a+z,\frac{1}{2})$ is $\frac{r+a+z}{2}$, it's mean. Since $\frac{r+a+z}{2} \leq \frac{r+2a+3z}{4}$ when $z \geq r$, the result follows.

Suppose now that $z < r$. We consider two case.

(a) Assume that $r - z$ is even. In that case we prove again that $\text{Med}(\Xi) \geq \frac{d-1}{2}$. Define $\Phi_1 := \text{Bin}(r,\frac{1}{4}) + \text{Bin}(a,\frac{1}{2}) + \text{Bin}(z,\frac{3}{4}) + \frac{r-z}{2}$. Then $\text{Med}(\Xi) \geq \frac{d-1}{2}$ if and only if $\text{Med}(\Phi_1) \geq \frac{3r+2a+z}{4}$. By the result of Hoeffding we have that $\text{Med}(\Phi_1) \geq \text{Med}(\text{Bin}(\hat{n},\hat{p}))$, where $\hat{p} = \frac{1}{r+a+z+\frac{r-z}{2}}(\frac{r}{4}+\frac{a}{2}+\frac{3z}{4}+\frac{r-z}{2}) = \frac{1}{2}$ and $\hat{n} = r+a+z+\frac{r-z}{2}$. Since $\hat{p} = 1/2$, we get that a median of $\text{Bin}(\hat{n},\hat{p})$ is its mean which in turn equals $\hat{n}\cdot\hat{p} = \frac{3r+2a+z}{4}$.

(b) Assume that $r-z$ is odd. In a similar way as above we show that $\text{Med}(\Xi) \geq \frac{d-2}{2}$. Define $\Phi_2 := \text{Bin}(r,\frac{1}{4})+\text{Bin}(a,\frac{1}{2})+\text{Bin}(z,\frac{3}{4})+\frac{r-z-1}{2}$. Then $\text{Med}(\Xi) \geq \frac{d-2}{2}$ if and only if $\text{Med}(\Phi_2) \geq \frac{3r+2a+z-2}{4} - \frac{2}{4}$. Again, by Hoeffding, we conclude that $\text{Med}(\Phi_2) \geq \text{Med}(\text{Bin}(\check{n},\check{p}))$, where $\check{p} = \frac{1}{r+a+z+\frac{r-z-1}{2}}(\frac{r}{4}+\frac{a}{2}+\frac{3z}{4}+\frac{r-z-1}{2})$ and $\check{n} = r+a+z+\frac{r-z-1}{2}$. Now the mean of $\text{Bin}(\check{n},\check{p})$ equals $\check{n}\cdot\check{p} = \frac{3r+2a+z-2}{4}$. It is known (see [7]) that the smallest uniform (with respect to both parameters) distance of the mean and a median of a Binomial distribution is $\leq \ln 2 \approx 0.69 < \frac{3}{4}$. This means that if $\check{n}\cdot\check{p}$ equals $\mu+\frac{1}{4}$, for some integer μ, then μ is a median of $\text{Bin}(\check{n},\check{p})$. If $\check{n}\cdot\check{p}$ equals $\mu+\frac{3}{4}$, for some integer μ, then $\mu+1$ is a median of $\text{Bin}(\check{n},\check{p})$ and if $\check{n}\cdot\check{p} = \mu+\frac{1}{2}$, then a median of $\text{Bin}(\check{n},\check{p})$ is $\geq \mu$. If the mean, $\check{n}\cdot\check{p}$ is an integer, then it is well known (see [9]) that mean and median coincide. In all cases a median is $\geq \check{n}\cdot\check{p} - \frac{1}{2}$ and the result follows.

$\square$

4.6 Final Comment

We have improved the previous estimates on the time to global equilibrium in the network coloring game by combining search games and combinatorial probability. The colored coin tossing problem which we considered is interesting in its own right. One open problem that deserves further study is the following: suppose you can color n biassed coins with n colors, all coins having the same bias. It is forbidden to color both sides of a coin with the same color, but all other colors are allowed. Let X be the number of different colors after a toss of the coins. In what way should you color the coins such that you maximize the median of X?

References

1. S. Alpern, D.J. Reyniers: Spatial Dispersion as a Dynamic Coordination Problem, *Theory and Decision*, **53**, p. 29–59, (2002).
2. K. Chaudhuri, F. Chung Graham , M. Shoaib Jamall: A Network Coloring Game, WINE '08, p. 522–530, (2008).

3. F. Chung Graham: Graph Theory in the Information Age, *Notices of AMS*, **57**, no. 6, p. 726–732, (July 2010).
4. B. Eisenberg: On the Expectation of the Maximum of IID Geometric Random Variables, *Statistics & Probability Letters*, **78**, p. 135–143, (2008).
5. B. Escoffier, L. Gourvés, J. Monnot: Strategic Coloring of a Graph, *Algorithms and Complexity, Lecture Notes in Computer Science*, 6087, Springer, Berlin, p. 155–166 (2010)
6. T.S. Ferguson: *Game Theory*, part II.
7. K. Hamza: The Smallest Uniform Upper Bound on the Distance Between the Mean and the Median of the Binomial and Poisson Distributions, *Statistics & Probability Letters*, **23**, p. 21–25, (1995).
8. W. Hoeffding: On the Distribution of the Number of Successes in Independent Trials, *An. Math. Statistics*, **27**, p. 713–721, (1956).
9. R. Kaas and J.M. Buhrman: Mean, Median and Mode in Binomial Distributions, *Statistica Neerlandica* **34**(1), p. 13–18, (1980).
10. M. Kearns, S. Suri, N. Montfort: An Experimental Study of the Coloring Problem on Human Suject Networks, *Science*, **313** (5788), p. 824–827, (2006)
11. T. L. Lai and H. Robbins: Maximally Dependent Random Variables, *Proc.Nat.Acad.Sci.USA*, **73**, No. 2, p. 286–288, February 1976, Statistics.
12. T. L. Lai and H. Robbins: A Class of Dependent Random Variables and their Maxima, *Z. Wahrscheinlichkeitstheorie verw. Gebiete*, **42**, p. 89–111, (1978).
13. M. Luby: Removing Randomness in Parallel Computation without a Processor Penalty, *FOCS*, p. 162–173, (1988).
14. P.N. Panagopoulou and P. G. Spirakis: A Game Theoretic Approach for Efficient Graph Coloring, *ISAAC 2008*, LNCS 5369, p. 183–195, (2008).
15. M. Shaked and J.G. Shanthikumar: *Stochastic Orders and their Applications*. Springer, New York (2007).

Chapter 5
Open Problems on Search Games

Robbert Fokkink, Leonhard Geupel, and Kensaku Kikuta

Abstract We discuss two classic search games: Isaacs' princess and monster game, and Dresher's high-low guessing game. Despite the fact that these games were introduced decades ago, there are still numerous open problems around them.

5.1 Introduction

Rufus Isaacs' princess and monster game, which was discussed in the first chapter by Shmuel Gal, is a classic search game. One could argue that this game, and its solution by Gal in 1979, started Search Games as an independent area of research. Gal essentially showed that the princess has the upper hand. From time to time she quickly moves to a new position, nullifying any progress that the monster has made in his search, so the time of capture becomes an exponential random variable. One could take the area that has been searched by the monster as a state variable. Every time the princess moves, she resets the state variable to zero. In Chap. 16, Rob Arculus analyzes the princess and monster game from this point of view, and applies it to predator-prey models in biology. It seems plausible that even if the princess is noisy and the monster is aware of the fact that she has just moved, it does not help him find the princess any more quickly.

R. Fokkink (✉)
Department of Applied Mathematics, TU Delft, PO Box 5031, 2600 GA, Delft, The Netherlands
e-mail: r.j.fokkink@tudelft.nl

L. Geupel
Department of Mathematics, TU München, Boltzmannstr 3, D-85747, Garching, Germany
e-mail: leonhardgeupel@web.de

K. Kikuta
Department of Strategic Management, University of Hyogo, 8-2-1 Gakuen nishi-machi,
Nishi-ku, Kobe-shi, 651-2197, Japan
e-mail: kikuta@biz.u-hyogo.ac.jp

S. Alpern et al. (eds.), *Search Theory: A Game Theoretic Perspective*,
DOI 10.1007/978-1-4614-6825-7__5, © Springer Science+Business Media New York 2013

The original problem posed by Isaacs, was to solve the game for a relatively small radius of detection. If the radius of detection is larger, and has an order of magnitude that is comparable to the size of the search space, then it is more convenient to model the game on a graph. The searcher then finds the hider once they are in the same place. This version of the princess and monster game was also proposed by Isaacs [11, p 349-350]. The game on a graph has only been solved for the circle (see Sect. 1.2 of this book), but it remains open for all other graphs. Even if the graph is an interval, although there is a conjectured solution in this case, which we will discuss below.

The princess and monster game is related to the game between the lion and a Christian, which was invented by Richard Rado: a lion and a Christian in a closed circular arena have equal maximum speeds. Can the lion catch the Christian in finite time? Surprisingly, the answer turns out to be NO [3, p 46], provided that the players are both represented by points, and the radius of detection is zero. So the lion only catches the Christian if the two points coincide and not any earlier, which in real life would require some divine intervention. This game was popularized by Littlewood in his Miscellany [14], and he added the following problem: can two lions catch a man in a bounded area with rectifiable lakes? According to Béla Bollobás [3], this problem remains unsolved. If there are two lions, then we have a multi-agent game that is similar to the game of cops and robbers [4], which has recently received a lot of attention. These are pursuit-evasion games: the players have visual contact. Contrary to search games, in which the players are blind. It is the difference between catching and finding.

Gal's 1980 book [9] is the first monograph on Search Games. Selmer Johnson's 1964 paper [13], which is aptly called 'A Search Game', may be the first serious publication on the topic, predating Isaacs' classic on Differential Games which is generally recognized as the work that initiated the field (see Gal's review in Chap. 1). Johnson writes:

> The following game was first suggested to the author by Melvin Dresher several years ago. Blue chooses h, an integer from the set 1 to n (a region to hide). Red guesses an integer from 1 to n, is told whether he is too high or too low, and repeats until he guesses h. The payoff to Blue is one unit for each guess.

This game had appeared earlier as an example in Dresher's monograph on Game Theory [5]. In that same year Rényi proposed to study a similar game. It is inspired by the parlor game 20 Questions, so it is a win-lose game with an upper bound on the number of questions. The original text is in Hungarian, and the following translation is taken from [15]:

> A thinks of something and B must guess it. B can ask questions which can be answered by 'yes' or 'no' and he must find out of what A had thought of [...] it is better to suppose that a given percentage of the answers are wrong (because A misunderstands the question or does not know certain facts).

The game is now commonly called Ulam's game or the Rényi-Ulam game, after Stanislaw Ulam popularized a similar problem in his autobiography [16]. Since the hider may give faulty feedback, the Rényi-Ulam game presents a more versatile problem than Dresher's guessing game, and the literature on the topic is extensive.

5.2 The Princess and Monster Game on Graphs

BLIND DATE

We consider the princess and monster game on the interval $[-1, 1]$. The maximum speed of the monster M is 1 and the maximum speed of the princess P is unbounded. If M moves at maximum speed, then we say that he runs. The positions of the players $m(t)$ and $p(t)$ vary continuously with t and $m(t)$ is Lipschitz of constant 1. The time that the monster finds the princess is $\min t_0 = \{t \colon m(t) = p(t)\}$, and the payoff to the princess is t_0 in this case. As a first attempt to solve this game, the following strategy is an obvious candidate solution: the monster flips a coin and start at either end of the interval equiprobably, and runs to the opposite end. We say that an M that moves like that is a sweeper. Against this sweeper strategy, it is optimal for the princess to initially hide in 0, wait until time $1 - \varepsilon$, flip a coin, and move to one of the end points equiprobably. The expected time that M finds P is $\frac{3}{2}$ in this case. This is not the solution of the game. P's strategy is optimal againt M's, but M's strategy is not optimal against P's. In [2] it has been conjectured that an optimal mixed strategy for M can be based upon the following pure strategies:

$\mathscr{M}_1$ Choose an arbitrary initial point and an arbitrary direction. M runs in that direction until the end, and then back until the other end.

$\mathscr{M}_2$ Again choose an arbitrary initial point and an arbitrary direction. But now M runs until he meets the sweeper coming from that direction, then turns around and runs until the end, joining the sweeper, and then back until the other end.

If the strategy space of the monster is restricted to these pure strategies, then the optimal mixed strategy of the princess is based upon the following pure strategies:

$\mathscr{P}$ Choose an infinitesimal $\varepsilon > 0$. Either hide at an end-point and remain immobile, or choose an arbitrary initial point in $[-1 + \varepsilon, -\varepsilon] \cup [\varepsilon, 1 - \varepsilon]$ and remain there until the sweeper that starts from the nearest end point is ε-close. Then run to the the middle until ε-close, turn, and run back until the end.

It is possible to show that there exists an optimal response in $\mathscr{M}_1 \cup \mathscr{M}_2$ against any mixed strategy that is based on $\mathscr{P}$. The conjecture in [2] is that conversely against any mixed strategy that is based on $\mathscr{M}_1 \cup \mathscr{M}_2$ there exists an optimal response that is in $\mathscr{P}$. If this conjecture is true, then the princess and monster game on an interval reduces to a standard optimization problem that can be solved numerically by discretization and linear programming. An initial computation of the value of the game under the assumption of the conjecture was carried out in [2], but it contained some inaccuracies that have been corrected in [10]. The table below gives the value of the game V_n against the number of grid points n of the discretization. In [2] it was stated that the value of the game could perhaps be $11/8$, but this table demonstrates that the value must be slightly lower: the value V_n for $n = 1,024$ in Table 5.1 is within 10^{-3} of the actual value of the game.

n	V_n
1	1
2	1.2667
4	1.3303
8	1.3547
16	1.3647
32	1.3689
64	1.3709
128	1.3719
256	1.3724
512	1.3726
1,024	1.3727

Table 5.1 Number of grid points versus value of the game

Since the solution of the game on an interval is already a hard problem, it may seem that the solution of the game on an arbitrary graph is impossible. However, one should observe the following nice confluency property of the princess' strategy : suppose that $p_1, p_2 \in \mathscr{P}$ and that $p_1(t_0) = p_2(t_0)$ for some t_0. Then $p_1(t) = p_2(t)$ for all $t_0 \geq t$ (in Chap. 14, Steve Alpern calls this 'sticky'). Even more so, a mixed strategy based on $\mathscr{P}$ can be described by a probability measure μ_t such that $\mu_t(A)$ is the probability that P is in A at time t. Perhaps it is possible to derive the optimal response of the monster from this property, and verify the conjecture. We conjecture more generally, that for the game on a tree there exists an optimal princess strategy that satisfies the confluency property.

5.3 High-Low Search Games

Dresher's guessing game has recently been solved asymptotically [8], but there is a very similar high-low search game that remains unsolved. It was proposed around the same time as Dresher's game, by Ed Gilbert [12]. As in Dresher's game, Blue chooses a secret number h and tells Red whether his guess is too high, too low, or correct. However, after doing this, Blue may change the secret number, but it has to be in accordance with the answers that have been given so far. Dresher's guessing game has an immobile hider, the secret number remains the same. In Gilbert's guessing game, the hider is mobile between consecutive guesses. Ordinarily, a search game with a mobile hider is a more difficult game to solve, but in this case that is not true. A guessing game is played over rounds and in Gilbert's game the players play the same game each round, over a reduced set of numbers. This recursion should make the game easier to solve. It certainly makes the value of the game easier to compute, and it is conjectured in [7] that the value of the game satisfies

$$\lim_{n \to \infty} V(n) - \log_2(n) = c$$

for some constant $c = 0.487 \ldots$.

There are many other versions of the high-low guessing game. The following continous version of the game was first proposed by Vic Baston and Fred Bostock: Blue chooses a secret number $h \in [0,1]$. Red repeatedly guesses it and is told whether the guess is too high or too low (the probability of guessing the exact number is zero). If g_n is the sequence of guesses, then Blue's payoff is equal to $\sum |g_n - h|$. Steve Alpern [1] found a pure minimax strategy for the searcher, see Fig 5.1. One should realize that a pure strategy in a guessing game is equal to a binary search tree. The next guess depends on the fact whether the previous guess is too high or too low. Alpern introduces a state variable that keeps track of the excess of guesses that were too low. After n guesses g_i, the state variable is $k = |\{i \le n: g_i < h\}| - |\{i \le n: g_i > h\}|$. There exists a unique number $0 < \lambda_k < 1$ for every integer k, which has the property that if the remaining interval that contains h is $[a,b]$ and if the searcher guesses $\lambda_k a + (1 - \lambda_k)b$ in the next round, then $\sum |g_n - h|$ is constant for every h, with the exception of h that are in the countable set of guesses. The searcher uses a single binary tree, which is illustrated in the figure, partly and for the first few guesses, that has equal payoff against almost every secret number. However, it is unknown if there exists a mixed strategy that performs better.

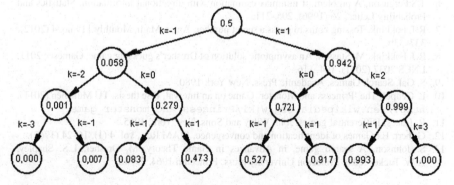

Fig. 5.1 A part of Alpern's minimax tree, rounded to three decimals

An optimal strategy for the hider has never been determined. One might try an algorithmic approach to this open problem, by giving the searcher a bounded number of questions, say 20 as in 20 questions, and taking $\sum_{n=1}^{20} |g_n - h|$ as the payoff to the hider.

One can modify the continuous guessing game following Gilbert, by allowing the hider to move from one secret number h_n to the next h_{n+1} between consecutive guesses. There are various choices for the payoff: it could be either $\sum |g_n - h_n|$ or $\sum |g_n - h_\infty|$, where h_∞ denotes the limiting value of the hider's secret numbers. Alpern's pure minimax strategy still works for the payoff $\sum |g_n - h_\infty|$, even though the hider has a larger strategy space. It should be easier to prove that the pure minimax strategy is optimal in this case. Instead of $\sum |g_n - h|$ one may also consider other norms or other payoffs. Tom Ferguson has studied a game with two guesses only, in which the payoff to the hider is $(g_2 - h)^2$. Despite its apparent simplicity, this game is non-trivial and the optimal strategies are not easy to find, see [6].

Finally, we would like to mention the following conjecture of Johnson: in Dresher's guessing game, the probability that Blue chooses the secret number 1 is equal to the probability that Blue chooses the secret number in $\{2, 3\}$, if $n > 4$. This problem has remained unsolved for almost 50 years now, and may very well be the longest standing open problem on search games.

References

1. S. Alpern, Search for point in interval, with high-low feedback, Math. Proc. Camb. Phil. Soc. 98 (1985), 569–578.
2. S. Alpern, R. Fokkink, R. Lindelauf, and G. J. Olsder. Numerical Approaches to the 'Princess and Monster' Game on the Interval, Annals Soc. Dynamic Games 10 (2009), 1–9.
3. B. Bollobás, The art of mathematics, coffee time in Memphis, Cambridge University Press, 2006.
4. A. Bonato, R.J. Nowakowski, The game of cops and robbers on graphs, American Mathematical Society, 2011.
5. M. Dresher, Games of Strategy: theory and application, Prentice Hall, 1961.
6. T.S.Ferguson, A problem of minimax estimation with directional information, Statistics and Probability Letters 26 (1996), 205–211.
7. R.J. Fokkink, Tossing coins to guess a secret number, Amer. Math. Monthly 119 no. 4 (2012), 337–339.
8. R.J. Fokkink, M. Stassen, An asymptotic solution of Dresher's guessing game, Gamesec 2011, LNCS 7037 (2011), 104–116.
9. S. Gal, Search Games, Academic Press, New York 1980.
10. L. Geupel, The 'Princess and Monster' Game on an Interval, BSc thesis, TU München (2011), http://en.wikipedia.org/wiki/Princess_and_monster_game
11. R. Isaacs, Differential games, John Wiley and Sons, New York 1965.
12. Gilbert, E.: Games of identification and convergence, SIAM Rev., vol. 4 (1), 16–24 (1962).
13. S. Johnson, A search game, in Advances in Game Theory, M. Dresher, L.S. Shapley, A.W. Tucker (eds), Princeton University press, Princeton 1964, 39–48.

14. J.E. Littlewood, A mathematician's miscellany, ed. B. Bollobás, Cambridge University Press, 1986.
15. A. Pelc, Searching games with errors - fifty years of coping with liars, Theoretical Computer Science 270 (2002), 71–109.
16. S. Ulam, Adventures of a Mathematician, Scribner, New York, 1976.

Part II
Geometric Games

Chapter 6
Some Cinderella Ruckle Type Games

Vic Baston

Abstract Nearly 30 years have elapsed since Ruckle's pioneering book (Ruckle WH (1983) Geometric games and their applications. Pitman, Boston) on geometric games was published; it was pioneering in the sense that it did not seek to detail a theory but dealt with a host of two-person zero-sum games which were easy to state and understandable to an intelligent layperson. Attractive features were the "toy" examples giving very idealized applications of the games and the list of open problems at the end of most chapters. Although many of the games had quite a lot in common, the professed aim was to provide "usable solutions" rather than a unified treatment. In fact Gal's book (Gal S (1980) Search games. Academic Press, New York) which developed a theory of search games had already appeared when Ruckle's book was published but, in the main, Ruckle's games fell outside its scope. Although there has been considerable activity in the theory of search games since then (see [2] and Chap. 9 of the book), the main emphasis has been on the development of the aspects covered in Gal's book. The aim of this chapter is to draw attention to some Ruckle games which the writer feels have not received the attention they deserve; hence the Cinderella in the title.

6.1 Introduction

Fraenkel divides games into two types, games people play (i.e. games that people buy and play) and games that mathematicians play or, in Peter Winkler's words, games people don't play. The selection of the games here fall very definitely into the latter category and the hope is that these games will lead to the development of

V. Baston (✉)
Faculty of Mathematical Studies, University of Southampton, Southampton, SO17 1BJ, UK
e-mail: vic.baston@btinternet.com

S. Alpern et al. (eds.), *Search Theory: A Game Theoretic Perspective*,
DOI 10.1007/978-1-4614-6825-7_6, © Springer Science+Business Media New York 2013

ideas which can be used to unify treatments for a variety of games, in other words, usable methods. Thus the choice has been strongly influenced by two factors. Firstly there needs to be a connection between the games and secondly each game has to provide a stimulus for further research; in most cases this means that the games are used as a starting off point from which attractive open questions can be generated. As a result Sects. 6.2 and 6.8 all include suggestions for further work of varying degrees of difficulty which are intended to encourage more researchers to take an interest in the games. However it does mean that other games associated with Ruckle such as lattice and accumulation games have been ignored. In keeping with the spirit of Ruckle's book, the games in this chapter are all two-person zero-sum ones and the players are called RED and BLUE with RED being the maximizer. The structure of the chapter is described in the following paragraphs.

The Several Intervals Game is played in the unit interval I with BLUE choosing a point of I and RED simultaneously selecting intervals of given lengths α_i in I; the payoff to RED is one if BLUE's point is in one of the intervals and zero otherwise. Although simply stated, this has proved to be an extremely difficult game to solve and no comprehensive solution has been found when RED can choose more than two intervals. Abbreviated details of the original (unpublished) approach used for the Two Intervals Game and some of the results on the Three Intervals case are given in Sect. 6.2.

Ruckle's greedy games have not attracted very much attention and are the subject of Sects. 6.3 and 6.4. Many games have the form that RED has the task of deciding where to hide a given amount of material but, in greedy games, RED has the additional decision of determining how much material to hide when facing the prospect that hiding more means a greater probability of discovery. Section 6.3 gives a formal definition of a greedy game and then concentrates attention on games in the unit interval whereas Sect. 6.4 looks briefly at greedy games on the unit square.

In the Number Hides Game RED and BLUE simultaneously choose subintervals of given length in an integer interval with RED getting a payoff equal to the number of integers the subintervals have in common. The game proved more tractable than the other games we consider and it was solved independently by three sets of researchers. Section 6.5 presents some natural variations of the game and Sect. 6.6 discusses the interesting generalization by Zoroa, Fernandez-Saez and Zoroa [7] in which BLUE has to hide a given quantity of objects in an integer subinterval of his choice with the stipulation that at least one object and at most c can be placed at each integer of the subinterval; as before RED chooses a subinterval and receives an amount equal to the number of objects in it. The solution of this game seems to be difficult so, as a first step, it is proposed that the solution of a couple of particular cases extending those of [7] be attempted.

The Hiding in a Disc Game is again easily stated. RED and BLUE simultaneously choose points in the unit disc and RED wins if and only the chosen points are at most a given distance c apart. It was already proving awkward 30 years ago as Ruckle demonstrated that an assertion concerning its value in the American Mathematical Monthly was false and, since then, there seems to have been little work done on it. Section 6.7 looks at its symmetrization and asks whether there are optimal strategies in this symmetrization which are probability distributions over a finite number of points.

Section 6.8 shows that some of our games can be thought of as special cases of a general game and Sect. 6.9 details come conclusions.

6.2 The Several Intervals Game

Probably the game in Ruckle's book which has since received the most attention in the literature is the Several Intervals Game. In it BLUE chooses a single point b in $[0,1]$ and RED chooses the union of several closed intervals $R = J_1 \cup \ldots \cup J_k$ where, for each i, the length of J_i is at most α_i. The payoff to RED is one if $b \in R$ and zero otherwise. Ruckle obtained solutions by trial and error for two special cases when $k = 2$ and, as a result, ventured that the solution for general k may be difficult; he therefore made a more modest proposal of solving it for $k = 2$ or 3. Even for this more limited objective progress has been slow and, to adapt Churchill's description of Attlee, workers on it have a lot to be modest about. However, after 20 years, Woodward in a doctoral thesis [5] managed to come up with what can be regarded as a complete solution for $k = 2$. We will indicate the processes that enabled him to arrive at this solution as they may provide ideas that can be used to solve the case $k = 3$ which, as we shall see, still presents a challenge.

In essence Woodward's approach was straightforward but the devil remained in the detail. Firstly it was shown that the game for general k is equivalent to a corresponding finite game meaning that a complete solution could be obtained for any given set of interval lengths using linear programming. Although theoretically useful, little immediate value was obtained from the computer results for $k = 2$ due to the sheer volume of data and the vast number of different strategies. In particular vastly different strategies could be produced for games which had the same game values and very similar α_1 and α_2. Also RED strategies proved particularly awkward as the computer did not find symmetric ones. This meant that the approach was throwing up interesting computing problems because it was important that a coherent set of solutions be produced for a theoretical analysis to be undertaken. By perturbing interval positions, swapping intervals from one strategy to another and other techniques, strategies were found which were valid for all cases with the same value; in many instances the resulting strategies showed very little resemblance to the original strategies generated by the computer. It was then possible to detect the pattern which enabled a theoretical analysis to be made. This analysis is contained in Woodward's thesis of almost 300 pages but recently new arguments (see Chap. 9) have been found which enable the treatment to be shortened.

Woodward also managed to generate a number of results for $k = 3$ by the same methods and we give an account of his findings. He produced expressions for the game value which cover all the cases when $1/3 \le \alpha_1 < 1/2$ and $\alpha_3 \ge 1/5$. It might have been expected that the justification of a RED optimal strategy would prove more troublesome than a BLUE one as it involves triplets of intervals played with certain probabilities whereas a BLUE strategy is simply a probability distribution on [0,1]. However the reverse was true because a RED optimal strategy could be easily verified by showing that each point of [0,1] meets its intervals with a certain probability. In many cases the RED optimal strategies could be derived using three appropriate inequalities of the form $x_{i1}\alpha_1 + x_{i2}\alpha_2 + x_{i3}\alpha_3 \geq 1$ $i = 1,2,3$ where the x_{ij} are non-negative integers; to be appropriate, a necessary condition is that there are positive integers p,q,r satisfying $px_{i1} + qx_{i2} + rx_{i3} = W$ for each i resulting in the value of the game being $(p+q+r)/W$. Note that the inequalities represent three different coverings of the unit interval by segments of lengths α_1, α_2 and α_3. This follows a similar pattern to that found in the Two Intervals Game where the RED optimal strategies can mostly be derived from two different coverings of the unit interval.

Although the examples give indications of how the case $k = 3$ might be treated in general, there are sufficient exceptions to suggest that further ideas are necessary if substantial progress is to be made. For example most have a BLUE optimal strategy which uses each of 0 and 1 with a probability equal to the game value. However no such BLUE strategy was found for the case $\alpha_1 = 1/3$, $\alpha_2 = 1/4$ and $\alpha_3 = 1/5$. Its value is 13/56 but the BLUE optimal strategy obtained used each of 0 and 1 with probability 10/56. This case is interesting in another way as the Red optimal strategy was derived from four covering inequalities rather than the usual three; in addition to the obvious $3\alpha_1 \geq 1$, $4\alpha_2 \geq 1$ and $5\alpha_3 \geq 1$, $\alpha_1 + 2\alpha_2 + \alpha_3 \geq 1$ is needed. It seems therefore that an interesting challenge on the way to solving the case $k = 3$ would be to answer the following open question.

Question 1. What is the value of the Three Intervals Game when the lengths of the intervals are of the form $\alpha_1 = 1/u, \alpha_2 = 1/v$ and $\alpha_3 = 1/w$ when u, v and w are positive integers satisfying $3 \leq u \leq v \leq w$?

To introduce a note of optimism, Woodward in his work on the Three Intervals Game has taken the length of the largest interval to be at least a third and it could be that the case when the longest interval has length less than a third will have a more unified treatment. After all, in the Two Intervals Game, there is a single expression for the value of the game when the length of the longest interval is less than a half but this expression does not always hold when it is greater than or equal to a half.

Woodward has also obtained some minor results for the n interval case. In particular he has shown that, when $\alpha_1 > 1/2$, the value of the game is one half of the value of the game with $n - 1$ intervals with lengths having values $\alpha_2/(1 - \alpha_1), \ldots, \alpha_n/(1 - \alpha_1)$. Also if there are n intervals all of the same length $1/k$ where $k > n$, the value of the game is $1 - k/n$.

6.3 Greedy Games

Consider the following scenario. RED wants to hide a quantity of arms or drugs within a given region which, if not detected within a given time, can be used to further RED's interests in the region. BLUE (the authorities within the region) can employ various measures in an attempt to find the hidden resource and frustrate RED's ambitions. The benefit to RED of successfully hiding the resource depends on the amount that has been hidden and circumstances dictate that the larger the amount RED tries to hide the greater the probability that it will be discovered by BLUE. This means that RED needs to balance two competing factors; RED would like to hide a large amount so that RED would derive a substantial benefit if it remains undetected but, at the same time, not so large that BLUE has a high probability of detecting it. In modelling this scenario as a game it seems reasonable to set the payoff to RED as q when an amount q is successfully hidden. The payoff when an amount q is detected by BLUE is not so clearcut as it may depend on how RED views the situation; we therefore introduce a parameter $\beta \geq 0$ and set the payoff to RED as $-\beta q$. If RED has very large global resources and the scenario is a comparatively minor one for it, the loss of resource would not be significant and a value of β near zero would be appropriate. On the other hand, if RED has little influence outside the region, a loss of a sizeable amount of resource could have major consequences for it so that a comparatively large value of β might be appropriate.

The above scenario provides the motivation for our definition of a *greedy game*.

It is a two-person zero-sum game which is played in a compact convex region S, with interior points, of n-dimensional Euclidean space. RED (the maximizer) chooses a member C from a given class $\mathscr{C}$ of measurable subsets of S and, without knowing RED's choice, BLUE (the minimizer) selects B from a given subset $\mathscr{B}$ of the power set of S. Letting A denote the measure of the C chosen by RED and $0 \leq \beta$, RED gets A if $B \cap C$ is empty and loses βA if it is not. Both players know S, $\mathscr{C}$, $\mathscr{B}$ and β.

In a number of ways this type of game is a mirror image of the type of game discussed in Sect. 6.2. Here, in a particular one-dimensional setting, an interval is being placed to *avoid* a point chosen by an opponent whereas, in the Several Interval Games, intervals are placed in an attempt to *include* a point chosen by an opponent.

Ruckle solved several greedy games which are played over the unit interval $I = [0,1]$ when $\beta = 0$. In the *length greedy game* RED can choose any measurable set and BLUE any set with at most k points whereas, in the *interval greedy game*, RED can choose any interval of I and BLUE an interval of length at most α. In both games BLUE has an optimal strategy which employs a uniform distribution. In the first BLUE chooses k points independently using the uniform distribution on I and, in the second, starts by choosing $t \in [0, 1 - \alpha]$ by the uniform distribution on $[0, 1 - \alpha]$ and then occupies the interval $[t, t + \alpha]$. RED has an ε-optimal strategy which involves a covering of I in both games. The basic idea underpinning the RED strategy for the length greedy game is that I is divided into an appropriately large number n of intervals and then a member is chosen at random from the set of unions of precisely m members of these intervals where m is defined in terms of α and n. For the interval greedy game the basic idea is simpler; RED divides I into an appropriate number of intervals and chooses one of them suitably modified.

When $\beta = 0$, the greedy length game in which $k = 1$ and the interval length game in which $\alpha = 0$ have a common solution so we now investigate the *point greedy interval game* Γ in which RED chooses an interval, BLUE chooses a point and $\beta \geq 0$. In common with Ruckle, we take the interval to be closed so that RED will in general have only ε-optimal strategies although it will be plain that RED has optimal strategies if open intervals are allowed. The next lemma generalises the strategies used by Ruckle in the game with $\beta = 0$ to obtain bounds on what the players can achieve in the more general game.

Lemma 1. *In the point greedy interval game, BLUE can restrict RED's expectation to at most $1/(4(1+\beta))$ whereas RED can guarantee an expectation of at least $\max\{(n-1-\beta)/n^2\} - \varepsilon$ where $\varepsilon > 0$ and the maximum is taken over all positive integers.*

Proof. If BLUE employs the uniform distribution on I, then BLUE has a probability of L of intersecting with a RED interval of length L giving RED an expectation of $(1-L)L - L\beta L = L(1 - (1+\beta)L)$. Hence the best that RED can do is to choose an interval of length $1/(2(1+\beta))$ and BLUE can restrict RED to a payoff of at most $1/(4(1+\beta))$.

For a positive integer n and $\eta > 0$ small, suppose RED plays one of the intervals $[k/n, (k+1)/n - \eta]$, $k = 0, 1, \ldots, n-1$ at random; any pure strategy of BLUE can intersect at most one of these intervals so RED can ensure an expectation of at least

$$(\frac{1}{n} - \eta)(\frac{n-1}{n}) - \beta\frac{1}{n}(\frac{1}{n} - \eta) = \frac{n-1-\beta}{n^2} - \frac{n-1-\beta}{n}\eta$$

and the lemma follows. $\square$

A consequence of the lemma is that, if there is a positive integer n such that $(n-1-\beta)/n^2 = 1/(4(1+\beta))$, then its common value is the value of the game. It is therefore easy to check that, for all positive integers $n \geq 2$, the game has value $1/(2n)$ when $\beta = (n-2)/2$.

The formulation of the game requires that the players know the value of β but, from a practical view, BLUE in particular may have little idea concerning its precise value. It can therefore be useful to know that a strategy is reasonably good for a range of values of β even it is not optimal, particularly if that strategy is fairly simple. The above analysis suggests the uniform distribution is such a strategy; in particular, it is likely to be effective if the loss of material has serious implications for RED, that is, when β is large.

For general values of $\beta \in [0, 1]$, the position is a good deal more complicated. First of all we should perhaps address the question of whether the games actually *have* a value; we do not wish to go into the details here but they do by an existence theorem of Alpern and Gal (see [5] Theorem A.1 on page 293). In the games in which a value has been found ($\beta = 2/(n-2)$), these RED strategies have been derived from coverings of I so it is difficult to see how they can be modified to improve the payoff for other values of β. On the other hand optimal alternatives to the uniform distribution for BLUE abound. For instance, when $\beta = 1/2$, the value is 1/6 so BLUE can afford to ignore all points in $[0, 1/6)$ and $(5/6, 1]$ and concentrate the distribution in an appropriate way in $[1/6, 5/6]$.

To illustrate the point we show that the value of the game is $(2-\beta)/9$ when $1/5 \le \beta \le 5/7$. Notice that $(n-1-\beta)/n^2$ equals 1/5 for $n = 2$ and 3 when $\beta = 1/5$ and equals 1/7 for $n = 3$ and 4 when $\beta = 5/7$. Thus $1/5 \le \beta \le 5/7$ is likely to be the maximum range of values of β giving the game value $(2-\beta)/9$ because not only does Lemma 1 tell us that RED can guarantee more (namely $(1-\beta)/4$) if $\beta < 1/5$ but it also suggests that RED cannot guarantee as much if $\beta > 5/7$. Let

$$F(x) = \begin{cases} 0 & \text{if } x < (2-\beta)/9, \\ 1/(1+\beta) - (2-\beta)/(9x(1+\beta)) & \text{if } (2-\beta)/9 \le x < 1/2, \\ (19\beta+7)/(18(1+\beta)) & \text{if } x = 1/2 \end{cases}$$

and $F(x) = 1 - F(1-x)$ for $1/2 < x \le 1$.

As $x \to 1/2-$, $F(x) \to (5+2\beta)/(9+9\beta) \le 1/2$ when $\beta \ge 1/5$. Thus $F(x)$ is a probability distribution over I which has a jump at 1/2 when $\beta > 1/5$ and is strictly concave in the interval $[(2-\beta)/9, 1/2)$.

Suppose BLUE employs the strategy $F(x)$. We first show that the properties of F mean that we only need to find the payoff of certain RED intervals in detail in order to find RED's best reply to F.

If $[a, a+x]$ and $[b, b+x]$ are two RED intervals with $F(a+x) - F(a) < F(b+x) - F(b)$, then $[a, a+x]$ gives a better payoff than $[b, b+x]$ so we need only consider $[a, a+x]$. Therefore any RED interval of the form $[a, a+x]$ with $0 < a \le (2-\beta)/9$ gives an inferior payoff than $[0, x]$ and so can be ignored.

Furthermore $F(a+x) - F(a) < F(b+x) - F(b)$ if $(2-\beta)/9 \le b < a < a+x < 1/2$ so intervals $[a, a+x] \subseteq [(2-\beta)/9, 1/2)$ have an inferior payoff to $[1/2-x, 1/2)$.

For $a < 1/2 < a+x$, $F(a+x) - F(a)$ has a minimum in a for fixed x at $a = (1-x)/2$ so, for intervals having 1/2 as an interior point, it is only necessary to consider those symmetric about 1/2.

Finally the symmetry of F means that we can assume a RED interval starts in $[0, 1/2)$.

Thus, in finding RED's best reply to F, the analysis is reduced to investigating three types of RED interval, namely (i) $[0, x]$, (ii) $[x, 1/2)$ where $x > (2-\beta)/9$ and (iii) $[1/2-x, 1/2+x]$.

(i) For $x < 1/2$, the payoff for $[0, x]$ is

$$(1 - F(x))x - \beta F(x)x = x(1 - (1+\beta)F(x)) = (2-\beta)/9.$$

For $1/2 < x \leq 1 - (2-\beta)/9$, it is

$$x(1 - (1+\beta)(1 - F(1-x))) = x(1 - \beta + \frac{2-\beta}{9(1-x)})$$

Routine calculations show that this expression has a minimum at $x = 1 - (1/3)\sqrt{(2-\beta)/(1-\beta)} \leq 1/2$ when $\beta \geq 1/5$; it is convex so, in $[1/2, (2-\beta)/9]$, its minimum occurs at $1/2$. The payoff is continuous on the right in $[0,1]$ so, for $x > 1/2$, it is at least that of the interval $[0,1/2]$ which is at least the payoff $(2-\beta)/9$ of $[0,1/2)$.

(ii) For $(2-\beta)/9 \leq x < 1/2$, the payoff of the interval $[x, 1/2)$ is

$$(1/2 - x)(1 - (1+\beta)(F(1/2-) - F(x))) = (1/2 - x)(1 - (2-\beta)(1 - 2x)/9x).$$

The two (main) brackets are decreasing functions of x so the maximum occurs at $x = (2-\beta)/9$ and is less than $(2-\beta)/9$.

(iii) Now $F((1+x)/2) - F((1-x)/2) = 1 - 2F((1-x)/2)$ so the expected payoff of $[(1-x)/2, (1+x)/2]$ is

$$x(1 - (1+\beta)(1 - 2F((1-x)/2))) = x(2-\beta)(1 - (4/(9(1-x)))).$$

This expression is concave and has a maximum of $(2-\beta)/9$ when $x = 1/3$.

Lemma 1 tells us that RED can ensure an expected payoff of at least $(2-\beta)/9$ so we have established the following theorem.

Theorem 1. *The value of the point greedy interval game is* $(2-\beta)/9$ *when* $1/5 \leq \beta \leq 5/7$.

This leads to the following conjecture.

Conjecture 1. The point greedy interval game has value $\max\{(n-1-\beta)/n^2\}$ where the maximum is taken over all positive integers.

Of course there are a legion of further challenges with an obvious one being the case when BLUE is allowed to select more than one point. An alternative approach would be to investigate other forms of payoff. We have introduced the idea of a cost to RED of being discovered so a natural extension would be to allow BLUE to choose the number of points to play but levy a cost on BLUE for them. BLUE would then have to make a judgement about the amount of resource to employ, thus creating a doubly greedy game. This might entail a movement away from zero-sum games but, on the basis that what is good for me is bad for my enemy and vice-versa, might realistically still remain in the zero-sum environment. As mentioned above Ruckle did frame his game in terms of BLUE having at most k points but, in the absence of a penalty for using more points, BLUE does not in fact have to make a judgement call. For the record, in the modified length greedy game Ruckle did introduce a modified payoff to RED of $a\alpha - bn$ where α is the length of RED's interval, n is the number of points in RED's interval and a and b are positive constants so this may also be setting off point for allied games.

6.4 Area Greedy Games

In the previous section we concentrated attention on greedy games played on the unit interval so we will now look at generalizations of the Area Greedy game which was given as a problem in Ruckle's book. In this two person zero-sum game played over the unit square S, RED (the maximizer) chooses a rectangular subset with edges parallel to S and BLUE (the minimizer) selects a path in S which is the graph of a continuous function from $[0,1]$ to $[0,1]$. BLUE wants the path to intersect RED's rectangle and, if it does, RED gets nothing. On the other hand RED gets the value of the area of the chosen rectangle if BLUE's path does not intersect it. The game has the flavour of a needle in a haystack game but, in this game, RED can choose the size of the rectangle as well as its position whereas the hider in a needle in a haystack game hides a needle of fixed length.

It is straightforward to see that Ruckle's game has a simple solution. By choosing the function f_n defined by

$$f_n(t) = \begin{cases} 2(nt - i) & \text{if } i \leq nt \leq i + 1/2 \\ 1 - 2(nt - i - 1/2)) & \text{if } i + 1/2 \leq nt \leq i + 1 \ \text{ for } i = 0, 1 \ldots, n - 1. \end{cases}$$

BLUE can ensure that every RED rectangle with horizontal length of at least $1/n$ is intersected so that RED's payoff is at most $1/n$. Thus, by choosing n sufficiently large, RED's payoff can be made arbitrarily small and the value of the game is therefore zero. Note that the same is true if RED has the freedom to choose any convex subset of S, not just rectangles.

To particularise the general definition of greedy game given in the previous section and stay within the spirit of the area greedy game of Ruckle, we define an *area greedy game* as a two person zero-sum game played over a compact region S, with interior points, of two-dimensional Euclidean space of the following type.

> RED (the maximizer) chooses a member from a class $\mathscr{C}$ of convex subsets of S and, without knowing RED's choice, BLUE (the minimizer) selects a path from a set of paths $\mathscr{P}$, all of length at most L, in S. Letting A denote the area of the set chosen by RED, RED gets A if BLUE's path does not intersect it and loses βA if it does. Both players know S, $\mathscr{C}$, $\mathscr{P}$, L and β.

Ruckle's original game is easy to solve because BLUE is allowed to choose a path that has no restrictions on its length. In fact every one of our area greedy games (and also some more general ones) has value zero if BLUE is allowed a completely free choice of path. This follows from Lemma 3.39 of Alpern and Gal [5] which ensures that, given $\varepsilon > 0$, there is a (closed) path in the two-dimensional compact convex set S such that every point of S has distance less than ε from the path. Hence a natural restriction to impose on BLUE's paths is that they should have length bounded by a positive real number, L say.

Notice that there are connections between the point greedy interval game analysed in the previous section and a number of area greedy games on the unit square when BLUE has to choose a path which is the graph of a constant function. In the latter games, the crucial factor that determines whether BLUE's path intersects the convex set C chosen by RED is the projection of C onto the y-axis. In particular, in this restricted form of Ruckle's area greedy game, in order to play optimally RED will choose a rectangle with base of length one so his choice is effectively to choose the height of the rectangle (which is a line segment) while BLUE is effectively choosing a point (the value of the constant). Perhaps the next natural step is to tackle the following problem.

Problem 1. Solve the Area Greedy Game on the unit square when BLUE can take any path of length one and RED can choose any compact convex set in the unit square.

6.5 The Numbers Hides Game

Like the Several Intervals Game the Number Hides Game is a two person zero-sum game which was posed as a problem in [4] with some special cases being solved. Its formulation is particularly simple.

> RED and BLUE choose sequences of p and q consecutive integers respectively between 1 and n. The payoff to RED is the number of integers in the intersection of the chosen intervals.

The game has proved more tractable than the Several Intervals Game. When Baston and Bostock submitted their solution to the proceedings of the American Mathematical Society, they were told that Ferguson had also solved it so a three author paper [3] was written in the style of Ferguson which the editors preferred. They subsequently learned that Zoroa, Zoroa and Ruiz had also found a solution [6].

The game is the discrete version of the Interval Overlap Game in which RED and BLUE choose intervals of lengths at most α and at least β respectively in the unit interval I, and RED has a payoff of the measure (length) of the intersection of the chosen intervals; this game was solved in [4]. We have seen in Sect. 6.2 that the Several Intervals Game is essentially equivalent to a corresponding finite game so it is natural to ask whether a similar situation pertains here. Although no formal justification has been given, the answer is probably yes as it was remarked in [3] that the ideas used to solve the Numbers Hide game carry over to the Interval Overlap Game; in fact these ideas enabled a fault in the analysis of BLUE's optimal strategies in [4] to be corrected. The games where the Number Hides Game is modified so that one or both players are permitted to choose an arbitrary set of integers rather than a set of consecutive integers have been solved in [3] and [7]. The game in which both players can choose an arbitrary set of integers is called the Simple Point Catcher Game. As pointed out in [3], its value is pq/n, when RED chooses at most p integers and BLUE chooses at least q but Ruckle [4] gave a more complicated expression, namely

$$\frac{{}^q C_1 \, {}^{n-q} C_{p-1} + 2 \, {}^q C_2 \, {}^{n-q} C_{p-2} + \cdots + r \, {}^q C_r \, {}^{n-q} C_{p-r} + \cdots + p \, {}^q C_p \, {}^{n-q} C_0}{{}^n C_p};$$

this expression can be rearranged to

$$\frac{pq/n}{{}^{n-1} C_{p-1}} W$$

where

$$W = {}^{n-q} C_{p-1} + {}^{q-1} C_1 \, {}^{n-q} C_{p-2} + \cdots + {}^{q-1} C_{r-1} \, {}^{n-q} C_{p-r} + \cdots + {}^{q-1} C_{p-1}$$

Because W is the coefficient of x^{p-1} in the expansion of $(1+x)^{n-q}(1+x)^{q-1}$ and the denominator is the coefficient of x^{p-1} in the expansion of $(1+x)^{n-1}$, they are equal and the expression simplifies to pq/n. Optimal strategies for RED and BLUE are to select the set of p, respectively q, integers from $\{1, 2, \ldots, n\}$ by simple random sampling.

We now introduce notation which enables us to give alternative optimal strategies in the Simple Point Catcher Game which are more useful as a guide for optimal strategies in the modified Interval Overlap games described below. In fact they are the optimal strategies Ruckle used to solve the Modified Number Hides Game in which the players can choose sequences modulo n. Let n be a fixed positive integer and, for each positive integer x, let $x = \lambda n + x^*$ where λ is an integer and $0 < x^* \leq n$. For positive integers $m < n$ and x, put

$$I_m(x) = \begin{cases} [x^*, x^* + m - 1] & \text{if } x^* + m - 1 \leq n, \\ [x^*, n] \cup [1, m + x^* - n - 1] & \text{if } x^* + m - 1 > n. \end{cases}$$

Thus the sets in

$$\mathscr{I}_m = \{I_m(x) : x = 1 + \mu m \text{ for } \mu = 0, 1, \ldots, n-1\}$$

cover the integer interval $[1, n]$ precisely m times and every integer $y \in [1, n]$ is in precisely m members of $\mathscr{I}_m$. Hence, if RED chooses one of the members of $\mathscr{I}_p$ at random and BLUE chooses any set of q integers in $[1, n]$, RED has an expectation of pq/n. Similarly, if BLUE chooses one of the members of $\mathscr{I}_q$ at random, RED has an expectation of pq/n whatever set of p integers in $[1, n]$ RED chooses. Thus, taking a member at random from $\mathscr{I}_p$ and $\mathscr{I}_q$ respectively are optimal strategies for RED and BLUE in the Simple Catcher Game. These optimal strategies and the ones mentioned earlier demonstrate that the game has the unusual property that, if the roles are reversed (so that RED loses the number of integers in the intersection of the chosen intervals), a player still has the same optimal strategy.

We now look at the analogous problems on the real interval $[0, 1]$ where one or both players can, instead of choosing an interval of length α, choose a set of (Lebesgue) measure α. This gives rise to the following three games over the unit interval where, in each case, the payoff to RED is the measure of the intersection of the sets chosen by the players:

$\Gamma_{MM}(\alpha,\beta)$. RED chooses a set of measure at most α and BLUE chooses a set of measure at least β.

$\Gamma_{IM}(\alpha,\beta)$. RED chooses an interval of length at most α and BLUE chooses a set of measure at least β.

$\Gamma_{MI}(\alpha,\beta)$. RED chooses a set of measure at most α and BLUE chooses an interval of length at least β.

The next result shows that Γ_{MM} is easy to solve using the ideas we have developed.

Proposition 1. *The value of* Γ_{MM} *is* $\alpha\beta$.

Proof. Firstly suppose both α and β are rational, say a_1/a_2 and b_1/b_2 respectively. For each real number x, let $x = m + x^*$ where m is an integer and $0 \leq x^* < 1$. Given $\gamma \in [0,1)$ and x a real non-negative number, $I_\gamma(x)$ is defined as the interval $[x^*, x^* + \gamma]$ if $x^* + \gamma \leq 1$ and the pair of intervals $[x^*, 1] \cup [0, \gamma + x^* - 1]$ if $x^* + \gamma > 1$. For rational $\gamma = c_1/c_2$ say, let

$$\mathscr{I}_\gamma = \{I_\gamma(x) : x = \mu\gamma \text{ for } \mu = 0, 1, \ldots, c_2 - 1\},$$

then $\mathscr{I}_\gamma$ covers $[0,1]$ precisely c_1 times. Thus, if RED chooses a member of $\mathscr{I}_\alpha$ at random, RED can guarantee an expectation of $\beta(c_1)/c_2) = \beta\alpha$ whatever set of measure β BLUE chooses. Similarly, if BLUE chooses a member of $\mathscr{I}_\beta$ at random, RED's expectation can be restricted to $\alpha\beta$. Hence the Proposition holds for rational α and β.

Clearly the expectation to RED does not decrease as the value of α increases or the value of β decreases. The result therefore follows for general α and β because, given any irrational number γ, there are rational numbers $r_1 < \gamma$ and $r_2 > \gamma$ arbitrarily close to γ. $\square$

The proof of Proposition 1 tells us that, provided a player can split the allowed measure between two intervals, he gets no benefit from being able to choose a measurable set.

Although, in general the games Γ_{IM} and Γ_{MI} are more difficult to analyse, some easy deductions can be made. Because BLUE's strategy space in Γ_{IM} contains BLUE's strategy space in the Interval Overlap Game and RED's strategy space is the same in both games, the value of Γ_{IM} is less than or equal to the value of the Interval Overlap Game. Furthermore the RED strategy in the proof of Proposition 1 ensures that the value of I_{MI} is at least $\alpha\beta$. By similar arguments, the value of Γ_{MI} is greater than or equal to the value of the Interval Overlap Game and Blue can ensure the value of I_{IM} is not more than $\alpha\beta$. Although one suspects that the following problem is not so difficult as many of the others in this chapter, it still may not be easy.

Problem 2. Solve Γ_{MI} and Γ_{IM}.

6.6 Relatives of the Number Hides Game

The Simple Point Catcher Game can be interpreted as BLUE having q objects to hide at integer points in $[1, p]$ with the constraint that exactly one object can be hidden at a point and RED being allowed to search p integer points in an attempt to find them. The payoff to RED is the number of objects found. A natural generalization of this game is for BLUE to have q objects to hide and be allowed to choose b points in which to hide them with the constraint that an amount between 1 and c must be placed in each of the chosen points. The game has an affinity with the greedy games covered in Sects. 6.3 and 6.4 because BLUE now has to balance two competing factors; whether to hide objects at a comparatively small number of integers meaning that it is relatively difficult for RED to find them but expensive if he does or to spread the objects over a comparatively large number of integers so that losses are more likely but less painful. N Zoroa, M. J. Fernández-Sáez and P Zoroa introduced these types of game involving capacities into the literature and have been in the forefront of research on them (see [7] and [8]); in particular an interesting general Point Catcher Game is solved in [8]. Although they have obtained many results, interesting and challenging problems remain open and we now detail some of them.

First consider the following generalization of the Numbers Hide Game; in [7] it is called the Hide and Seek Game with Capacities equal to c, but we will call it the Integer Number Hides Game with Capacities to emphasize not only that it has a strong relationship with the Numbers Hide Game but also that it is not an isolated game but one that forms part of a coherent body of work.

> BLUE has q indivisible objects to hide in the integer interval $L = [1, n]$ and must choose an interval B of L to do so under the restriction that between 1 and c objects must be allocated to each point of the chosen interval. Simultaneously RED picks an interval R of length p and gets a payoff equal to the number of objects that BLUE allocated to the points of A.

Two results for this game are given in [7]. Firstly, if p is a divisor of N, then the value of the game is pq/N. Secondly, let $n = \lambda p + r$ where λ is a positive integer and $0 \le r < p$, then, provided $q \le (c-1)r + p$, the value of the game is $q/(\lambda + 1)$ if $q \le rc$ and $(q(\lambda + 1) - cr)/(\lambda(\lambda + 1))$ if $q > rc$. In addition two particular examples are given which show that the value of the game can equal $(q - c)/\lambda$. As they had found other examples which had the same expression for the value and in which all RED optimal strategies have a similar structure, they suggested that there may be a general structure for games with value $(q - c)/\lambda$. Very modest progress on this front when $n = \lambda p + 2$ and $c = 2$ is detailed below.

Lemma 2. Let $n = \lambda p + 1$ or $n = \lambda p + 2$ for some integer λ and $q \le n - 2 + c$, then RED can ensure a payoff of at least $(q - c)/\lambda$.

Proof. Let $J_j(i) = [ip + j, (i+1)p + j - 1]$. First suppose $n = \lambda p + 1$ and RED chooses a member of $\{J_1(i) : i = 0, 1, \ldots, \lambda - 1\} \cup \{J_2(i) : i = 0, 1, \ldots, \lambda - 1\}$ at random. The probability of an x satisfying $2 \le x \le \lambda p$ occurring in the RED strategy is $1/\lambda$ whereas each of 1 and $\lambda p + 1$ occur with probability $1/(2\lambda)$. Hence in a best reply BLUE will place as many objects as possible at the points 1 and $n = \lambda p + 1$. As $q \le n - 2 + c$, BLUE can put a total of c objects at 1 and n because an object must be placed at each point of $[2, n-1]$ if objects are put in both 1 and n. Thus a minimum of $q - c$ objects must be put in $[2, n-1]$ so that RED is assured of a payoff of at least $(q - c)/\lambda$. □

Now suppose $n = \lambda p + 2$ for some integer λ and RED chooses a member of $\{J_2(i) : i = 0, 1, \ldots, \lambda - 1\}$ at random. The probability of an x satisfying $2 \le x \le \lambda p + 1$ occurring in the RED strategy is $1/\lambda$ whereas each of 1 and n occur with probability 0. Hence, as before, in a best reply BLUE will put as many objects as possible at the points 1 and n but be forced to put at least $q - c$ in $[2, n-1]$ because $q \le n - 2 + c$. Thus RED's strategy ensures a payoff of at least $(q - c)/\lambda$.

The structure of our BLUE strategies is much more complicated so we first look at a particular example to illustrate the general case.

Example 1. The game in which $n = \lambda p + 2 = 32$, $p = 5$, $c = 2$ and $q = 28$ has value $26/6 = (q - c)/\lambda$.

Proof. Consider the BLUE strategy which chooses one of the following pure strategies at random.

$$B_1 = (2,2,2,1,1,2,2,2,2,1,2,2,2,2,1,2,\overbrace{0,\ldots,0}^{16 \text{ times}}), \quad \overleftarrow{B_1},$$

$$B_2 = (2,2,2,2,1,2,2,2,1,1,2,2,2,2,1,2,\overbrace{0,\ldots,0}^{16 \text{ times}}), \quad \overleftarrow{B_2},$$

$$B_3 = (2,2,2,2,1,2,2,2,2,1,2,2,2,1,1,2,\overbrace{0,\ldots,0}^{16 \text{ times}}), \quad \overleftarrow{B_3}$$

where $\overleftarrow{(x_1,\ldots,x_n)} = (x_n,\ldots,x_1)$. The points in $\{1,2,3,6,7,8,11,12,13,16\}$ each have an expected capacity of 6/6, those in $\{4,9,14\}$ each have an expected capacity of 5/6 while those in $\{5,10,15\}$ each have an expected capacity of 3/6. Thus every five successive points in $[1,16]$ have an expected capacity of $(18 + 5 + 3)/6 = 26/6 = (q - 2)\lambda$. By symmetry the same holds for every five successive points in $[17,32]$. Furthermore every 5 consecutive points containing 16 and 17 must contain at least 1 of 15 and 18 and at least 1 of 14 and 18 so has an expected capacity of at most 26/6. Thus the value of the game is at most 26/6. □

By introducing some notation we can see that the example has a structure which will be useful in the general case. Let g_i $(i = 1, \ldots, m)$ denote sequences of lengths $\alpha_1, \ldots, \alpha_m$ respectively, then we use $g_1 \oplus g_2 \oplus \cdots \oplus g_m$ to denote the sequence of length $\alpha_1 + \cdots + \alpha_m$ given by the members of g_1 in order, followed by the members of g_2 in order and so on, finishing up with the members of g_m in order. Thus, putting $J_5(1) = (2,2,2,1,2)$, $J_5(2) = (2,2,1,1,2)$ and letting m_t denote the sequence of t m's, B_1 in our example can be written as $2_1 \oplus J_5(2) \oplus J_5(1) \oplus J_5(1) \oplus 0_{16}$. It is then clear that B_2 and B_3 can be obtained from B_1 by suitably permuting the J_5's in B_1.

More generally let $J_p(w)$ denote the sequence $(x_1, \ldots, x_p)$ with $x_i = 1$ for $p - w \leq i \leq p - 1$ and $x_i = 2$ otherwise; in particular $J_p(0) = 2_p$. For w_i satisfying $0 \leq w_i \leq p - 2$, let

$$H_p(w_1, \ldots, w_\mu) = 2_1 \oplus J_p(w_1) \oplus \cdots \oplus J_p(w_\mu) \oplus 0_{\mu p + 1}.$$

Thus, in the example, $B_1 = H_5(2, 1, 1)$, $B_2 = H_5(1, 2, 1)$ and $B_3 = H_5(1, 1, 2)$. Note that $H_p(w_1, \ldots, w_\mu)$ represents a BLUE strategy in the game in which $n = 2\mu p + 2$, $c = 2$ and $q = 2(\mu p + 1) - \sum_{i=1}^{\mu} w_i$ as does $H_p(w_{\sigma(1)}, \ldots, w_{\sigma(\mu)}) = H_p(\sigma w)$ (abusing notation) where σ is any permutation of $\{1, \ldots, \mu\}$. Given $w = (w_1, \ldots, w_\mu)$, let

$$\mathscr{H}_p(w) = \{H_p(\sigma w) : \sigma \in \mathscr{I}(\mu)\} \cup \{\overleftarrow{H_p(\sigma w)} : \sigma \in \mathscr{I}(\mu)\}$$

where $\mathscr{I}(\mu)$ denotes the set of permutations of $\{1, 2, \ldots, \mu\}$

Theorem 2. *For* $p > 2$, $n = 2\mu p + 2$, $c = 2$ *and* q *satisfying* $\mu(p+2) + 2 \leq q \leq 2\mu p + 1$, *the value of the game is* $(q - c)/(2\mu)$.

Proof. RED can ensure an expectation $(q - c)/(2\mu)$ by Lemma 2 so we only need to show that BLUE can restrict RED to that expectation. Take any $w = (w_1, \ldots, w_\mu)$ satisfying $0 \leq w_i \leq p - 2$ and $\sum_{i=1}^{\mu} w_i = 2(\mu p + 1) - q$; such a w exists because $1 \leq 2(\mu p + 1) - q \leq \mu(p - 2)$. Suppose BLUE adopts the strategy which picks a member of $\mathscr{H}_\mu(w)$ at random, then x_1, $x_2 \in [1, \mu p + 1]$ satisfying $x_1 - x_2 = 0$ (mod p) have the same expected allocation. Thus, if the RED strategies starting at $1, \ldots, p$ all have expectation at most $(q - c)/(2\mu)$, then so does every RED strategy contained in $[1, \mu p + 1]$. Put $\rho_w(m) = |\{i : w_i \geq m\}|$, then the expected allocation of $j \in [1, p]$ is $(2 - \rho_w(p + 1 - j)/\mu)/2$ which is a decreasing function of j. Hence every RED strategy contained in $[1, \mu p + 1]$ has an expected allocation of $p - \sum_{j=1}^{p} \rho_w (p + 1 - j)/(2\mu)$. Let $t_m = |\{j : w_j = m\}|$, then

$$2(\mu p + 1) - q = \sum_{i=1}^{\mu} w_i = \sum_{m=1}^{p} m t_m = \sum_{m=1}^{p} m(\rho_w(m) - \rho_w(m + 1)) = \sum_{m=1}^{p} \rho_w(m).$$

Thus every RED strategy contained in $[1, \mu p + 1]$, and by symmetry, every RED strategy contained in $[\mu p + 2, n]$, has an expected allocation of $(q - 2)/(2\mu)$.

Suppose a RED strategy contains both $\lambda p + 1$ and $\lambda p + 2$. By symmetry the expected allocations of $\lambda p + 1 - j$ and $\lambda p + 2 + j$ are the same for $j = 1, \ldots, p - 2$ and, from the above, we know that these allocations are increasing functions of j. Hence every RED strategy containing both $\lambda p + 1$ and $\lambda p + 2$ has an expected allocation of at most that of $[(\mu - 1)p + 2, \mu p + 1]$ which has the same expected allocation as $[2, p + 1]$. Thus RED has an expectation of at most $(q - c)/(2\lambda)$. $\square$

Apart from the extreme cases $q = 2\mu p + 1$ and $q = \mu(p + 2) + 2$, there are several possibilities for the choice of $(w_1, \ldots, w_\mu)$ in the proof of the previous theorem so there are in general a number of optimal strategies for BLUE. However the examples in [7] show that there are BLUE optimal strategies which do not follow our structure, illustrating that it may not be easy to home in on particular BLUE optimal structures for other cases. It is probably over-optimistic to hope for a complete solution of the game without further inroads into special subcases being made first. Zoroa, Fernández-Sáez and Zoroa have solved the game when q is relatively small so it would seem that the two subcases that present the best chance of progress on an analytical front are:

Problem 3. Solve The Integer Number Hides Game with Capacities for comparatively large q.

Problem 4. Solve The Integer Number Hides Game with Capacities for $c = 2$.

Like the Several Intervals Game, one feels that a comprehensive solution of this game may need the insight given by computer generated solutions where the computer has been programmed to target certain types of solution.

A natural variation of the above game in which BLUE has an amount q, not necessarily an integer, of divisible material to hide was introduced by Zoroa, Fernández-Sáez and Zoroa in [8]. It can be formulated as follows.

> BLUE has an amount q, not necessarily an integer, of divisible material to hide in the integer interval I and must choose a subinterval B of I with length at most b in which to do so under the restriction that an amount of at most c is allocated to each point of B. Simultaneously RED picks a subinterval R of length r and gets a payoff equal to the amount of material that BLUE allocated to the points of R.

It is not easy to give a summary of the theorems obtained in [8] which does justice to them without involving detailed notation so the reader is encouraged to read the paper itself. Although a complete solution appears to be extremely difficult, many open questions regarding partial results suggest themselves. Note that, in this game, BLUE is allowed to put an amount zero at some of the chosen points so it is not totally obvious that there is a close connection between this game and the previous one. Thus it is of interest that [8] points out that there are similarities between the two in some cases.

6.7 Hiding in a Disc Game

The Hiding in a Disc game is very simple to state and easy to understand but seems difficult to solve. It can be described as follows.

> Without knowing each other's choices, RED and BLUE choose points r and b in a disc D with centre O and radius one. The payoff to RED in this zero-sum game is one if $|r - b| \leq c$ and zero otherwise.

When $1/\sqrt{2} \leq c < 1$, the value is the ratio of the length of the arc whose chord has length c to the circumference of D; optimal strategies for BLUE and RED are to choose a point according to a uniform distribution on the circumference of D and on the circumference of a circle with centre O and radius $\sqrt{1 - c^2}$ respectively.

What intrigued me when I first read Ruckle's book was that the solution for $1/2 < c < 1/\sqrt{2}$ was still open, particularly so because the value when $c = 1/2$ is known. Ruckle showed that, in the range, RED can guarantee a payoff of at least $(1/\pi)\arccos(1/2c)$ demonstrating that a solution communicated to the American Mathematical Monthly was incorrect. The RED strategy which ensured this payoff seemed an intuitively natural one so all that was needed was to produce a BLUE strategy showing RED could do no better. However the "all" proved elusive and, after prolonged efforts, I gave the problem best. On re-reading Ruckle's book recently I was curious whether progress had been made on the problem but I have been unable to find references to it.

It is natural to wonder whether a symmetry argument which is standard in the literature might be of use for this game; for a formal group theoretic justification of the following process see [1]. Let $\mathscr{A}$ denote the set of all rotations about the centre and let Γ denote the symmetized version of the game in which, after RED and BLUE choose strategies r and b respectively, a random (equiprobable) member γ is selected from $\mathscr{A}$ and the payoff $P(\gamma r, b) = P(r, \gamma^{-1}b)$ is assigned to RED. Observe that either player can ensure that Γ is played by applying a random automorphism to his own strategy so its value must be the same as that of the original game. Hence the Hiding in a Disc game is solved once Γ is solved. We may therefore regard mixed strategies of the players in Γ as distributions over the equivalence classes of $\mathscr{A}$ so that the strategy spaces are represented by the unit interval. The optimal strategies of the game given in [4] can all be expressed in Γ as probability distributions over a finite number of points in the unit interval. Unfortunately the payoff of the symmetrized game is much more complicated than that for the original game so there may be few practical benefits of symmetrising this particular game. However it does highlight a question that is of interest.

Question 2. In the symmetrized Hiding in a Disc game, do there always exist optimal strategies for the players which are probability distributions over a finite number of points in the unit interval?

6.8 A General Ruckle-Type Game

Ruckle proposed the problem of solving the Hiding in a Disc Game played on a set S more general than the circular disc. As that game appears to be still unsolved, it might seem somewhat bizarre to give a game formulation of which it is a special case. However the game we now introduce does show that a number of win-lose Ruckle-type games (payoff 0 or 1) do have a common structure and that there may be interesting research to be done on them in topologies other than the Euclidean one. The reader is reminded that a closed ball with centre c and radius r is the set of points which are at a distance less than or equal to r from c.

Let $\Gamma_S(b; r_1, \ldots, r_k)$ denote the following two-person zero-sum game played on a convex compact subset S of R^n endowed with a topology from a metric. BLUE chooses a closed ball B of radius b and RED closed balls $R_1, \ldots, R_k$ of radii $r_1, \ldots, r_k$ where all the closed balls have their centres in S. The payoff (to RED) is 1 if $S \cap B \cap \bigcup_{i=1}^{k} R_i \neq \emptyset$ and 0 otherwise.

Note that the Hide in a Disc Game is equivalent to one in which BLUE chooses a closed disc of radius r_B and RED a closed disc of radius $c - r_B$. Thus, when the topology is given by the Euclidean metric, special cases of $\Gamma_S(b; r_1, \ldots, r_k)$ include:

- *Hide in a Disc Game* where S is the unit disc, $b = r_b, k = 1$ and $r_1 = c - r_B$;
- *Several Intervals Game* where S is the unit interval, the closed balls are closed intervals and $b = 0$;
- *Several Intervals Game Variation* in which BLUE chooses an interval of length b instead of a point.

So far research on Ruckle-type games has almost exclusively concerned itself with problems in which the Euclidean topology is employed but, from a games that people do not play standpoint, there is no reason why other topologies should be ignored. In particular the Euclidean topology is a special case ($p = 2$) of the topology given by the distance function

$$||x - y||_p = \left(\sum_{i=1}^{n} (|x_i - y_i|)^p \right)^{1/p}$$

where $x = (x_1, \ldots, x_n)$, $y = (y_1, \ldots, y_n)$ and $p \geq 1$. When $p = 1$, we have the Manhattan, or taxicab, topology whereas the limit as $p \to \infty$ gives the Chebyshev topology which is represented by the distance function $||x - y||_\infty = \max_{1 \leq i \leq n} |x_i - y_i|$. Note that closed balls in R^2 takes the shape of a square in the Chebyshev topology and the shape of a diamond in the Manhattan topology. A closed ball in R^1 is a linear segment for all p.

In contrast to the Euclidean version, the Chebyshev Hide in a Disc Game, even in its n-dimensional form, is easy to solve; it can be stated as follows:

Play takes place in I^n where I denotes the unit interval. BLUE chooses a point and RED a n-cube of side $2r_1$ and the payoff to RED is 1 if BLUE's point is in RED's cube and zero otherwise.

Let $\mathcal{G}$ be a minimum cover of I^n by cubes of side $2r_1$ then, if RED chooses a member of $\mathcal{G}$ at random, RED's expectation is at least $1/|\mathcal{G}|$. Let m be the positive integer such that $2mr_1 < 1 \leq 2mr_1 + 1$, δ satisfy $0 < \delta < (1 - 2mr_1)/m$ and $P(i) = 2i(r_1 + \delta)$. If BLUE selects one of the points $\mathcal{B} = \{(P(i_1), \ldots, P(i_n)) : i_j = 0, 1 \ldots, m$ for $j = 1, \ldots, n\}$ at random, then

$$||(P(i), P(j)) - (P(s), P(t))||_1 = \max\{|2(s - i)(r_1 + \delta)|, |2(t - j)(r_1 + \delta)|\}$$
$$\geq 2(r_1 + \delta).$$

Thus any closed ball of radius r_1 contains at most one point of $\mathcal{B}$ so BLUE can restrict RED's expectation to at most $1/(m + 1)^2$. But, taking $S(i, j)$ to denote the closed ball with centre at the point $((2i + 1)r_1, (2j + 1)r_1 + 1)$ and radius r_1, $\{S(i, j) : 0 \leq i, j < m\}$ is a cover of $I \times I$ containing $(m + 1)^2$ members so RED can expect at least $1/(m + 1)^2$ and the game is solved.

In addition to problems in the topology given by $||x - y||_p$, readers who relish the more esoteric problems may like to investigate problems in the topology arising from the distance function $d(x,y) = \sum_{i=1}^{n} |x_i - y_i|^p$ where $0 < p < 1$; in this topology the closed ball in R^2 is not convex.

6.9 Conclusions

In this chapter we have investigated only a few of the games proposed by Ruckle in his book but they indicate how the apparently simple games there can provide a challenge in themselves or the foundation for significant generalisations. A common thread running through most of the games is that there are optimal strategies which involve, in some way, coverings of the set the game is played on. Research problems are the lifeblood of any mathematical discipline and it is hoped that it has been shown that Ruckle's problems are in rude health. However one can also expect the games to evolve in different directions. With the current global financial crisis there is a much greater questioning as to whether projects are affordable so a natural direction would be to incorporate costs into many of Ruckle's games. For instance the several intervals game has been interpreted as a game in which a defender puts detecting devices (intervals) across a channel in an attempt to detect an infiltrator but little interest has so far been shown in creating scenarios in which the defender has a limited budget and the more efficient the device (the larger the length of the interval) the greater the cost of deployment. Be that as it may, the important aspect of Ruckle's games from my viewpoint is that they provide one with intellectual fun. His book even includes a game on a Möbius band; definitely a game that people don't play.

References

1. S. Alpern and M. Asic: *The Search Value of a Network.* Networks **15** 229–238 (1985).
2. S. Alpern and S. Gal: *The Theory of Search Games and Rendezvous.* Kluwer, Boston (2003).
3. V. J. Baston, F.A. Bostock and T. S. Ferguson: *The Numbers Hides Game*, Proc. American Math. Soc. **107** 437–447 (1989).
4. W. H. Ruckle: *Geometric Games and Their Applications.* Pitman, Boston (1983).
5. I. Woodward, I: *Cable Laying Ambush Games,* Ph.D. thesis, University of Southampton (2002).
6. P. Zoroa, N. Zoroa and J. M. Ruiz: *Juego de Intersección de Intervalos Finitos,* Revista de la Real Academia de Ciencias Exactas, Fisicas y Naturales, Madrid, Spain, **82** 469–481 (1988).
7. N. Zoroa, M. J. Fernández-Sáez and P. Zoroa: *A Game Related to the Number Hides Game,* JOTA (**103**) 457–473 (1999).
8. N. Zoroa, M. J. Fernández-Sáez, M. J., and P. Zoroa: *Search and Ambush Games with Capacities,* JOTA (**123**) 431–450 (2004).

Chapter 7
The Cardinality of the Sets Involved in Lattice Games

Noemí Zoroa, María-José Fernández-Sáez, and Procopio Zoroa

Abstract Lattice games were introduced by Ruckle in (Geometric games and their applications. Pitman Advanced Publishing Program, 1983). These are games on the Lattice games where at least one of the players can move only from one point to an adjacent lattice point. This restriction on the movements of the player is realistic because it expresses that his movements are difficult. Although different results have been obtained for games on the lattice since the book of Ruckle, the work on lattice games is very scarce, and none of the problems set up there has been totally solved. In this chapter we obtain the cardinalities of the sets of strategies for the players of lattice games, this is the first of the problems proposed by Ruckle, and we hope, as does he that it will be of value in attacking such games.

7.1 Introduction

In this chapter we deal with two-person zero-sum games on the lattice

$$L = \{1, 2, \ldots, n\} \times \{1, 2, \ldots, m\}$$

in which, one of the sets of strategies for the players is the set of all the functions from $\{1, 2, \ldots, n\}$ into $\{1, 2, \ldots, m\}$ such that $f(i+1)$ equals one of the three values $f(i)$, $f(i+1)$, or $f(i-1)$ or a subset of it. Ruckle, in his interesting book [5] on geometric games, includes some games of these kind and calls them lattice games. In Chap. 3 we considered algorithms for solving search games on a lattice. In this chapter, we consider the cardinality of the strategy space.

N. Zoroa (✉) • M.-J. Fernández-Sáez • P. Zoroa
Faculty of Mathematics, Department of Statistics and Operational Research,
University of Murcia, Campus of Espinardo, 30071, Murcia, Spain
e-mail: zaroa@um.es; majose@um.es; procopio@um.es

S. Alpern et al. (eds.), *Search Theory: A Game Theoretic Perspective*,
DOI 10.1007/978-1-4614-6825-7_7, © Springer Science+Business Media New York 2013

Let $\mathscr{F}_{n,m}$ denote the set of all the functions from $\{1,2,\ldots,n\}$ to $\{1,2,\ldots,m\}$, that is $\mathscr{F}_{n,m}$ consists of all subsets of L which have a single point in each column, and let $\mathscr{F}_{n,m}^0$ denote the subset of $\mathscr{F}_{n,m}$ containing the functions such that $f(i+1)$ equals one of the three values $f(i)$, $f(i+1)$, or $f(i-1)$. The lattice games presented by Ruckle in [5] are the Lattice Ambush Game (LAG), the Lattice Search Game (LSG), the Lattice Penetration Game (LPG), and those obtained from these on laying out the lattice on a cylinder, the Cylindrical Ambush Game (CAG), the Cylindrical Search Game (CSG) and the Cylindrical Penetration Game (CPG).

In the LAG the set of strategies for player I is $\mathscr{F}_{n,m}^0$, the set of strategies for player II is $\mathscr{F}_{n,m}$ and the payoff to player I is 1 if both players do not meet, and zero otherwise. In the LSG the sets of strategies for both players are the same as in the LAG and the payoff to player I is equal to 0 if they do not meet and 1 if they meet. In the third game, the LPG, player I receives a payoff equal to his degree of penetration in the lattice if he is not intercepted, and 0 otherwise. The set of strategies of player I is $\mathscr{F}_{1,m}^0 \cup \mathscr{F}_{2,m}^0 \cup \cdots \cup \mathscr{F}_{n,m}^0$ and the set of strategies for player II is $\mathscr{F}_{n,m}$. The CAG, CSG and CPG are games that have the same set of strategies for player II and the payoff functions are equal to those of the LAG, LSG and LPG respectively, but player I is allowed to pass from one edge of the lattice to another. For all these games Ruckle obtains results for some special cases, but the general solutions are not obtained.

Although there has been activity in the study of games on the lattice [1, 6–8], there are few results on lattice games. The CAG is studied in [7] where it is called ambush game over time on a cyclic set; it is solved for the cases $n=2$, $n=3$, $m > 3(n-1)$ and for some cases when $m = 3(n-1)$. Bounds for the value of the game are also obtained, but the general solution is not obtained. Patrolling games on different graphs are studied in [1], the games studied when the graph is a line graph are similar to the LSG and the CSG, but in these games the attacker attack just one node along a period of time of length k. Properties for the optimal strategies for the players and the solution for some particular cases are obtained. Ruckle remarks that, among the games on the lattice, the hardest to handle are the lattice games, because of the combinatorial difficulties involved. In an attempt to shed some light on lattice games we have studied the sets of strategies of the players and we have obtained the cardinalities of all the sets involved in these games.

A two-person zero-sum game will be expressed by $G = (X,Y,M)$ where X, Y are the sets of pure strategies for players I and II, respectively, and

$$M : X \times Y \to \quad \mathbb{R} \tag{7.1}$$

is the payoff function which represents the winnings of player I and the losses of player II. Player I chooses a strategy $A \in X$, player II chooses a strategy $B \in Y$ and these choices determine the payoff $M(A,B)$ to player I and $-M(A,B)$ to player II.

Throughout this chapter X and Y are finite sets, therefore a probability distribution on X, that is to say, a mixed strategy for player I, can be written as a function

$$x : X \to \mathbb{R}$$

such that $x(C) \geq 0$ for all $C \in X$ and $\sum_{C \in X} x(C) = 1$. Similarly, a mixed strategy for player II will be given by a function

$$y : Y \to \mathbb{R}$$

such that $y(C) \geq 0$ for all $C \in Y$ and $\sum_{C \in Y} y(C) = 1$. When the players use their mixed strategies x and y, the payoff $M(x,y)$ is the expected value of $M(A,B)$.

A very simple lattice game $G = (X,Y,M)$ is the following, $X = Y \subset \mathscr{F}_{n,m}^0$ and

$$M(A,B) = \begin{cases} 1 & \text{if } A = B, \\ 0 & \text{if } A \neq B. \end{cases} \tag{7.2}$$

It is easy to see that an optimal strategy for both players is the uniform distribution on the set of their pure strategies and the value of the game v is given by

$$v = \frac{1}{|X|},$$

but, to completely solve this game we have to know the cardinality of the set X.

Given $A \in \mathscr{F}_{n,m}$ we denote its increments by $\Delta A(i)$, $\Delta A(i) = A(i+1) - A(i)$ for $i = 1,2,\ldots,n-1$ and $\Delta A(n) = A(n) - A(1)$. We consider the following sets:

$$\mathscr{F}_{n,m}^0 = \{A \in \mathscr{F}_{n,m} : \Delta A(i) \in \{0,1,-1\}, i = 1,\ldots,n-1\}, \tag{7.3}$$

$$\mathscr{F}_{n,m}^1 = \{A \in \mathscr{F}_{n,m} : \Delta A(i) \in \{0,1,-1,m-1,1-m\},$$
$$i = 1,\ldots,n-1\}, \tag{7.4}$$

$$\mathscr{F}_{n,m}^2 = \{A \in \mathscr{F}_{n,m} : \Delta A(i) \in \{0,1,-1\}, i = 1,\ldots,n\}, \tag{7.5}$$

$$\mathscr{F}_{n,m}^3 = \{A \in \mathscr{F}_{n,m} : \Delta A(i) \in \{0,1,-1,m-1,1-m\}, i = 1,\ldots,n\} \tag{7.6}$$

Clearly $\mathscr{F}_{n,m}^2 \subset \mathscr{F}_{n,m}^0 \subset \mathscr{F}_{n,m}$ and $\mathscr{F}_{n,m}^3 \subset \mathscr{F}_{n,m}^1 \subset \mathscr{F}_{n,m}$. The elements of the set $\mathscr{F}_{n,m}^2$ can be interpreted as paths on the cylinder, that is paths on the lattice $\{1,2,\ldots,n+1\} \times \{1,2,\ldots,m\}$ where the points $(1,j)$ and $(n+1,j)$ are considered to be the same point. In a similar way the elements of the set $\mathscr{F}_{n,m}^1$ can be interpreted as paths which can surround a cylinder one or more times and the elements of $\mathscr{F}_{n,m}^3$ as paths which can surround a cylinder or as paths on a torus. Figure 7.1 shows a representation of the element $\{(1,6),(2,7),(3,8),(4,9),(5,10),(6,1),(7,2)(8,2),$ $(9,3),(10,4),(11,3),(12,4),\ (13,5)\} \in \mathscr{F}_{13,10}^3 \subset \mathscr{F}_{13,10}^1 \subset \mathscr{F}_{13,10}$ on the lattice $L = \{1,2,\ldots,13\} \times \{1,2,\ldots,10\}$ and on the torus. The sets $\mathscr{F}_{n,m}^0, \mathscr{F}_{n,m}^1, \mathscr{F}_{n,m}^2$ and $\mathscr{F}_{n,m}^3$ appear in many situations in which the path of a person needs to be described.

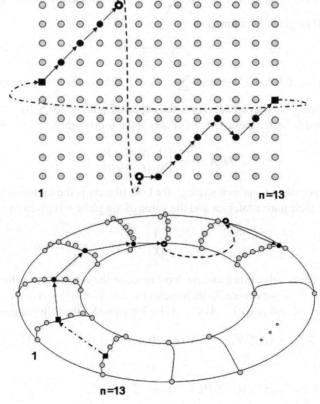

Fig. 7.1 Paths on the torus

To obtain the cardinality of these sets we will use the tridiagonal Toeplitz matrices. Other games where these matrices appear can be seen in [2].

7.2 The Cardinality of $\mathscr{F}_{n,m}^0$ and $\mathscr{F}_{n,m}^2$

Let us consider the set $\mathscr{F}_{n,m}^0$ defined by (7.3). Let a pair r, s be given, $1 \leq r \leq m, 1 \leq s \leq m$, and let us denote by $a_{rsm}^{(n)}$ the number of paths of $\mathscr{F}_{n,m}^0$ satisfying $A(1) = r$ and $A(n) = s$, since m is fixed, we omit it for simplification, so

$$a_{rs}^{(n)} = \left| \{A \in \mathscr{F}_{n,m}^0 : A(1) = r, A(n) = s\} \right|$$

and hence

$$|\mathscr{F}_{n,m}^0| = \sum_{r,s} a_{rs}^{(n)}, r = 1, \ldots, m, s = 1, \ldots, m. \tag{7.7}$$

Let B denote the square matrix of order m

$$B = \begin{bmatrix} 1 & 1 & 0 & 0 & \dots & 0 \\ 1 & 1 & 1 & 0 & \dots & 0 \\ 0 & 1 & 1 & 1 & \dots & 0 \\ \cdot & \cdot & \cdot & \cdot & \cdot & \cdot \\ 0 & 0 & 0 & 0 & \dots & 1 \end{bmatrix},$$ (7.8)

that is the matrix having zeros as elements everywhere, with the exception of the elements of the principal diagonal and that immediately above and below which are equal to 1.

If we call $M^{(n)}$ the square matrix of order m whose elements are the $a_{rs}^{(n)}$, $r = 1, \dots, m$, $s = 1, \dots, m$

$$M^{(n)} = \left[a_{rs}^{(n)} \right],$$ (7.9)

it is clear that

$$M^{(2)} = B,$$

$$M^{(n)} = M^{(n-1)}B,$$

and so

$$M^{(n)} = \left[a_{rs}^{(n)} \right] = B^{n-1}.$$

The following lemma and proposition are known, they give us the tools to obtain the cardinalities of the sets $\mathscr{F}_{n,m}^0$ and $\mathscr{F}_{n,m}^2$.

Lemma 1. *The eigenvalues of the square matrix B of order m, given by (7.8) are*

$$\lambda_k = 1 - 2\cos\frac{k\pi}{m+1}, \quad k = 1, \dots, m,$$ (7.10)

and the components of the eigenvector $c_{.k}$ (column vector) corresponding to λ_k are

$$c_{hk} = (-1)^{h+k}\sqrt{\frac{2}{m+1}}\sin\left(hk\frac{\pi}{m+1} \right), \quad h = 1, \dots, m.$$ (7.11)

Let us denote by C the square matrix of order m whose elements are the c_{hk}, $C = [c_{hk}]$. Then C is an orthogonal and symmetric matrix, that is to say,

$$C^T = C, \quad CC = I, \quad C\Lambda C = B$$ (7.12)

is satisfied, where Λ is the diagonal matrix whose elements of the principal diagonal are the eigenvalues λ_k.

Proof. Let T be the square matrix

$$T = \begin{bmatrix} b & c & 0 & 0 & \cdots & 0 \\ a & b & c & 0 & \cdots & 0 \\ 0 & a & b & c & \cdots & 0 \\ \cdot & \cdot & \cdot & \cdot & \cdots & \cdot \\ 0 & 0 & \cdots & a & b & c \\ 0 & 0 & \cdots & 0 & a & b \end{bmatrix},$$

where $a \neq 0$ and $b \neq 0$. These are the tridiagonal Toeplitz matrices, these matrices are among the few nontrivial structures that admit formulas for their eigenvalues and eigenvectors. Matrix B given by (7.8) is the particular case of matrix T where $a = b = c = 1$, therefore it is known that its eigenvalues are given by (7.10) and the components of the eigenvector $c_{.k}$ (column vector) corresponding to λ_k are given by (7.11). See [4].

Since the m eigenvalues are all different, it follows that the eigenvectors are orthogonal, that is to say

$$c_{.k}^T c_{.k'} = 0 \quad \text{if} \quad k \neq k'.$$

With the constants chosen in the elements c_{hk} the equalities

$$c_{.k}^T c_{.k} = \sum_{h=1}^{m} c_{hk}^2 = 1, \quad k = 1,\ldots,m$$

are satisfied, this can be proved directly or by applying Lemma 2, which we will see below. Therefore (7.12) is proved. $\square$

Proposition 1. *The number of elements of $\mathscr{F}_{n,m}^0$ satisfying $A(1) = r$ and $A(n) = s$ is given by $a_{rs}^{(n)}$, and is equal to*

$$a_{rs}^{(n)} = (-1)^{r+s} \frac{2}{\pi} \sum_{k=1}^{m} \frac{\beta_k}{k} \, (1 - 2\cos\beta_k)^{n-1} \sin(r\beta_k)\sin(s\beta_k)$$

where

$$\beta_k = \frac{k\pi}{m+1}. \tag{7.13}$$

Proof. A proof of this result can be found in [3]. We can provide the following proof; let C and Λ be the given in Lemma 1. $CC = I$ and $C\Lambda C = B$ is satisfied, therefore

$$M^n = B^{n-1} = C\Lambda^{n-1}C \tag{7.14}$$

is fulfilled. From the above equality it follows that

$$a_{rs}^{(n)} = \sum_{k=1}^{m} \lambda_k^{n-1} c_{rk} c_{sk}$$

$$= (-1)^{r+s} \sum_{k=1}^{m} (1 - 2\cos\beta_k)^{n-1} \frac{2}{m+1} \sin(r\beta_k)\sin(s\beta_k)$$

$$= (-1)^{r+s} \frac{2}{\pi} \sum_{k=1}^{m} \frac{\beta_k}{k} \, (1 - 2\cos\beta_k)^{n-1} \sin(r\beta_k)\sin(s\beta_k) \tag{7.15}$$

where β_k is given by (7.13), and the proof is complete. $\square$

Theorem 1. *The cardinality of the set $\mathscr{F}_{n,m}^0$ is given by the expression*

$$\left|\mathscr{F}_{n,m}^0\right| = \sum_{k=1}^{m} (1 - 2\cos\beta_k)^{n-1} \frac{1 + (-1)^{m+k}}{(m+1)} tg^2 \frac{\beta_k}{2}, \qquad (7.16)$$

where β_k is given by (7.13).

Proof. We can express the cardinality of $\mathscr{F}_{n,m}^0$ as

$$\left|\mathscr{F}_{n,m}^0\right| = \sum_{h,r} a_{hr}^{(n)} = \sum_{k=1}^{m} \lambda_k^n \sum_{h,r} c_{hk} c_{rk} = \sum_{k=1}^{m} \lambda_k^n \left(\sum_{h=1}^{m} c_{hk}\right)^2. \qquad (7.17)$$

First we will compute the sum

$$\sum_{h=1}^{m} c_{hk} = (-1)^k \sqrt{\frac{2}{m+1}} \sum_{h=1}^{m} (-1)^h \sin(h\beta_k) = (-1)^k \sqrt{\frac{2}{m+1}} S. \qquad (7.18)$$

The value of S is obtained from the imaginary part of the expression

$$\sum_{h=0}^{m} (-1)^h \exp(ih\beta_k)$$

$$= \frac{1 + (-1)^m e^{i(m+1)\beta_k}}{1 + e^{i\beta_k}}$$

$$= \frac{1 + (-1)^m e^{ik\pi}}{1 + e^{i\beta_k}} = \frac{1 + (-1)^m \cos(k\pi)}{1 + e^{i\beta_k}}$$

$$\frac{1 + (-1)^{m+k}}{e^{i\frac{\beta_k}{2}}(e^{-i\beta_k/2} + e^{i\beta_k/2})} = \frac{e^{-i\frac{\beta_k}{2}}(1 + (-1)^{m+k})}{2\cos\frac{\beta_k}{2}},$$

and the imaginary part of this value is equal to

$$\frac{\sin(-\frac{\beta_k}{2})(1 + (-1)^{m+k})}{2\cos\frac{\beta_k}{2}} = -tg\frac{\beta_k}{2}\frac{1 + (-1)^{m+k}}{2} = S. \qquad (7.19)$$

Therefore,

$$\sum_{h=1}^{m} c_{hk} = (-1)^k \sqrt{\frac{2}{m+1}} S = (-1)^{k+1} \sqrt{\frac{2}{m+1}} tg\frac{\beta_k}{2}\frac{1 + (-1)^{m+k}}{2}$$

and substituting the last value into (7.17) we conclude that

$$|\mathscr{F}_{n,m}^0| = \sum_{k=1}^{m} \lambda_k^{n-1} \left(\sum_{h=1}^{m} c_{hk} \right)^2 = \sum_{k=1}^{m} \lambda_k^{n-1} \frac{2}{m+1} \frac{(1+(-1)^{m+k})^2}{4} tg^2 \frac{\beta_k}{2}$$

$$= \sum_{k=1}^{m} (1 - 2\cos\beta_k)^{n-1} \frac{1+(-1)^{m+k}}{(m+1)} tg^2 \frac{\beta_k}{2},$$

which proves the theorem. □

Lemma 2. *The sum $\sum_{r=1}^{p} \sin^2 \frac{nr\pi}{p+1}$ is equal to 0 if n is a multiple of $(p+1)$, and to $\frac{p+1}{2}$ otherwise.*

Proof. It is clear that, if n is a multiple of $(p+1)$ the equality

$$\sum_{r=1}^{p} \sin^2 \frac{nr\pi}{p+1} = 0$$

is satisfied. Let us suppose that n is not a multiple of $(p+1)$, then

$$\sum_{r=0}^{p} \sin^2 \frac{nr\pi}{p+1} = \sum_{r=0}^{p} \frac{1}{2}(1 - \cos \frac{2nr\pi}{p+1}) = \frac{p+1}{2} - \frac{1}{2} \sum_{r=0}^{p} \cos \frac{2nr\pi}{p+1}. \qquad (7.20)$$

The last sum of the above expression is the real part of $\sum_{r=0}^{p} \exp \frac{i2nr\pi}{p+1}$, and with the notation

$$z = \exp \frac{i2n\pi}{p+1} \neq 1$$

we have

$$\sum_{r=0}^{p} \exp \frac{i2nr\pi}{p+1} = \sum_{r=0}^{p} z^r = \frac{1 - z^{p+1}}{1-z} = 0$$

and (7.20) gives the desired conclusion, which finishes the proof. □

Theorem 2. *The cardinality of $\mathscr{F}_{n,m}^2$ is equal to*

$$|\mathscr{F}_{n,m}^2| = \sum_{k=1}^{m} (1 - 2\cos\beta_k)^n$$

where β_k is given by (7.13).

Proof. We can express the cardinality of $\mathscr{F}_{n,m}^2$ in the way

$$|\mathscr{F}_{n,m}^2| = a_{11}^{(n)} + a_{12}^{(n)} + \sum_{r=2}^{m-1}(a_{rr-1}^{(n)} + a_{rr}^{(n)} + a_{rr+1}^{(n)}) + a_{mm-1}^{(n)} + a_{mm}^{(n)}$$

$$= a_{11}^{(n+1)} + \sum_{r=2}^{m-1} a_{rr}^{(n+1)} + a_{mm}^{(n+1)} = Trace \ of \ M^{(n+1)} = Trace \ of \ B^n$$

$$= \sum_{k=1}^{m}(1 - 2\cos\beta_k)^n \frac{2}{m+1} \sum_{h=1}^{m} \sin^2(h\beta_k)$$

and from Lemma 2 it follows

$$|\mathscr{F}_{n,m}^2| = \sum_{k=1}^{m}(1 - 2\cos\beta_k)^n, \qquad (7.21)$$

which establishes the formula. $\square$

7.3 The Cardinality of $\mathscr{F}_{n,m}^1$ and $\mathscr{F}_{n,m}^3$

Let us consider now the sets $\mathscr{F}_{n,m}^1$ and $\mathscr{F}_{n,m}^3$. The elements of $\mathscr{F}_{n,m}^1$ are the functions $A \in \mathscr{F}_{n,m}$ satisfying $\Delta A(i) \in \{0, 1, -1, m-1, 1-m\}$, for $i = 1, 2, \dots, n-1$, and $\mathscr{F}_{n,m}^3$ is the subset of $\mathscr{F}_{n,m}^1$ set up by the lattice paths satisfying the additional condition

$$\Delta A(n) = A(1) - A(n) \in \{0, 1, -1, m-1, 1-m\},$$

so the relation $\mathscr{F}_{n,m}^3 \subset \mathscr{F}_{n,m}^1 \subset \mathscr{F}_{n,m}$ is satisfied.

Theorem 3. *The cardinality of the set $\mathscr{F}_{n,m}^1$ is equal to $3^{n-1}m$ if $m \geqslant 3$, and to 2^n for $m = 2$.*

Proof. First let us assume $m \geqslant 3$. For every $A \in \mathscr{F}_{n,m}^1$ we define the n values $a, x_1, x_2, \dots, x_{n-1}$ to be

$$a = A(1), \ x_i = \begin{cases} \Delta A(i), & \text{if} \quad |\Delta A(i)| \leq 1, \\ -1, & \text{if} \quad \Delta A(i) = m-1, \\ 1, & \text{if} \quad \Delta A(i) = 1-m, \end{cases}$$

so

$$a \in \{1, 2, \dots, m\}$$

and

$$x_i \in \{0, -1, 1\}, \ i = 1, 2, \dots, n-1.$$

There is a bijection between the set of sequences $a, x_1, x_2, \dots, x_{n-1}$ and $\mathscr{F}_{n,m}^1$, and clearly the number of different possibilities for the sequences $a, x_1, x_2, \dots, x_{n-1}$ is equal to $m3^{n-1}$, which is the desired conclusion.

Now let $m = 2$, this part of the proof is similar to the previous one, but we have to bear in mind that, in this case, it is the same consider an increment equal to 1 than an increment equal to -1. For every $A \in \mathscr{F}_{n,2}^1$, we can define the n values $a, x_1, x_2, \ldots, x_{n-1}$ to be

$$a = A(1), \quad x_i = \begin{cases} 1, & \text{if } |\Delta A(i)| = 1; \\ 0, & \text{if } \Delta A(i) = 0; \end{cases}$$

then there is a bijection between the set of sequences $a, x_1, x_2, \ldots, x_{n-1}$ and $\mathscr{F}_{n,2}^1$, and clearly the number of different possibilities for the sequences $a, x_1, x_2, \ldots, x_{n-1}$ is equal to 2^n as it was stated, and the proof finishes. $\quad\square$

It is clear that $\mathscr{F}_{n,2}^3 = \mathscr{F}_{n,2}^1$ and $\mathscr{F}_{n,3}^3 = \mathscr{F}_{n,3}^1$. Therefore

$$\left|\mathscr{F}_{n,2}^3\right| = 2^n \quad \text{and} \quad \left|\mathscr{F}_{n,3}^3\right| = 3^n.$$

In the next theorem the cardinality of the rest of the sets $\mathscr{F}_{n,m}^3$ is obtained.

Theorem 4. *Let $m > 3$. The cardinality of the set $\mathscr{F}_{n,m}^3$ is given by*

$$\left|\mathscr{F}_{n,m}^3\right| = m \sum_{h=0}^{\left[\frac{n}{2}\right]} \binom{n}{2h}\binom{2h}{h}$$

$$+ 2m \sum_{r=1}^{\left[\frac{n}{m}\right]} \sum_{h=0}^{\left[\frac{n-rm}{2}\right]} \binom{n}{rm+2h}\binom{rm+2h}{h}. \quad (7.22)$$

Proof. For every $A \in \mathscr{F}_{n,m}^3$ we define the $n+1$ values $a, x_1, x_2, \ldots, x_n$ to be

$$a = A(1), \quad x_i = \begin{cases} \Delta A(i), & \text{if } |\Delta A(i)| \leq 1, \\ -1, & \text{if } \Delta A(i) = m-1, \\ 1, & \text{if } \Delta A(i) = 1-m, \end{cases}$$

clearly the sequence $a, x_1, x_2, \ldots, x_n$ determines the function A, so there is a bijection between the set of sequences $a, x_1, x_2, \ldots, x_n$ and the set $\mathscr{F}_{n,m}^3$. But now the values x_i satisfy the following relation

$$\sum x_i = rm, \quad -\left[\frac{n}{m}\right] \leq r \leq \left[\frac{n}{m}\right]. \quad (7.23)$$

The number of functions $A \in \mathscr{F}_{n,m}^3$ for which $r = 0$, clearly is given by

$$m \sum_{h=0}^{\left[\frac{n}{2}\right]} \binom{n}{2h}\binom{2h}{h}$$

and, for every r, $0 < r \leq \left[\frac{n}{m}\right]$, the number of $A \in \mathscr{F}_{n,m}^3$ for which $\sum x_i = rm$ or $\sum x_i = -rm$ is equal to

$$2m \sum_{h=0}^{\left[\frac{n-rm}{2}\right]} \binom{n}{rm+2h} \binom{rm+2h}{h}. \tag{7.24}$$

Combining (7.23) and (7.24) we can assert that the cardinality of $\mathscr{F}_{n,m}^3$ is given by (7.22), which finishes the proof. □

Remark 1. (a) Theorem 4 and its proof are also valid when $m = 3$, therefore in this case the value of expression (7.22) is equal to 3^n.

(b) Due to the symmetry it is clear that the number of functions belonging to $\mathscr{F}_{n,m}^1$ for which $A(1) = r$ is equal to the cardinality of $\mathscr{F}_{n,m}^1$ divided by m. In a similar way, the number of functions belonging to $\mathscr{F}_{n,m}^3$ for which $A(1) = r$ is equal to the cardinality of $\mathscr{F}_{n,m}^3$ divided by m.

Example 1. In a warlike situation the guerrilla wants to place n mines around the perimeter of a zone. There is a weak point to penetrate in the ring where the mines are going to be placed and it will be used by the guerrilla to do the incursion. The ring is discretized in a grid with n columns and m rows, considering column 1 as next to column n, and the weak point situated in the midst of them. The grid can be represented by

$$L = \{1, 2, \ldots, n\} \times \{1, 2, \ldots, m\},$$

and column i by the set $L_i = \{i\} \times \{1, 2, \ldots, m\}$. The guerrilla will set a mine in each column. In L_1 he will set the mine at any of its m points, but he moves with difficulty therefore if one mine is set at point (i, j) in column L_i the next mine has to be placed in column L_{i+1} at one of the points $(i+1, j-1)$, $(i+1, j)$, $(i+1, j+1)$. Bearing in mind that column 1 is next to column n, it follows that, if the first mine was placed in the point $(1, j)$ the last mine has to be placed at one of the points $(n, j-1)$, (n, j), $(n, j+1)$. Furthermore, the perimeter is protected by the army, which tries to deactivate as many set mines as it can, beginning by the the first column and ending in column L_n. The payoff to the army is maximum if it deactivates all the mines. In the rest of the cases the payoff is equal to the number of deactivated mines from column L_1 to the first column containing a mine that has not been deactivated, because the risk of an accident in this column is very big. This situation can be modeled by the two-person zero-sum game (X, Y, M) where player I is the army, player II the guerrilla, $X = Y = \mathscr{F}_{n,m}^2 = \{A \in \mathscr{F}_{n,m} : \triangle A(i) \in \{0, 1, -1\}, i = 1, \ldots, n\}$ and the payoff function

$$M(A, B) = \begin{cases} h & \text{if } A(i) = B(i) \quad \text{for} \quad i = 1, 2, \ldots, h, \quad A(h+1) \neq B(h+1), \\ kn & \text{if } A(i) = B(i) \quad \text{for} \quad i = 1, 2, \ldots, n. \end{cases}$$

where, k is a constant, $k > 1$.

Figure 7.2 shows a representation of the strategy

$$\{(1,4), (2,3), (3,2), (4,3), (5,4), (6,3), (7,2), (8,3), (9,4), (10,5), (11,6),$$
$$(12,5)(13,4)(14,5)\}$$

on the lattice $L = \{1, 2, \ldots, 14\} \times \{1, 2, \ldots, 8\}$ and on the ring. In this game an optimal strategy for both players is the uniform distribution over the set of their pure strategies

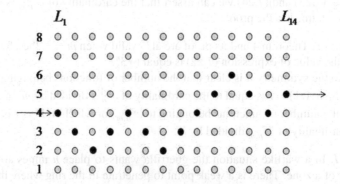

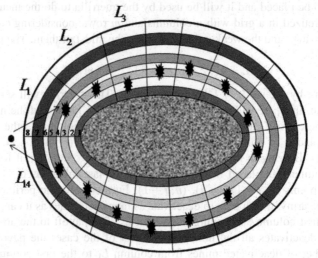

Fig. 7.2 Strategy on the lattice and on the ring

$$\alpha(A) = \frac{1}{|X|} = \frac{1}{|Y|} = \frac{1}{\left|\mathscr{F}_{n,m}^2\right|}.$$

To obtain the value of the game we have to compute

$$M(A, \alpha) = M(\alpha, B) = \frac{1}{\left|\mathscr{F}_{n,m}^2\right|} \sum_{A \in X} M(A, B)$$

let X_h the subset of X defined by

$$X_h = \{A \in X : A(i) = B(i) \quad \text{for} \quad i = 1, 2, \ldots, h, \quad A(h+1) \neq B(h+1)\}$$

then we can write

$$M(A,\alpha) = \frac{1}{|\mathscr{F}_{n,m}^2|}\left(\sum_{A\in X_1}1 + \sum_{A\in X_2}2 + \ldots + \sum_{A\in X_{n-1}}(n-1) + kn\right)$$

$$= \frac{\left(2\cdot 3^{n-2} + 2\cdot 3^{n-3}\cdot 2 + 2\cdot 3^{n-4}\cdot 3 + \ldots + 2\cdot 3\cdot(n-2) + 2(n-1) + kn\right)}{|\mathscr{F}_{n,m}^2|}$$

$$= \frac{2}{|\mathscr{F}_{n,m}^2|}\left(\sum_{h=1}^{n-1}h\,3^{n-h-1}\right) + \frac{kn}{|\mathscr{F}_{n,m}^2|}$$

$$= 2\frac{3^{n-1}}{|\mathscr{F}_{n,m}^2|}\left(\sum_{h=1}^{n-1}h\,3^{-h}\right) + \frac{kn}{|\mathscr{F}_{n,m}^2|} = 2\frac{3^{n-1}}{|\mathscr{F}_{n,m}^2|}\left(\frac{3^n - 2n - 1}{3^{n-1}2^2}\right) + \frac{kn}{|\mathscr{F}_{n,m}^2|}$$

$$= \frac{3^n - 2n(1-k) - 1}{2|\mathscr{F}_{n,m}^2|}$$

and this is the value of the game.

7.4 Further Results

Lattice games were introduced by Ruckle in [5]. These are games on the lattice where at least one of the players can move only from one point to an adjacent lattice point. This restriction on the movements of the player is realistic because it expresses that his movements are difficult. Although different results have been obtained for games on the lattice since the book of Ruckle, the work on lattice games is very scarce, and none of the problems set up there has been totally solved. In this chapter we obtain the cardinality of the sets of strategies for the players of lattice games, this is the first of the problems proposed by Ruckle, and we hope, as him, that it will be of value in attacking such games.

References

1. Alpern, S., Morton, A. and Papadaki, K.: Patrolling games. Oper. Res. 59, 1246–1257 (2011)
2. Garnaev, Y.: A remark on a helicopter and submarine game. Naval Res. Logistics 40, 745–753 (1993)
3. Krattenthaler, C. and Mohanty, S. G.: Lattice path combinatorics - applications to probability and statistics. http://www.mat.unuvie.ac.at/~kratt/artikel/encystat.html
4. Meyer, K.: Matrix Analysis and Applied Linear Algebra. Published electronically at http://matrixanalysis.com/DownloadChapters.html

5. Ruckle, W.: Geometric games and their applications. Pitman Advanced Publishing Program (1983)
6. Zoroa, N. and Zoroa, P.: Some games of search on a lattice. Naval Res. Logistics 40, 525–541 (1993)
7. Zoroa, N., Fernández-Sáez, M.J., Zoroa, P.: Patrolling a perimeter. Eur. J. Oper. Res. 222, 571–582 (2012)
8. Zoroa, N., Zoroa, P., Fernández-Sáez, M.J.: Weighted search games. Eur. J. Oper. Res. 195, 394–411 (2009)

Chapter 8
Effective Search for a Naval Mine with Application to Distributed Failure Detection

Jun Kiniwa, Kensaku Kikuta, and Toshio Hamada*

Abstract We consider a kind of reconnaissance problem which has an application to distributed failure detection. The problem can be considered as a multistage two-person zero-sum game. The two-person, player A and player B, consists of a transport ship and a terrorist, respectively, where the ship is equipped with an unmanned reconnaissance boat. The ship circulates ports again and again and the terrorist may lay a naval mine on the shipping route. For safety, the ship dispatches the unmanned reconnaissance boat and removes the risk of a mine. However, it is very rare that the terrorist lays the mine, while the circulation of the reconnaissance boat is very costly. So, we introduce a mine-preparing probability, represented by geometric distribution, preceding the terrorist's strategy. The ship has to determine when it should dispatch the boat so that it can maximize its expected payoff. First, we assume that the mine is laid at each beginning of a stage and investigate two cases, a game continuation case and a game termination case, after the ship has been broken by a mine. Next, we assume that the mine may be laid at any timing of a stage and investigate two methods, dispatching two boats and dispatching one boat, for the game continuation case. Finally, we state that the problem can also be applied to a failure detection problem in a distributed system if we regard the ship as a token and the terrorist as an adversary who causes a failure.

* Deceased.

J. Kiniwa (✉)
Department of Applied Economics, University of Hyogo, 8-2-1 Gakuen nishi-machi, Nishi-ku, Kobe-shi, 651-2197, Japan, Phone: +81-78-794-5844, Fax: +81-78-794-6166
e-mail: kiniwa@econ.u-hyogo.ac.jp

K. Kikuta
Department of Strategic Management, University of Hyogo, 8-2-1 Gakuen nishi-machi, Nishi-ku, Kobe-shi, 651-2197 Japan, Phone: +81-78-794-5829, Fax: +81-78-794-6166
e-mail: kikuta@biz.u-hyogo.ac.jp

S. Alpern et al. (eds.), *Search Theory: A Game Theoretic Perspective*,
DOI 10.1007/978-1-4614-6825-7_8, © Springer Science+Business Media New York 2013

8.1 Introduction

We consider a reconnaissance problem which has an application to distributed failure detection. The problem is described as follows. A transport ship circulates a route containing distinct n ports again and again. A terrorist aims to attack the ship by a naval mine. The ship can avoid the attack by dispatching an unmanned reconnaissance boat before starting from one port to another. However, it is costly to dispatch the boat at every starting time because the ship has to wait until the circulation of the boat. So, the ship uses a mixed strategy which maximizes the expected payoff in two-person zero-sum game, where the ship is considered as player A and the terrorist as player B [13].

The topic of this chapter is motivated by a distributed failure detection/repair problem. In distributed systems, e.g., the Internet, it is difficult to detect/repair a failure. So, it is useful to consider a self-mending system in particular for a *transient failure*, a kind of memory corruption. Self-stabilizing *mutual exclusion* systems [2, 3], in which at most one process is allowed to obtain a *privilege* to some resource, e.g., a shared printer, tolerate the transient failure. One of the systems uses a *token* circulation mechanism [14] and is carefully reinforced with a *sub-token*, called a *superstabilization* [8, 10]. The token and the sub-token in a distributed system correspond to the transport ship and the reconnaissance boat in our reconnaissance problem, respectively. In addition, the failure caused by an *adversary* is considered as the naval mine laid by a terrorist. The great advantage is that our problem enables us to omit the detailed implementation of the distributed system and to extract an essential point. Thus, we interpret the solution of the reconnaissance problem as that of the token circulation without proving the correctness of a protocol.

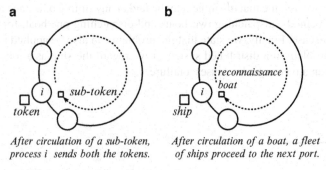

After circulation of a sub-token,   After circulation of a boat, a fleet
process i sends both the tokens.   of ships proceed to the next port.

Fig. 8.1 Superstabilizing protocol vs. mine reconnaissance

Figure 8.1a illustrates the superstabilizing protocol. The feature of the method is that the process holding a token always dispatches a sub-token before acquiring a privilege. When the sub-token meets a failure, it is corrected to a legitimate state. After the circulation of the sub-token, the process holding a token acquires a privilege. Figure 8.1b illustrates our reconnaissance problem, where the same strategies of the ship as those of the token can be discussed. Kiniwa and Kikuta [11] improved the superstabilizing protocol by taking a mixed strategy into consideration in two-person zero-sum game. In their chapter, the process holding a token issues a sub-token with some probability. However, their model is so simple and ignores empirical probability of failures. It follows that the adversary always aims to bring about a failure. Thus, an unlikely failure may frequently occur against our intuition.

In this chapter, we incorporate an empirical failure probability into our model, and adapt our idea to a practical case. We consider that the terrorist cannot make a decision until the naval mine is prepared, where the *mine-preparing probability* is empirical. We interpret an interval between failures as a state that a mine is not prepared and is followed by a state that a mine has been prepared but not laid yet. By adjusting the mine-preparing probability, we can use our idea in practice.

Our topic is related to a well-studied problem, called a sequential inspection game, consisting of an inspector and an inspectee, or customs and a smuggler, respectively [1, 5–7, 9, 12, 15]. It was originally started by Dresher[4] as arms control and disarmament. That is, the customs patrols in order to stop the smuggler attempting to ship a cargo of perishable contraband across a strait. The customs has limited resources, i.e., the number of boats, and can patrol only during k of n nights. The smuggler ships the cargo of contraband during l of n nights, where $l = 1$ in its original work. When the smuggling coincides with the patrol, the smuggler is captured with probability q, where originally $q = 1$. The modeling and the technique of this chapter is due to the inspection game.

The difference between our problem and the inspection game is as follows. First, in the inspection game, the inspectee may be captured, while in our problem not. On the contrary, the inspector in our problem may suffer severe damage. Second, the inspection game plans the number of the inspectee's illegal actions in advance, while our problem does not and the number of mine-laying is very rare. Third, the inspection game has the upper bound of the number of patrols, while our problem considers it as a cost.

The rest of this chapter is organized as follows. Section 8.2 states our model. Section 8.3 includes one stage case which presents our fundamental idea and an idea of mine-preparing probability. Then, for the iterated circulation of a ship, we have several interpretations. First, Sect. 8.3.1 considers that player A corresponds to a fleet of ships and our game continues after a ship suffers damage. Second, Sect. 8.3.2 considers that player A corresponds to only one ship and our game terminates when the ship suffers damage. Third, Sect. 8.4 changes the assumption adopted in previous sections and compares two boats case with one boat case. Fourth, Sect. 8.5 shows our application to distributed failure detection. Then, Sect. 8.6 concludes the chapter. Finally, Appendix refers to our distributed failure detection protocol.

8.2 Model

A fleet of ships transport (or a ship transports) goods/materials on a circular route, a *ring*, again and again. Since there is a risk that a terrorist may lay a naval mine on the route, it is sometimes necessary to take a scout along the route by using an unmanned reconnaissance boat. If the boat finds a naval mine, a crew on the ships can remove it by remote control. On the shipping route, there are n ports $(n > 1)$,[1] where the crew on the ships makes a decision whether or not it dispatches the reconnaissance boat. Notice that the ships cannot move while the boat circulates a ring if it is dispatched once.

It is assumed to be very rare that a terrorist lays the naval mine. Therefore, it is too costly to dispatch the boat every arrival time to each port. So we have two choices, to dispatch the boat and not to dispatch the boat. The actual decision of the dispatch is modeled by the mixed strategy in a two-person zero-sum game. An interval while the transport ships move one port to another is called a *stage*, which may contain a circulation of the reconnaissance boat or not. We consider different multistage games $\Gamma(k)$, where the maximum number of stages is assumed to be $k = m$. Let α be the reward that the reconnaissance boat finds and handles the naval mine, and let β be the loss that the transport ships suffer damage without dispatching the boat. Let $-n + 1$ be the cost n of circulating the boat plus the reward 1 of moving the ships to the next port. The ships safely proceed to the next port without circulating the boat with reward 1. We assume $1 < \alpha < \beta$ because the removal of risk does not make money, while the crash of a ship incurs high cost. Since the crews' purpose is to proceed the ships, we do not consider the successful ship moving as a cost but as a reward.

Furthermore, we make the following assumptions.

1. The mine laid by a terrorist is surely removed by a reconnaissance boat if it is dispatched. The ship without dispatching a boat surely suffers damage if the mine is laid.
2. The mine may be laid on the route at each beginning of a stage. So, we assume that the mine is not laid immediately after the boat has passed.
3. The number of stages m is sufficiently larger than the interval during which failures occur.
4. Any distance between neighboring ports is almost the same for simplicity. So, it takes n units time for a boat to circulate the ring.

Notice that the assumption 2 is used only in the games in Sect. 8.3. It is replaced by another assumption in Sect. 8.4.

8.3 Two-Person Zero-Sum Game

In this section, we first consider a game with $m = 1$, and then consider two multi-stage games containing mine preparation. The transport ships, player A, use a mixed strategy $a = (p, 1 - p)$ such that he dispatches the reconnaissance boat with probability p and does not dispatch it with probability $1 - p$. On the other hand, the terrorist, player B, uses a mixed strategy $b = (q, 1 - q)$ such that he lays a naval mine with probability q and does not lay it with probability $1 - q$.

---

[1] In Sect. 8.4, we assume $n > 2$

	A mine	No mine
Reconnaissance	α	$1-n$
No reconnaissance	$-\beta$	1

Table 8.1 Payoff matrix for a stage

Table 8.1 shows the payoff matrix of our game. In the left column, where a terrorist lays a mine, if a ship dispatches a reconnaissance boat, the reward of the ship is $\alpha > 0$. Otherwise, the reward of the ship is $-\beta$. In the right column, where a terrorist does not lay a mine, if a ship dispatches a reconnaissance boat, the reward of the ship is $1 - n$ because it takes n steps for the boat to circulate a ring and then the ship proceeds to the next port.

Let $E(a, \text{Mine})$ be the expected payoff of player A when player B lays a naval mine. On the contrary, let $E(a, \neg\text{Mine})$ be the expected payoff of player A when player B does not lay a naval mine.

$$E(a, \text{Mine}) = p \cdot \alpha + (1 - p) \cdot (-\beta) = p(\alpha + \beta) - \beta$$
$$E(a, \neg\text{Mine}) = p \cdot (1 - n) + (1 - p) \cdot 1 = 1 - pn$$

Since two straight lines intersect in $0 \le p \le 1$, player A's optimal strategy a^* is

$$a^* = \left(\frac{\beta + 1}{\alpha + \beta + n}, \frac{\alpha + n - 1}{\alpha + \beta + n} \right).$$

The value of the game is

$$\frac{\alpha + \beta(1 - n)}{\alpha + \beta + n}. \tag{8.1}$$

Let $E(\text{Recon}, b)$ be the expected payoff of player B when player A dispatches a reconnaissance boat. On the contrary, let $E(\neg\text{Recon}, b)$ be the expected payoff of player B when player A does not dispatch a reconnaissance boat.

$$E(\text{Recon}, b) = (1 - q) \cdot (1 - n) + q \cdot \alpha = q(n + \alpha - 1) - n + 1$$
$$E(\neg\text{Recon}, b) = (1 - q) \cdot 1 + q \cdot (-\beta) = 1 - q(\beta + 1)$$

Since two straight lines intersect in $0 \le q \le 1$, player B's optimal strategy b^* is

$$b^* = \left(\frac{n}{\alpha + \beta + n}, \frac{\alpha + \beta}{\alpha + \beta + n} \right).$$

Notice that if we use the player A's optimal strategy as it is, it means that the adversary always aims to bring about a failure. So, an unlikely failure may frequently occur against our intuition. To avoid it, we consider the empirical probability of a terrorist from now. Let s be a mine-preparing probability for each stage. We assume that the terrorist succeeds in preparing the mine by geometric distribution. Then, he can prepare it with probability

$$r_k = (1-s)^{k-1}s \quad (1 \le k \le m)$$

at the k-th stage. It is assumed that the terrorist has material only to prepare a mine, therefore the terrorist can prepare a mine on stage k only if he has not been able to prepare one on the previous stages $(1, 2, \ldots, k-1)$. After preparing the mine, he makes a decision to lay it for each stage. The following table shows the payoff matrix where the mine-preparing probability is considered. It is essentially the same one as Table 8.1 except for the probability and the value, where v_{k-1} is the value of the game $\Gamma(k-1)$. In Sects. 8.3.1 and 8.3.2, we consider two cases, a game continuation case and a game termination case, after the ship is broken by a mine.

	A mine	No mine
Reconnaissance	$r_k(\frac{\alpha n}{n+1} - \frac{\beta}{n+1}) + v_{k-1}$	$1 - n + v_{k-1}$
No reconnaissance	$r_k(-\beta) + v_{k-1}$	$1 + v_{k-1}$

Table 8.2 Payoff matrix for $\Gamma(k)$ with mine-preparation

8.3.1 Mine-Preparing Probability: Game Continuation Case

In this section, we consider that a fleet of ships can continue to sail and thus the game continues if one of them is broken by a mine. Then, the game $\Gamma(m)$ for $m > 1$ can be expressed as follows:

$$\Gamma(m) = \begin{bmatrix} r_m \alpha + \Gamma(m-1) & 1 - n + \Gamma(m-1) \\ r_m(-\beta) + \Gamma(m-1) & 1 + \Gamma(m-1) \end{bmatrix},$$

where $\Gamma(k)$ $(k < m)$ is the first k stages of the game $\Gamma(m)$. The game value v_m for the game $\Gamma(m)$, where $v_1 = (r_1(\alpha + \beta(1-n)))/(r_1(\alpha + \beta) + n)$, can be solved as follows.

$$
\begin{aligned}
v_m &= val \begin{bmatrix} r_m \alpha + v_{m-1} & 1 - n + v_{m-1} \\ r_m(-\beta) + v_{m-1} & 1 + v_{m-1} \end{bmatrix} \\
&= v_{m-1} + val \begin{bmatrix} r_m \alpha & 1 - n \\ r_m(-\beta) & 1 \end{bmatrix} \\
&= v_{m-1} + \frac{r_m(\alpha + \beta(1-n))}{r_m(\alpha + \beta) + n} \\
&= \sum_{h=1}^{m} \frac{r_h(\alpha + \beta(1-n))}{r_h(\alpha + \beta) + n}.
\end{aligned}
$$

The optimal strategies of player A and player B for the k-th stage ($1 \le k \le m$), denoted by a_k^* and b_k^*, are

$$a_k^* = \left(\frac{r_k \beta + 1}{r_k(\alpha + \beta) + n}, \; \frac{r_k \alpha + n - 1}{r_k(\alpha + \beta) + n} \right)$$

and

$$b_k^* = \left(\frac{n}{r_k(\alpha + \beta) + n}, \; \frac{r_k(\alpha + \beta)}{r_k(\alpha + \beta) + n} \right),$$

respectively.

If k is very large or $\alpha < \beta \ll n$ holds, we have

$$a_k^* \simeq \left(\frac{1}{n}, \frac{n-1}{n} \right), \; b_k^* \simeq (1,0).$$

On the other hand, if $n \ll \alpha < \beta$ holds, we have

$$a_k^* \simeq \left(\frac{\beta}{\alpha + \beta}, \; \frac{\alpha}{\alpha + \beta} \right), \; b_k^* \simeq (0,1).$$

If $n \simeq r_k \alpha$ and $\beta \simeq \alpha$ hold, we have

$$a_k^* \simeq \left(\frac{n+1}{3n}, \frac{2n-1}{3n} \right), \; b_k^* \simeq (1/3, 2/3).$$

8.3.2 Mine-Preparing Probability: Game Termination Case

In this section, we consider that only one transport ship sails and the game terminates if it is broken by a mine. Then, the game $\Gamma(m)$ is determined by an auxiliary game $\Gamma'(k,m)$, i.e., the last k stages of the game $\Gamma(m)$:

$$\Gamma(m) = \begin{bmatrix} r_1 \alpha + \Gamma'(m-1,m) & 1 - n + \Gamma'(m-1,m) \\ r_1(-\beta) & 1 + \Gamma'(m-1,m) \end{bmatrix}$$

and

$$\Gamma'(k,m) = \begin{bmatrix} r_{m-k+1} \alpha + \Gamma'(k-1,m) & 1 - n + \Gamma'(k-1,m) \\ r_{m-k+1}(-\beta) & 1 + \Gamma'(k-1,m) \end{bmatrix}$$

for $(k = m-1, m-2, \ldots, 2)$, where $\Gamma'(1,m)$ is the game with payoff matrix

$$\begin{bmatrix} r_m \alpha & 1-n \\ r_m(-\beta) & 1 \end{bmatrix}.$$

If the ship does not dispatch a reconnaissance boat when a naval mine is laid, the ship is sunk and the game terminates with an expected cost $r_1(-\beta)$. The value of the game is represented by

$$v_m = val \begin{bmatrix} r_1\alpha + v'_{m-1,m} & 1 - n + v'_{m-1,m} \\ r_1(-\beta) & 1 + v'_{m-1,m} \end{bmatrix}$$

and

$$v'_{k,m} = val \begin{bmatrix} r_{m-k+1}\alpha + v'_{k-1,m} & 1 - n + v'_{k-1,m} \\ r_{m-k+1}(-\beta) & 1 + v'_{k-1,m} \end{bmatrix}$$

for $(k = m - 1, m - 2, \ldots, 2)$, where $v'_{k,m}$ is the value of $\Gamma'(k,m)$, and

$$v'_{1,m} = val \begin{bmatrix} r_m\alpha & 1 - n \\ r_m(-\beta) & 1 \end{bmatrix} = \frac{r_m(\alpha + \beta(1 - n))}{r_m(\alpha + \beta) + n}.$$

For the game $\Gamma'(k,m)$, the expected payoffs of player A's mixed strategy $a' = (p', 1 - p')$ against player B's pure strategies are

$$E(a', \text{Mine}) = p'(r_{m-k+1}\alpha + v'_{k-1,m}) + (1 - p')r_{m-k+1}(-\beta)$$
$$E(a', \neg\text{Mine}) = p'(1 - n + v'_{k-1,m}) + (1 - p')(1 + v'_{k-1,m}). \tag{8.2}$$

Then, the intersection of them is

$$p'(r_{m-k+1}\alpha + v'_{k-1,m}) + (1 - p')r_{m-k+1}(-\beta) = p'(1 - n + v'_{k-1,m})$$
$$+ (1 - p')(1 + v'_{k-1,m})$$

$$p' = \frac{r_{m-k+1}\beta + v'_{k-1,m} + 1}{r_{m-k+1}(\alpha + \beta) + v'_{k-1,m} + n} = 1 - \frac{r_{m-k+1}\alpha + n - 1}{r_{m-k+1}(\alpha + \beta) + v'_{k-1,m} + n}. \tag{8.3}$$

Since the value of the game is $1 + v'_{k-1,m} - np'$ from (8.2), we have

$$v'_{k,m} = 1 + v'_{k-1,m} - n + n \cdot \frac{r_{m-k+1}\alpha + n - 1}{r_{m-k+1}(\alpha + \beta) + v'_{k-1,m} + n}.$$

Then, the value of $v'_{k,m}$ can be approximated by

$$v'_{k,m} \simeq r_{m-k+1}\left(\frac{\alpha}{n - 1} - \beta\right). \tag{8.4}$$

Since $v_m = v'_{m,m}$, we have

$$v_m \simeq r_1\left(\frac{\alpha}{n - 1} - \beta\right).$$

From (8.3) and (8.4), the probability p' for the k-th from the last stage is

$$p' = 1 - \frac{r_{m-k+1}\alpha + n - 1}{r_{m-k+1}g_{n,s} + n},$$

where

$$g_{n,s} = \frac{\alpha(n-s)}{n-1} + s\beta.$$

Since the k-th from the last stage corresponds to the $(m-k+1)$-th (from the first) stage, the player A's optimal strategy for the k-th stage ($1 \le k \le m$) is

$$a_k^* \simeq \left(1 - \frac{r_k\alpha + n - 1}{r_k g_{n,s} + n}, \frac{r_k\alpha + n - 1}{r_k g_{n,s} + n}\right).$$

On the other hand, the expected payoffs of player B's mixed strategy $b = (q, 1-q)$ against player A's pure strategies are

$$E(\text{Recon}, b) = q(r_{m-k+1}\alpha + v'_{k-1,m}) + (1-q)(1 - n + v'_{k-1,m})$$
$$E(\neg\text{Recon}, b) = q r_{m-k+1}(-\beta) + (1-q)(1 + v'_{k-1,m}). \qquad (8.5)$$

Since we obtain

$$q = \frac{n}{r_{m-k+1}(\alpha+\beta) + v'_{k-1,m} + n} \simeq \frac{n}{r_{m-k+1}g_{n,s} + n}$$

from (8.4) and (8.5), similar to a_k^*, the player B's optimal strategy $b_k^* = (q, 1-q)$ for the k-th stage ($1 \le k \le m$) is

$$b_k^* \simeq \left(\frac{n}{r_k g_{n,s} + n}, 1 - \frac{n}{r_k g_{n,s} + n}\right).$$

For simplicity, we assume $g_{n,s} \simeq \alpha + s\beta$. Then, the approximation can be derived as follows. If k is very large or $\alpha < \beta \ll n$ holds, we have

$$a_k^* \simeq \left(\frac{1}{n}, \frac{n-1}{n}\right), \ b_k^* \simeq (1,0).$$

On the other hand, if $n \ll \alpha < \beta$ holds, we have

$$a_k^* \simeq \left(\frac{s\beta}{\alpha + s\beta}, \frac{\alpha}{\alpha + s\beta}\right), \ b_k^* \simeq (0,1).$$

If $n \simeq r_k\alpha$ and $\beta \simeq \alpha$ hold, we have

$$a_k^* \simeq \left(\frac{sn+1}{(2+s)n}, \frac{2n-1}{(2+s)n}\right), \ b_k^* \simeq \left(\frac{1}{2+s}, \frac{1+s}{2+s}\right).$$

8.4 Another Assumption

If assumption 2 described in Sect. 8.2 is changed as follows, a new argument is possible. We call the time interval that a boat/ship moves from one port to another a *slot*.

New Assumption 1 *The mine may be laid on the route at each beginning of a* **slot**, *where every slot is selected equally likely.*   □

It means that the mine can be laid between the two ports immediately after the boat has passed across the ports. In what follows, we compare two boats case with one boat case for a fleet of ships (i.e., the game continues after the attack in both cases) and for $n > 2$ ports.

Notice that it takes $n + 1$ slots, consisting of reconnaissance (n slots) and the move of ships (1 slot), to proceed to the next stage if the reconnaisance boat is dispatched. Therefore, the terrorist has $n + 1$ chances to lay a mine in such a case.

8.4.1 Two Boats Case

To deal with the new assumption, we consider a new method, that is, the transport ships dispatch two reconnaissance boats in both clockwise direction and counter-clockwise direction. Then, the mine laid immediately after a boat has passed can be found by another boat (see Fig. 8.2). It seems that the method can be replaced by dispatching one boat in the counterclockwise direction. However, the two boats method is meaningful in the distributed failure detection because no spurious tokens are allowed in the global network as well as in front of the true token. Since the two boats method does not work for $n = 2$ ports, we assume $n > 2$ in this section.

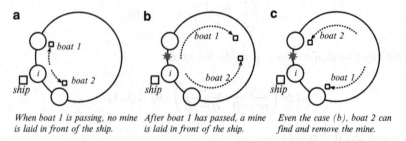

a   b   c

When boat 1 is passing, no mine   After boat 1 has passed, a mine   Even the case (b), boat 2 can
is laid in front of the ship.   is laid in front of the ship.   find and remove the mine.

Fig. 8.2 Usefulness of two boats for the new assumption

By using this method, we have the following payoff matrix for the game continuation case. Notice that even the two boats method cannot prevent the ships from suffering damage if the mine is laid after a circulation of them.

	A mine	No mine
Reconnaissance	$r_k \alpha + v_{k-1}$	$1 - n + v_{k-1}$
No reconnaissance	$r_k(-\beta) + v_{k-1}$	$1 + v_{k-1}$

Table 8.3 Payoff matrix for $\Gamma(k)$—two boats case

Table 8.3 differs from Table 8.2 in the left upper part. If the ships dispatch reconnaissance boats, a stage contains n slots for the circulation of a boat and 1 slot for the move of ships. Even if a boat missed a mine, it would be found by another boat except for the interval while 1 slot move of the ships. Thus, the reward of the ships is $\alpha n/(n+1)$. On the contrary, one of the ships strikes the mine if it is laid after the boats circulate the ring (and before the ships move). Thus, the loss of the ships is $\beta/(n+1)$.

Next, the game $\Gamma(m)$ for $m > 1$ can be expressed as follows.

$$\Gamma(m) = \begin{bmatrix} r_m(\frac{\alpha n - \beta}{n+1}) + \Gamma(m-1) & 1 - n + \Gamma(m-1) \\ r_m(-\beta) + \Gamma(m-1) & 1 + \Gamma(m-1) \end{bmatrix}$$

The game value v_m for $\Gamma(m)$, where $v_1 = 1 - (r_1\beta + 1)/(r_1(\alpha + \beta)/(n+1) + 1)$, can be solved as follows.

$$\begin{aligned} v_m &= val \begin{bmatrix} r_m(\frac{\alpha n - \beta}{n+1}) + v_{m-1} & 1 - n + v_{m-1} \\ r_m(-\beta) + v_{m-1} & 1 + v_{m-1} \end{bmatrix} \\ &= v_{m-1} + val \begin{bmatrix} r_m(\frac{\alpha n - \beta}{n+1}) & 1 - n \\ r_m(-\beta) & 1 \end{bmatrix} \\ &= v_{m-1} + \left(1 - \frac{r_m\beta + 1}{r_m(\alpha + \beta)/(n+1) + 1} \right) \\ &= m - \sum_{k=1}^{m} \frac{r_k\beta + 1}{r_k(\alpha + \beta)/(n+1) + 1}. \end{aligned}$$

Similar to the argument for $m = 1$, the optimal strategies of player A and player B for the k-th stage ($1 \le k \le m$), denoted by a_k^* and b_k^*, are

$$a_k^* = \left(\frac{(r_k\beta + 1)(n+1)}{r_k(\alpha + \beta)n + n(n+1)}, \frac{r_k(\alpha n - \beta) + n^2 - 1}{r_k(\alpha + \beta)n + n(n+1)} \right)$$

and

$$b_k^* = \left(\frac{n+1}{r_k(\alpha + \beta) + n + 1}, \frac{r_k(\alpha + \beta)}{r_k(\alpha + \beta) + n + 1} \right),$$

respectively.

If k is very large or $\alpha < \beta \ll n$ holds, we have

$$a_k^* \simeq \left(\frac{1}{n}, \frac{n-1}{n} \right), \; b_k^* \simeq (1, 0).$$

On the other hand, if $n \ll \alpha < \beta$ holds, we have

$$a_k^* \simeq \left(\frac{\beta}{\alpha+\beta}, \frac{\alpha}{\alpha+\beta} \right), \ b_k^* \simeq (0,1).$$

If $n \simeq r_k \alpha$ and $\beta \simeq \alpha$ hold, we have

$$a_k^* \simeq \left(\frac{(n+1)^2}{n(3n+1)}, \frac{(n-1)(2n+1)}{n(3n+1)} \right), \ b_k^* \simeq \left(\frac{n+1}{3n+1}, \frac{2n}{3n+1} \right).$$

8.4.2 One Boat Case

If the ship can dispatch only one boat, the payoff matrix is shown in Table 8.4.

	A mine	No mine
Reconnaissance	$r_k(\frac{\alpha}{n+1} - \frac{\beta n}{n+1}) + v_{k-1}$	$1 - n + v_{k-1}$
No reconnaissance	$r_k(-\beta) + v_{k-1}$	$1 + v_{k-1}$

Table 8.4 Payoff matrix for $\Gamma(k)$—one boat case

Table 8.4 also differs from Table 8.2 in the left upper part. Similar to the explanation above, if the ships dispatch a reconnaissance boat, a stage contains $n+1$ slots. Since there is only one boat, it can find a mine only if it is laid at the first slot of the stage. Thus, the reward of the ships is $\alpha/(n+1)$. On the contrary, one of the ships strikes the mine if it is laid at other slots. Thus, the loss of the ships is $\beta n/(n+1)$.

Next, the game $\Gamma(m)$ for $m > 1$ can be expressed as follows.

$$\Gamma(m) = \begin{bmatrix} r_m(\frac{\alpha-\beta n}{n+1}) + \Gamma(m-1) & 1 - n + \Gamma(m-1) \\ r_m(-\beta) + \Gamma(m-1) & 1 + \Gamma(m-1) \end{bmatrix}$$

The game value v_m for $\Gamma(m)$, where $v_1 = 1 - n(r_1\beta+1)/(r_1(\alpha+\beta)/(n+1)+n)$, can be solved as follows.

$$\begin{aligned}
v_m &= val \begin{bmatrix} r_m(\frac{\alpha-\beta n}{n+1}) + v_{m-1} & 1 - n + v_{m-1} \\ r_m(-\beta) + v_{m-1} & 1 + v_{m-1} \end{bmatrix} \\
&= v_{m-1} + val \begin{bmatrix} r_m(\frac{\alpha-\beta n}{n+1}) & 1 - n \\ r_m(-\beta) & 1 \end{bmatrix} \\
&= v_{m-1} + \left(1 - \frac{n(r_m\beta+1)}{r_m(\alpha+\beta)/(n+1)+n} \right) \\
&= m - \sum_{k=1}^{m} \frac{n(r_k\beta+1)}{r_k(\alpha+\beta)/(n+1)+n}.
\end{aligned}$$

Similar to the argument for $m = 1$, the optimal strategies of player A and player B for the k-th stage ($1 \leq k \leq m$), denoted by a_k^* and b_k^*, are

$$a_k^* = \left(\frac{(r_k \beta + 1)(n+1)}{r_k(\alpha + \beta) + n(n+1)}, \frac{r_k(\alpha - \beta n) + n^2 - 1}{r_k(\alpha + \beta) + n(n+1)} \right)$$

and

$$b_k^* = \left(\frac{n(n+1)}{r_k(\alpha + \beta) + n(n+1)}, \frac{r_k(\alpha + \beta)}{r_k(\alpha + \beta) + n(n+1)} \right),$$

respectively.

If k is very large or $\alpha < \beta \ll n$ holds, we have

$$a_k^* \simeq \left(\frac{1}{n}, \frac{n-1}{n} \right), \; b_k^* \simeq (1,0).$$

On the other hand, if $n \ll \alpha < \beta$ holds, we have

$$a_k^* \simeq \left(\frac{\beta}{\alpha + \beta}, \frac{\alpha}{\alpha + \beta} \right), \; b_k^* \simeq (0,1).$$

If $n \simeq r_k \alpha$ and $\beta \simeq \alpha$ hold, we have

$$a_k^* \simeq \left(\frac{(n+1)^2}{n(n+3)}, \frac{n-1}{n(n+3)} \right), \; b_k^* \simeq \left(\frac{n+1}{n+3}, \frac{2}{n+3} \right).$$

8.5 Application

We briefly state how our reconnaissance problem can be applied to a self-stabilizing mutual exclusion, when there is at most one (major) token in a system. The process holding the (major) token is given a privilege to do some task, called a *critical section*. As shown in Fig. 8.1, our protocol works for ring networks, in which a token, called a *major token*, and a sub-token, called a *minor token*, are used. Our protocol is the combination of a *base state protocol* and a *superstabilizing protocol* [11].

In the base state protocol, we predefine $H > n$ integer states $I_H = \{0, 1, \ldots, H - 1\}$, called *bases*, such that every correct state of a major token takes in a domain $R_H = [0, H) = \{x \mid 0 \leq x < H\}$. If some major token has a non-base state, called a *non-base falut*, it is reset to a neighboring base state. To avoid a deadlock in which every major token has a non-base state, process 0 sends a deadlock-breaking token, called dtoken. When the process 0 receives the dtoken, it resets to some base state independently because such a situation means the deadlock occurs.

In the superstabilizing protocol, we can also define the correct states of a major token as I_H. In addition, we can recognize a process has a major token if it has a different state from its neighboring state(s). Notice that there may be other spurious major tokens even if every process has a base state. In such a case, we cannot restore a non-error state by the base state protocol but by the superstabilizing protocol. That is, when the minor token encounters the spurious major token, the erroneous state can be corrected. Since it is very costly to send the minor token at every moving the major token, we considered a mixed strategy of sending a minor token (ST, for short) with probability p and not sending it (NS, for short) with probability $1 - p$.

Here is our protocol. There are two descriptions for $i = 0$ and $i \neq 0$, but we omit the one for $i = 0$.

Our protocol for process $i \neq 0$

if (i has a non-base fault) **then**
```
reset to a neighboring base (if any);
send dtoken if i receives it
```
else if (i has both major and minor tokens) **then**
if (i has just received the major token) **then**
```
choose ST with probability p and NS with probability 1−p
```
if (i has been waiting for the minor token) or (i chooses NS) **then**
```
perform critical section;
send major and minor tokens to i+1
```
else if (i has a spurious major token) **and** (i has a minor token) **then**
```
eliminate spurious major token
```
fi
```
send a minor token to i+1 (if any)
```

8.6 Conclusion

In this chapter we considered an effective search strategy for a naval mine by using a two-person zero-sum game. The problem is motivated by distributed failure detection/repair as shown in the appendix. The advantage of our topic is to simplify the original problem and to show its wide application.

We can find the feature of a method in some typical cases. For example, suppose that $n \simeq r_k \alpha$ and $\beta \simeq \alpha$ hold. First, the reconnaissance probabilities in the optimal strategies are $(n + 1)/(3n)$ for the game continuation case and $1/n$ for the game termination case. This means the probability for the termination case is smaller than that for the continuation case for $n > 2$. Next, the reconnaissance probabilities in the optimal strategies are $(n + 1)^2/(3n^2 + n)$ for the two boats case and $(n + 1)^2/(n^2 + 3n)$ for the one boat case, where a detailed timing assumption is used. This means the probability for the two boats case is smaller than that for the one boat case.

Our future work includes another application of reconnaissance problem to distributed protocols.

Acknowledgements The authors would like to thank the anonymous referees for their helpful comments. This work was partially supported by Grant-in-Aid for Scientific Research ((C)23510177) of the Ministry of Education, Science, Sports, and Culture of Japan.

References

1. V.Baston and F.Bostock, "A generalized inspection game," *Naval Research Logistics*, **38**, (1991) 171–182.
2. E.W.Dijkstra, "Self-stabilizing systems in spite of distributed control," *Communications of the ACM*, **17**, 11 (1974) 643–644.
3. S.Dolev, "Self-stabilization," The MIT Press, (2000).
4. M.Dresher, "A sampling inspection problem in arms control agreements: a game-theoretic analysis," Memorandum RM-2972-ARPA, The RAND Corporation, Santa Monica, California, 1962.
5. T.Ferguson and C.Melolidakis, "On the inspection game," *Naval Research Logistics*, **45**, (1998) 327–334.
6. A.Garnaev, "A remark on the customs and smuggler game," *Naval Research Logistics*, **41**, (1994) 287–293.
7. A.Garnaev, "Search games and other applications of game theory," Heidelberg, New York, Springer, (2000).
8. T.Herman, "Superstabilizing mutual exclusion," *Distributed Computing*, **13**, 1 (2000) 1–17.
9. R.Hohzaki, "A compulsory smuggling model of inspection game taking account of fulfillment probabilities of players' aims," *Journal of the Operations Research Society of Japan*, **49**, 4 (2006) 306–318.
10. Y.Katayama, E.Ueda, H.Fujiwara and T.Masuzawa, "A latency optimal superstabilizing mutual exclusion protocol in unidirectional rings," *Journal of Parallel and Distributed Computing*, **62**, 5 (2002) 865–884.
11. J.Kiniwa and K.Kikuta, "Analysis of an intentional fault which is undetectable by local checks under an unfair scheduler," In *Proceedings of the 11th International Symposium on Stabilization, Safety, and Security of Distributed Systems (SSS)*, LNCS 5873, (2009) 443–457.
12. M.Maschler, "A price leadership method for solving the inspection's non-constant-sum game," *Naval Research Logistics Quarterly*, **13**, (1966) 11–33.
13. R.B.Myerson, "Game theory: analysis of conflict," Harvard University Press (1991).
14. L.Rosaz, "Self-stabilizing token circulation on asynchronous uniform unidirectional rings," in *Proceedings of the 19th Annual ACM Symposium on Principles of Distributed Computing* (2000) 249–258.
15. M.Thomas and Y.Nisgav, "An infiltration game with time dependent payoff," *Naval Research Logistics Quarterly*, **23**, (1976) 297–302.

Chapter 9
The Value of the Two Cable Ambush Game

Vic Baston and Ian Woodward

Abstract This chapter finds the value of Ruckle's much studied two-person zero-sum ambush game which involves an Infiltrator attempting to traverse a channel undetected when the channel is protected by two lengths of electronic cable. The value is obtained by exploiting the structure of a Defender optimal strategy without explicitly constructing an optimal strategy for either player. It is expressed as the minimum of a finite number of simple expressions, each of which is easily calculated from the values of the cable lengths. This result identifies ranges of values for the lengths of the two cables over which the games not only have the same value but also have Defender optimal strategies with the same basic structure.

9.1 Introduction

In [10, 11] Ruckle introduced a number of search and ambush games. Some of the games he described have generated a considerable amount of interest and one game in particular has been the subject of a number of papers. It can be described as follows.

Infiltrator wishes to travel down a channel without being detected by Defender who has two electronic cables of lengths a and b which can be placed in the channel. Infiltrator knows the lengths of the cables but has no means of determining where they have been laid and he will be detected if he crosses them. As it is natural for Defender to place the cables across the channel where its width is minimal, the problem can be formulated mathematically by letting Infiltrator choose a point x in the interval $[0,1]$ and Defender choose closed intervals $[y, y+a]$ and $[z, z+b]$ in $[0,1]$.

V. Baston (✉) • I. Woodward
Faculty of Mathematical Studies, University of Southampton, Southampton SO17 1BJ, UK
e-mail: vic.baston@btinternet.com; ian.woodward@advantage-business.co.uk

S. Alpern et al. (eds.), *Search Theory: A Game Theoretic Perspective*,
DOI 10.1007/978-1-4614-6825-7_9, © Springer Science+Business Media New York 2013

The situation can then be considered as a two person zero-sum game $\Gamma(a,b)$ by choosing an appropriate payoff function to Infiltrator; the one favoured is 1 if Infiltrator's point is not in either of the intervals chosen by Defender and 0 otherwise. More formally $\Gamma(a,b)$ can be regarded as the two person zero-sum game with strategy spaces $[0,1]$ and $[0,1] \times [0,1]$ with payoff function f to Infiltrator given by

$$f(x,(y,z)) = \begin{cases} 0 & \text{if } x \in [y, y+a] \cup [z, z+b], \\ 1 & \text{otherwise.} \end{cases}$$

This formulation makes use of the fact that the position of an interval is determined by describing where its left-hand endpoint is located.

Ruckle [11] found a complete solution for $\Gamma(a,b)$ when $a = b$ and also for some particular numerical values of a and b. Baston and Bostock [2] solved the game for the case $b \geq 1/2$ and Lee [8] extended their results to cover all cases for which $a \leq b$ and $1/3 \leq b < 1/2$ as well as producing some bounds for other cases. Further, less easy to describe, results have been obtained by Zoroa, Zoroa and Fernández-Sáez [15]. On a slightly different tack, Garnaev [6] considered a discrete version of the game in which play takes place over an integer interval $[1, n]$; further details of this and similar games can be found in Garnaev [7].

In the chapter mentioned above, the approach was to nominate explicit strategies for the players and show that they were optimal. However Woodward [13, 14] has adopted a different approach by investigating the structure of the more general continuous game in which Defender has n barriers. He demonstrated that it is equivalent to a finite game in the sense that the games have the same value and that optimal strategies in the finite game remain optimal when transferred to the continuous game. This enabled him to use linear programming for the continuous game and, in [14], he solved the three barrier game for lengths $a_1 \geq a_2 \geq a_3 \geq 1/4$ and some additional cases where $a_3 \geq 1/5$. Woodward has also shown that his finite game is equivalent to Garnaev's discrete game generalized to n barriers. On a slightly different tack, Baston and Kikuta have considered two variations of the n barrier continuous game. In [3] Infiltrator may send more than one agent down the channel and the players have limited information available to them whereas, in [4], Infiltrator has a positive width and a proportion of the width needs to be detected for a positive identification.

In this paper we develop the approach used by Woodward [14], concentrating on the structure of the game $\Gamma(a,b)$ and, in particular, on the structure of its optimal Defender strategies in the finite game which Woodward showed is equivalent to $\Gamma(a,b)$. This enables us to find the value of $\Gamma(a,b)$ for all values of a and b. The result was first obtained by the second author in his Ph.D. thesis [14] but the methods of proof here are radically different and much shorter than those in the thesis. An interesting aspect of the proofs is that, unlike previous work on the game, we do not at any stage nominate explicit strategies for the players and show that they are optimal. The value of $\Gamma(a,b)$ is given in terms of the minimum of a finite number of terms; loosely speaking the number of terms increase as the length of the longer interval decreases. Applying our result to the case $a \leq b$ and $1/3 \leq b \leq 1/2$, we provide a much simpler expression for the value of $\Gamma(a,b)$ than that of Lee [8].

The structure is as follows. In Sect. 9.2 we introduce the notation we require and give a summary of Woodward's results for the two barrier game $\Gamma(a,b)$. After analysising an illustrative example in Sect. 9.3, these results are used in Sect. 9.4 to obtain a lower bound for the value of $\Gamma(a,b)$. This is achieved by proving that Defender has an optimal strategy of a particular form and then showing by linear programming that a strategy of this form cannot restrict Infiltrator to less than a certain quantity. A second example is introduced in Sect. 9.5 to illustrate the ideas used in Sect. 9.6 to demonstrate that a Defender strategy can be constructed which achieves the lower bound obtained in Sect. 9.4. In the final section we comment on and give some consequences of our result.

9.2 Notation and Woodward's Results

We will always assume that a and b are positive real numbers with $a \leq b < 1$, $\Gamma(a,b)$ is the game defined in the Introduction and $v(a,b)$ is the value of $\Gamma(a,b)$. Apart from the final section we will also assume that $b < 1/2$; as we shall see, our expression for the value of $\Gamma(a,b)$ applies for $b < 1/2$ but there are some cases with $b \geq 1/2$ for which it does not. As Baston and Bostock [2] have completely solved the case $b \geq 1/2$, this exception is not a problem, just something of a curiosity.

As mentioned previously, Woodward [13] has shown that $\Gamma(a,b)$ is equivalent to a finite game $\Gamma_F(a,b)$ which is similar to $\Gamma(a,b)$ but one in which the strategy spaces of the players are finite subsets of $[0,1)$. In $\Gamma_F(a,b)$ Defender's strategy space is $Z \times Z$ where

$$Z = \{\mu b + \rho a \in [0,1) : \mu \text{ and } \rho \text{ non-negative integers}\}$$

so that $(w_1, w_2) \in Z \times Z$ represents the strategy that places the left-hand endpoints of the intervals with lengths a and b at the points w_1 and w_2. This result is crucial to the proof of our theorem in Sect. 9.4 but, for technical reasons, we take the Defender strategy space as $Z^* \times Z^*$ where $Z^* = Z \cup \{1\}$. Let

$$z_0 = 0, z_1 = a, z_2, \ldots, z_{|Z|-1}, z_{|Z|} = 1 \tag{9.1}$$

be the elements of Z^* in increasing order. We do not need to consider the strategy space for Infiltrator in $\Gamma_F(a,b)$ in detail but, for the information of the reader, it comprises of a set of points obtained by taking, for each $z \in Z$, a point just to the right of z, the distance past z being larger the farther z is to the right of the interval.

In Sect. 9.6 a Defender strategy in $\Gamma_F(a,b)$ will be defined using a number of coverings of $[0,1]$ with intervals of lengths a and b and it is reasonable to expect that, for an optimal strategy, these coverings will be minimal in some sense. This is the motivation for the following notation.

Given a and b, define $\lambda_{\lceil 1/b \rceil} = 0$, where $\lceil 1/b \rceil$ denotes the least integer greater than or equal to $1/b$, and, for $i = 0, 1, \ldots, \lceil 1/b \rceil - 1$, λ_i by the integer such that

$$\frac{1-ib}{a} \leq \lambda_i < \frac{1-ib}{a} + 1. \tag{9.2}$$

Thus a covering of $[0,1]$ having precisely i intervals of length b must have at least λ_i intervals of length a.

Since our (mixed) Defender strategy in $\Gamma_F(a,b)$ will involve the same "number" of intervals of length a as of length b, the associated coverings will be a balance of two types;

(i) Those that have more intervals of length b than intervals of length a
   and

(ii) Those that have at least as many intervals of length a as those of length b.

As a consequence the sets Λ^+ and Λ^- defined by

$$\Lambda^+ = \Lambda^+(a,b) = \{i : i > \lambda_i\} \quad \text{and} \quad \Lambda^- = \Lambda^-(a,b) = \{i : i \leq \lambda_i\} \tag{9.3}$$

will play a prominent role and feature in our expression for the game value.

9.3 An Illustrative Example

In the next section we prove a theorem that gives a lower bound for the value of our game. Its proof takes an arbitrary optimal strategy for Defender and shows that it must have a certain structure. The arguments to do this are somewhat technical so, in this section, we provide an example which illustrates them.

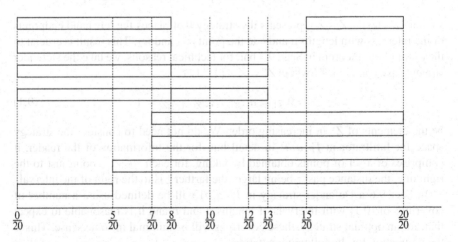

Fig. 9.1 An optimal Defender strategy when $a = 5/20$ and $b = 8/20$

Consider the case when $a = 5/20$ and $b = 8/20$. Its value is 3/8 (see [8]) so it is easy to check that an optimal strategy for Defender is given in Fig. 9.1 in which each (horizontal) line of rectangles represents a pure Defender strategy played with probability 1/8. Notice that lines 3 and 4 represent the same strategy as do lines 5 and 6. Most vertical lines intersect five rectangles although some intersect more; in particular vertical lines from points between 7/20 and 8/20 meet seven rectangles so we could move the rectangles of length 5/20 in rows 5 and 6 which start at 7/20 to the right so that they start at 8/20 without affecting optimality. Doing this means that they overlap the rectangles of length 8/20 on the same line; with regard to the proof of the theorem, it is undesirable for rectangles on the same line to intersect except possibly at a point. Thus the rectangles of length 8/20 in rows 5 and 6 are moved to the right so that they start at 13/20. From a strictly pedantic standpoint this is not permissible as they end beyond 1 but this is not a problem because we are at an interim stage and, when the final stage is reached, they can be moved to the left so that the intervals lie in [0,1]. Having moved the rectangles on lines 5 and 6, we see that the vertical lines between 12/20 and 13/20 still intersect seven rectangles so we perform a similar process to the above on lines 7 and 8 to arrive at the position in Fig. 9.2. in which every vertical line between 0 and 1 not only intersects precisely five rectangles but also precisely one rectangle of each shading. Thus rectangles with the same shading cover the interval [0,1].

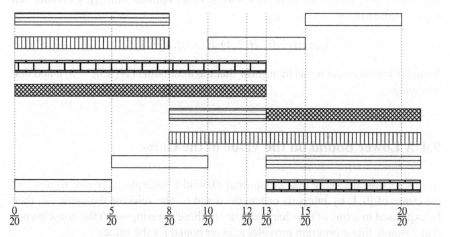

Fig. 9.2 The modified Defender strategy when $a = 5/20$ and $b = 8/20$

Notice that the rectangles with no shading is a minimal cover of [0,1] by rectangles of length 5/20. The other rectangles with a particular shading all comprise two rectangles of length 8/20 and one of length 5/20 and so provide minimal covers of [0,1] when precisely two rectangles of length 8/20 can be used; although these covers are different, this is not significant in the analysis and they will all be regarded as the same type, say $C(1,2)$, where 1 and 2 represent the number of rectangles of length a and b respectively in the cover.

The proof of Theorem 1 shows that this process can be applied to every game $\Gamma(a,b)$ demonstrating that any optimal Defender strategy in $\Gamma(a,b)$ gives rise to two types of minimal cover of [0,1], one, $C(\lambda_{j+}, j^+)$, with a $j^+ \in \Lambda^+$ and the other, $C(\lambda_{j-}, j^-)$, with a $j^- \in \Lambda^-$ where Λ^+ and Λ^- are given by (9.3) and the λ s by (9.2). The number of each type of cover that arises is easy to calculate because, in the optimal strategy, each pure strategy contains precisely one rectangle of each length. Hence the number of pure strategies equals the number of rectangles of length a which equals the number of length b. If there are x of type $C(\lambda_{j+}, j^+)$ and y of type $C(\lambda_{j-}, j^-)$, the number of rectangles of length a is $x\lambda_{j+} + y\lambda_{j-}$ while the number of rectangles of length b is $xj^+ + yj^-$. Equating them gives $x(j^+ - \lambda_{j+}) = y(\lambda_{j-} - j^-)$; in practice we can always take $x = \lambda_{j-} - j^-$ and $y = j^+ - \lambda_{j+}$. Thus the number of pure strategies is $j^+\lambda_{j-} - j^-\lambda_{j+}$. Furthermore the number of covers is $x + y = \lambda_{j-} - j^- + j^+ - \lambda_{j+}$ so every vertical line meets at least this number of rectangles. Thus any choice of a point in [0,1] by Infiltrator will also intersect at least this number of pure strategies. It follows that Defender can restrict Infiltrator to at most

$$H(j^+, j^-) = 1 - \frac{\lambda_{j-} - j^- + j^+ - \lambda_{j+}}{j^+\lambda_{j-} - j^-\lambda_{j+}}.$$

In our example $\Lambda^+ = \{2,3\}$, $\Lambda^- = \{0,1\}$. so $\lambda_0 = 4$, $\lambda_1 = 3$, $\lambda_2 = 1$, $\lambda_3 = 0$. Theorem 1 says that, without the knowledge of an optimal strategy, Defender can restrict Infiltrator to

$$\min\{H(2,0), H(2,1), H(3,0), H(3,1)\}.$$

Straightforward calculations then show that the minimum is $H(2,0) = 3/8$ and that it is unique.

9.4 A Lower Bound on the Value of the Game

In this section we show that an optimal Defender strategy gives rise to a set of coverings of $[0,1]$ by intervals of lengths a and b. The value of the game can then be expressed in terms of the characteristics of these coverings with the consequence that a simple linear program provides a lower bound for the value.

For $i \in \Lambda^+$, $j \in \Lambda^-$, where Λ^+ and Λ^- are defined by (9.3), let

$$G(i,j) = (i - \lambda_i + \lambda_j - j)/(i\lambda_j - j\lambda_i) \qquad (9.4)$$

and

$$G = \max\{G(i,j) : i \in \Lambda^+, j \in \Lambda^-\}. \qquad (9.5)$$

Theorem 1. *The value of* $\Gamma(a,b)$ *is bounded below by* $1 - G$ *where* G *is given by* (9.5).

Proof. From Lemmas 1 and 2 and Theorem 3 in [13], Defender and Infiltrator have optimal strategies contained in finite strategy spaces which are subsets of the strategy spaces of $\Gamma(a,b)$. As stated in Sect. 9.2 we need only be specific about the Defender strategy space which, for technical reasons, we take to be $Z^* \times Z^*$ where $Z^* = Z \cup \{1\}$ and Z is given by (9.1). The only feature of the finite Infiltrator strategy space that is relevant here is that it can be taken so that it is contained in $[0,1)$. As the payoffs in the finite game are rational, a theorem of Weyl (see [12]) ensures that the game value is rational and that each player has an optimal strategy in which the probability of playing any pure strategy is also rational. Hence we may make the following assumptions.

Assumption 1 *There is a Defender optimal strategy S_0^* made up of k (not necessarily distinct) pure strategies in $Z^* \times Z^*$ each played with probability $1/k$.*

Assumption 2 *If the intervals in one of the pure strategies in S_0^* intersect within $[0,1]$, then they do so only at a point of Z^*. For instance, if a pure strategy places the two intervals $[w_1, w_1 + a]$ and $[w_2, w_2 + b]$ such that $w_1 < w_2$ and $w_1 + a \in (w_2, w_2 + b)$ then it can be replaced by the strategy placing the intervals at $[w_1, w_1 + a]$ and $[w_1 + a, w_1 + a + b]$ because every Infiltrator strategy which is intercepted by the former pair of intervals is intercepted by the latter pair.*

Assumption 3 *In the set of all strategies satisfying Assumptions 1 and 2 S_0^* is one with a maximal number of pure strategies having the interval of length b preceding the interval of length a.*

Because there are precisely k Defender strategies, it is clear that the value of the game $v(a,b)$ must take the form $(k - m)/k$ for some positive integer m. For each $i = 1, \ldots, |Z|$ Infiltrator can choose a point of (z_{i-1}, z_i) so there must be at least m of the k members of S_0^* which contain (z_{i-1}, z_i).

Note also that the intervals in the pure strategies of S_0^* all start at a point of Z^* and thus terminate at a point in $Z^* \cup (1, 1 + b]$.

CLAIM 1: An optimal strategy S^* can be obtained from S_0^* using an inductive argument with the property that, for $i = 1, \ldots, |Z|$, S^* has precisely m pure strategies having intervals which contain (z_{i-1}, z_i).

The inductive hypothesis is stated as follows. Suppose, for some j satisfying $0 \le j < |Z|$, S_j^* is an optimal Defender strategy which selects at random a member from a multiset K_j of k pure strategies in $Z^* \times Z^*$ and which satisfies the following conditions:

(α) the intervals in a member of K_j intersect in $[0,1]$ (if at all) only at a point of Z^*,

(β) for each $i = 1, \ldots, |Z|$, there are at least m members of K_j which have intervals containing (z_{i-1}, z_i),

(γ) for each $i = 1, \ldots, j$ there are precisely m members of K_j which have intervals containing (z_{i-1}, z_i).

We construct a set S_{j+1}^* from S_j^* which has the same properties as S_j^*.

If there are precisely m members of K_j containing (z_j, z_{j+1}), put $K_{j+1} = K_j$ and define $S_{j+1}^* = S_j^*$. Clearly S_{j+1}^* satisfies the inductive conditions.

If there are $m_j > m$ members of K_j containing (z_j, z_{j+1}), then, because there are precisely m members of K_j containing (z_{j-1}, z_j), there are at least $m_j - m$ members of K_j which have intervals starting at z_j. We can therefore modify precisely $m_j - m$ of them by moving the interval starting at z_j so that it starts at z_{j+1}. If, in a modified pure strategy, the interval I which has been moved to start at z_{j+1} has interior points in common with the other interval J, we further modify the strategy by moving J to the right until it starts at w, where w is the termination point t of I if $t < 1$ and 1 otherwise; note that w is a point of Z^* by the definitions of Z and Z^*. We now define K_{j+1} as the set of the modified strategies together with the unmodified strategies of K_j. Taking S_{j+1} to be the strategy which chooses a member of K_{j+1} at random it is easy to see that S_{j+1} satisfies the inductive conditions.

Clearly S_0^* satisfies the above conditions on S_j with (γ) being satisfied vacuously. Thus, by induction we have established Claim 1.

CLAIM 2: S^* is a strategy in $Z \times Z$.

Suppose $K_{|Z|}$ has a pure strategy W_r^* which has an interval X_r of length x starting at the point 1 and let Y_r denote the other interval in W_r^*.

If Y_r also starts at 1, the intervals of W_r^* do not contain any points in Infiltrator's finite strategy space so the strategy given by randomly choosing a member of $K_{|Z|} \setminus \{W_r^*\}$ would give a value strictly less than $(k - m)/k$ which is a contradiction.

We can therefore assume that Y_r covers some interval $[y_1, y_2]$ where $y_1 \in Z$. Let $x^{(1)}$ and $x^{(2)}$ be the number of closed intervals of length x needed to cover the intervals $[0, y_1]$ and $[y_2, 1]$ respectively. Let K be the multiset comprising of $x^{(1)} + x^{(2)} - 1$ copies of $K_{|Z|}$ together with an extra W_r^*, then K contains $x^{(1)} + x^{(2)}$ copies of W_r^*. From these copies of W_r^*, define $x^{(1)} + x^{(2)}$ pure strategies in $Z \times Z$ by repositioning the $x^{(1)} + x^{(2)}$ intervals X_r so that they cover $[0, y_1]$ and $[y_2, 1]$ but do not have y_1 or y_2 as an interior point and let K^* denote the multiset obtained from K by substituting these strategies for the $x^{(1)} + x^{(2)}$ copies of W_r^*. The Defender strategy which selects a member of K^* at random restricts Infiltrator to at most

$$1 - \frac{m(x^{(1)} + x^{(2)} - 1) + 1}{k(x^{(1)} + x^{(2)} - 1) + 1} < 1 - mk = v(a, b)$$

which is a contradiction. Hence S^* is a strategy in $Z \times Z$. We have established Claim 2.

Let T denote the multiset of intervals in the members of $K_{|Z|}$. By the construction of S^*, if there are j members of T which terminate at $z_i > 0$, then there are j members of T which start at z_i. We can therefore obtain a "chain" of intervals by selecting a sequence of intervals $I_0, I_1, \ldots, I_r$ such that I_0 starts at $z_0 = 0$, I_i starts at the point where I_{i-1} terminates $(i = 1, \ldots, r)$ and I_r contains the point 1 but starts before it.

Deleting the intervals of this chain from T, we can form a second chain and so on until we obtain m chains in all.

CLAIM 3: A chain which has precisely i intervals of length b where $0 \leq i \leq \lceil 1/b \rceil$ has exactly λ_i intervals of length a where λ_i satisfies (9.2).

Suppose there is a chain $I_1, I_2, \ldots, I_r$ with precisely i intervals of length b for which the assertion is false then, using the definition of λ_i in (9.2), there must be at least $\lambda_i + 1 \geq 1$ intervals of length a in the chain and the termination point w of I_r satisfies $w \geq 1 + a$. Because the interval I_r starts before the point 1, I_r must be an interval of length b. Let $s = \max\{j : |I_j| = a\}$ where $|I|$ denotes the length of I, then $s < r$. Let $T_{s+1}, \ldots, T_r$ denote the pure strategies containing $I_{s+1}, \ldots, I_r$ respectively; for $j = s+1, \ldots, r$ let T_j^- denote the pure strategy obtained from T_j by moving I_j a distance a to the left.

The analysis now divides into two parts (I) and (II).

(I) T_j^- has overlapping intervals.

If there is a T_j^- which has overlapping intervals, let T_g^- denote the one for which j is maximal, then the strategy S^- obtained from S^* by replacing T_j with T_j^- for $j \geq g$ is clearly still optimal. Note that, in T_g, the interval of length a precedes the interval of length b because $|I_g| = b$. The intervals of T_g^- overlap so we can take a pure strategy T_g^{-*} whose intervals do not overlap and cover the points covered by the intervals of T_g^- and in which the interval of length b precedes the interval of length a. Replacing T_g^- by T_g^{-*} in S^- now gives rise to an optimal strategy and this optimal strategy has more pure strategies having the interval of length b preceding the interval of length a than S^*. However the construction of S^* ensures that S^* and S_0^* have the same number of pure strategies having the intervals of length b preceding the interval of length a so we have a contradiction to Assumption 3 at the beginning of the proof.

(II) No T_j^- has overlapping intervals

If no T_j^- has overlapping intervals, replacing T_j by T_j^- for $j = s+1, \ldots, r$ in S^* and moving the interval I_s so that it starts at the point 1 would still be an optimal strategy and the argument above that proves S^* is a strategy in $Z \times Z$ shows that this cannot be the case. Thus we have established the claim, a chain which has precisely i intervals of length b where $0 \leq i \leq \lceil 1/b \rceil$ has exactly λ_i intervals of length a where λ_i satisfies (9.2).

For each integer $i = 0, 1, \ldots, \lceil 1/b \rceil$, there is a non-negative integer β_i of chains that have precisely i intervals of length b and hence, by what we have just proved, each of these chains must have precisely λ_i intervals of length a. The chains have a total of $\sum_{i=0}^{\lceil 1/b \rceil} i\beta_i$ intervals of length b and $\sum_{i=0}^{\lceil 1/b \rceil} \lambda_i \beta_i$ intervals of length a. Since no member of $K_{|Z|}$ has an interval starting at 1 or more,

$$\sum_{i=0}^{\lceil 1/b \rceil} i\beta_i = \sum_{i=0}^{\lceil 1/b \rceil} \lambda_i \beta_i = k \quad \text{and} \quad \sum_{i=0}^{\lceil 1/b \rceil} \beta_i = m$$

so

$$v(a,b) = 1 - \left(\sum_{i=0}^{\lceil 1/b \rceil} \beta_i / \sum_{i=0}^{\lceil 1/b \rceil} i\beta_i \right).$$

Thus, putting $g_j = (\beta_j / \sum_{i=0}^{\lceil 1/b \rceil} i\beta_i)$, a lower bound for $v(a,b)$ is $1 - H$ where

$$H = \max \sum_{j=0}^{\lceil 1/b \rceil} g_j$$

subject to

$$\sum_{j=0}^{\lceil 1/b \rceil} \lambda_j g_j = \sum_{j=0}^{\lceil 1/b \rceil} j g_j = 1, \quad \text{and} \quad g_j \geq 0 \quad (j = 0, 1, \ldots, \lceil 1/b \rceil).$$

This is a linear program with dual

$$\min x + y$$

subject to

$$\lambda_i x + i y \geq 1, \quad x, y \text{ unrestricted} \quad (i = 0, 1, \ldots, \lceil 1/b \rceil).$$

Let $s \in \Lambda^+$ and $q \in \Lambda^-$ be such that $G(s,q) = G$, where $G(i,j)$ and G are given by (9.4) and (9.5) respectively. It is routine to check that, for $i, \in \Lambda^-$,

$$0 \leq G(s,q) - G(s,i) = \frac{(s - \lambda_s)(i(\lambda_q - \lambda_s) + \lambda_i(s - q) - (s\lambda_q - q\lambda_s))}{(s\lambda_q - q\lambda_s)(s\lambda_i - i\lambda_s)}$$

and that, for $i \in \Lambda^+$,

$$0 \leq G(s,q) - G(i,q) = \frac{(\lambda_q - q)(i(\lambda_q - \lambda_s) + \lambda_i(s - q) - (s\lambda_q - q\lambda_s))}{(s\lambda_q - q\lambda_s)(s\lambda_i - i\lambda_s)}.$$

It is now straightforward to verify by the duality theorem that $H = G(s,q)$ by taking $g_q = G(s,q) - g_s = (s - \lambda_s)/(s\lambda_q - q\lambda_s)$, $g_i = 0$ otherwise and $x = G(s,q) - y = (s - q)/(s\lambda_q - q\lambda_s)$. $\square$

9.5 A Second Illustrative Example

In the previous section we showed that an optimal Defender strategy gives rise to a number of chains of intervals and used this fact to obtain a lower bound for $v(a,b)$. However the type of chains and the number of each type that give this lower bound can be determined from the values of the lengths a and b without knowing the optimal Defender strategy. Thus it is natural to ask whether a knowledge of these chains enables us to construct an optimal Defender strategy. In the next section we will demonstrate that this is always possible when $b \le 1/2$ but, as in the case of Theorem 1, the justification is somewhat technical so we first analyse a particular example.

Consider the case when $a = 6/21$ and $b = 7/21$, then $\lambda_0 = 4, \lambda_1 = 3, \lambda_2 = 2$, $\lambda_3 = 0$ so that $\Lambda^+ = \{3\}$, $\Lambda^- = \{0, 1, 2\}$. Routine calculations give [see (9.5)]

$$G = \max\{G(\lambda_{j+}, \lambda_{j-}) : \lambda_{j+} \in \Lambda^+, \ \lambda_{j-} \in \Lambda^-\} = G(3, 0) = 7/12$$

giving a lower bound for the game of 5/12 by Theorem 1. Hence, to find a Defender optimal strategy, we are led to consider two chains, one, $C(0, 3)$, which has precisely three intervals of length b and none of length a and the other, $C(4, 0)$, with no intervals of length b and precisely four intervals of length a. To be contained in $[0, 1]$, this second chain would have to contain overlapping intervals so we can reduce the value of a to $1/4$ (which we do) and still have a cover of $[0, 1]$. To ensure that the number of intervals of length b equals the number of intervals of length a, we need four $C(0, 3)$ chains and three $C(4, 0)$ chains. Represent the intervals in the four $C(0, 3)$ chains by $B_i(j)$ where, for $i = 1, 2, 3$ and $j = 1, \ldots, 4$, $B_i(j) = [(i-1)b, ib]$ is in the j-th $C(0, 3)$ chain and the intervals in the three $C(4, 0)$ chains by $A_i(j)$ where, for $i = 1, \ldots, 4$ and $j = 1, 2, 3$, $A_i(j) = [(i-1)a, ia]$ is in the j-th $C(4, 0)$ chain. We want to construct 12 pure Defender strategies from these intervals with each strategy containing an interval of length a and an interval of length b which intersect in at most one point.

For each $B_i(j)$ construct the set $\mathscr{B}_i(j)$ of intervals of length a which have at most one point in common with it so that, for $j = 1, 2, 3$,

$$\mathscr{B}_1(j) = \{A_i(k) : i = 1, 2, 3, \ k = 3, 4\}, \quad \mathscr{B}_2(j) = \{A_i(k) : i = 1, 2, 3, \ k = 1, 4\}$$

and

$$\mathscr{B}_3(j) = \{A_i(k) : i = 1, 2, 3, \ k = 1, 2\}.$$

Provided we can take four members from each of these sets in such a way that each one of the $A_i(k), i = 1, \ldots, 4, \ k = 1, 2, 3$ is chosen, we will have 12 Defender pure strategies. One possibility is shown in Fig. 9.3; clearly an optimal Defender strategy for the original problem in which $a = 6/21$ and $b = 7/21$ can be obtained by increasing the lengths of the shorter intervals appropriately. Notice that Λ^+ and Λ^- remain the same for all $a \in [1/4, 1/3]$ when $b = 1/3$ so we have effectively found Defender optimal strategies for all these values. Furthermore, if $a = (1 - 3\varepsilon)/3$ where $\varepsilon > 0$ is very small, by appropriate lengthening of the shorter intervals in Fig. 9.3 all vertical lines except those at 0, 1 and those just either side of 1/3 and 2/3 can be made to intersect more than seven intervals. Thus it is reasonable to expect Infiltrator to choose points from the set $\{0, (1 - 2\varepsilon)/3, (1 + \varepsilon)/3, (2 - \varepsilon)/3, (2 + 2\varepsilon)/3, 1\}$ so that an interval of length 1/3

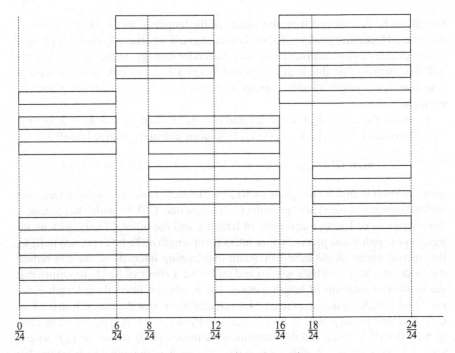

$$\frac{0}{24} \qquad \frac{6}{24} \quad \frac{8}{24} \qquad \frac{12}{24} \qquad \frac{16}{24} \quad \frac{18}{24} \qquad\qquad \frac{24}{24}$$

Fig. 9.3 A Defender optimal strategy when $a = 6/24$ and $b = 8/24$

or $(1 - 3\varepsilon)/3$ can contain at most two of these points; in addition, if an interval of length $(1 - 3\varepsilon)/3$ contains two of them, the two points must be either $(1 - 2\varepsilon)/3$ and $(1 + \varepsilon)/3$ or $(2 - \varepsilon)/3$ and $(2 + 2\varepsilon)/3$. Using symmetry and a little intuition, it is not too difficult to deduce that an optimal Infiltrator strategy is given by playing the points $0, (1 - 2\varepsilon)/3, (1 + \varepsilon)/3, (2 - \varepsilon)/3, (2 + 2\varepsilon)/3, 1$ with probabilities $10/36, 3/36, 5/36, 5/36, 3/36, 10/36$ respectively.

9.6 The Value of the Game

In Sect. 9.4 we showed that an optimal Defender strategy gives rise to a number of chains of intervals and used this fact to obtain a lower bound for $v(a, b)$. The lower bound involved only two non-negative integers, s and q, which satisfy $s \in \Lambda^+$, $q \in \Lambda^-$ and $G(s, q) = G$. This suggests that, if the lower bound is attained, an optimal Defender strategy might be constructible from a number, x_s, of chains having precisely s intervals of length b and a number, x_q, of chains having precisely q intervals of length b. The total number of intervals of length b in these chains is $sx_s + qx_q$ whereas the total number of intervals of length a is $\lambda_s x_s + \lambda_q x_q$. From the proof of Theorem 1, we would expect these two quantities to be equal which implies $x_s = \alpha(\lambda_q - q)$ and $x_q = \alpha(s - \lambda_s)$ for some positive integer α. For most cases $\alpha = 1$

would be sufficient for our purposes but, to make full use of symmetry, we will take $\alpha = 2$. We will be able to restrict attention to two particular types of chain, one in which all the intervals of length b precede all the intervals of length a and the other in which all the intervals of length a precede all the intervals of length b.

As the number of intervals of length b equals the number of intervals of length a, there are $(1-1)$ correspondences between them. Each correspondence f can be used to define a Defender strategy $S(f)$ by choosing one of the intervals of length b at random together with the interval of length a corresponding to it. We shall see in the proof of the next theorem that, by setting up an appropriate correspondence f, $S(f)$ restricts Infiltrator to at most $1 - G(s,q)$ where $G(s,q)$ is given by (9.4). To prove the existence of an appropriate correspondence, we need Hall's theorem ([1] or see [5] p. 89) on systems of distinct representatives.

A *system of distinct representatives* of not necessarily distinct subsets $C_1, C_2, \ldots, C_n$ of a set X is an n-tuple $(c_1, c_2, \ldots, c_n)$ such that $c_i \in C_i$ for $i = 1, \ldots, n$ and $c_i \neq c_j$ if $i \neq j$.

For a subset $J \subseteq \{1, 2, \ldots, n\}$, define $C(J) = \cup_{j \in J} C_j$ where $C(\emptyset) = \emptyset$.

Theorem 2 (Hall's Theorem). *The family $C_1, C_2, \ldots, C_n$ has a system of distinct representatives if and only if $|C(J)| \geq |J|$ for all subsets J of $\{1, 2, \ldots, n\}$.*

For positive integers i and j let $A_i^{\rightarrow}(j) = [(i-1)a, ia]$, $B_i^{\rightarrow}(j) = [(i-1)b, ib]$, $B_i^{\leftarrow}(j) = [1 - ib, 1 - (i-1)b]$ and $A_i^{\leftarrow}(j) = [1 - ia, 1 - (i-1)a]$. Although $B_i^{\rightarrow}(j_1)$ and $B_i^{\rightarrow}(j_2)$ are equal as sets, we will want to look on them as different entities when $j_1 \neq j_2$. Similar remarks apply to the other entities. To avoid unwieldy notation we effectively treat these entities as multisets and adopt the following convention.

Convention. For $X, Y \in \{B^{\rightarrow}, A^{\leftarrow}, B^{\leftarrow}, A^{\rightarrow}\}$, we consider $X_{i_1}(j_1) = Y_{i_2}(j_2)$ if and only if $X = Y$, $i_1 = i_2$ and $j_1 = j_2$.

If Defender can restrict Infiltrator's payoff to at most v with intervals of lengths b and a, he can clearly restrict Infiltrator's payoff to at most v with intervals of lengths b' and a' where $b' \geq b$ and $a' \geq a$. We will use this fact for some particular cases in the proof of the next theorem which gives an upper bound for $v(a,b)$ in terms of an $s \in \Lambda^+(a,b)$ and a $q \in \Lambda^-(a,b)$. These cases use the following remarks.

Remark 1. When $\lambda_s = 0$ and $qb + \lambda_q a > 1$, we can decrease the value of a to a^* where $qb + \lambda_q a^* = 1$ and just prove the theorem for the case when $\lambda_s = 0$ and $qb + \lambda_q a = 1$. Furthermore, if we also have $q = 0$, we can decrease b to b^* where $sb^* = 1$ and just prove the theorem for $b^* = 1/s$ and $a = 1/\lambda_q$.

Remark 2. When $q = 0$ and $sb + \lambda_s a > 1$ where $\lambda_s > 0$, we can decrease b to $b^- = \max\{b_1, b_2\}$ where $sb_1 + \lambda_s a = 1$ and $b_2 = a$; if $b^- = b_2$, we can then further decrease the values of b^- and a together until they equal b^* where $sb^* + \lambda_s b^* = 1$ (note that $(s + \lambda_s - 1)a \leq sb + (\lambda_s - 1)a < 1 \leq \lambda_0 a$ so $s + \lambda_s \leq \lambda_0$ giving $\lambda_0 b^* \geq 1$). Hence when $q = 0$ and $\lambda_s > 0$ we need only prove the theorem for the case when $q = 0$ and $sb + \lambda_s a = 1$.

Although the technical details are somewhat involved, the basic idea in the proof of the next theorem is quite simple. Two sets of intervals $\mathscr{X}$ and $\mathscr{Y}$ are defined with all the members of $\mathscr{X}$ having length b and all the members of $\mathscr{Y}$ having length a. For each interval J in $\mathscr{X}$ the set of all those sets in $\mathscr{Y}$ which are disjoint from J is denoted by $\mathscr{B}_J$. It is shown that $\mathscr{B}_J$ has a set of distinct representatives so that the representative of $\mathscr{B}_J$ together with J can be interpreted as a Defender pure strategy. The strategy which chooses one of these pure strategies at random enables the result to be established.

Theorem 3. *Let $2b < 1$, $s \in \Lambda^+(a,b)$ and $q \in \Lambda^-(a,b)$, then Defender can restrict Infiltrator to at most $1 - G(s,q)$ where $G(s,q)$ is given by (9.4).*

Proof. Let $2b < 1$, $s \in \Lambda^+$, and $q \in \Lambda^-$. Define $\mathscr{X}^\to$ and $\mathscr{Y}^\leftarrow$ by

$$\mathscr{X}^\to = \{B_i^\to(j) : 1 \le i \le q, \ 1 \le j \le s - \lambda_s + \lambda_q - q\}$$
$$\cup \{B_i^\to(j) : q+1 \le i \le s, \ 1 \le j \le \lambda_q - q\}$$

and

$$\mathscr{Y}^\leftarrow = \{A_i^\leftarrow(j) : 1 \le i \le \lambda_s, \ 1 \le j \le s - \lambda_s + \lambda_q - q\}$$
$$\cup \{A_i^\leftarrow(j) : \lambda_s + 1 \le i \le \lambda_q, \ 1 \le j \le s - \lambda_s\}$$

Definitions for $\mathscr{X}^\leftarrow$ and $\mathscr{Y}^\to$ have the directions of the arrows reversed. It is easy to see that

$$|\mathscr{X}^\to| = |\mathscr{X}^\leftarrow| = s\lambda_q - q\lambda_s = |\mathscr{Y}^\to| = |\mathscr{Y}^\leftarrow|.$$

Note that, for $1 \le j \le \lambda_q - q$, $B_1^\to(j), \ldots, B_s^\to(j), A_{\lambda_s}^\leftarrow, A_{\lambda_{s-1}}^\leftarrow(j), \ldots, A_1^\leftarrow(j)$ is not in general a chain in the sense of Sect. 9.4 because the definition of λ_s in (9.2) means that $B_s^\to(j)$ and $A_{\lambda_s}(j)$ can have interior points in common. However it is easy to generate a chain from it by moving the $A_i^\leftarrow(j)$s an appropriate distance to the right. Similar comments apply to $B_1^\to(j), \ldots, B_q^\to(j), A_{\lambda_q}^\leftarrow, A_{\lambda_{q-1}}^\leftarrow(j), \ldots, A_1^\leftarrow(j)$. This motivates the definitions for $\mathscr{B}_q^\to(j)$ and $\mathscr{B}_s^\to(j)$ below. Putting $\mathscr{Y}^\to \cup \mathscr{Y}^\leftarrow = \mathscr{Y}$ and denoting the interior of an interval I by int I, let, for each $B_i^\to(j) \in \mathscr{X}^\to$,

$$\mathscr{B}_i^\to(j) = \begin{cases} \{A \in \mathscr{Y} : B_i^\to(j) \cap \operatorname{int} A = \emptyset\} & \text{if } i \notin \{q,s\} \\ \{A \in \mathscr{Y} : A = A_{\lambda_i}^\leftarrow(j) \text{ or } B_i^\to(j) \cap \operatorname{int} A = \emptyset\} & \text{if } i \in \{q,s\}. \end{cases} \quad (9.6)$$

The $\mathscr{B}_i^\leftarrow(j)$ are defined by reversing the directions of the arrows.

We will obtain the existence of a mixed optimal strategy by taking pure strategies comprising $B_i^\to(j)$ (respectively $B_i^\leftarrow(j)$) and a member of $\mathscr{B}_i^\to(j)$ (respectively $\mathscr{B}_i^\leftarrow(j)$). To do this, we will show that, for most cases, the family $\mathscr{B}$ of all the $\mathscr{B}_i^\to(j)$ and $\mathscr{B}_i^\leftarrow(j)$ has a set of distinct representatives; by Remarks 1 and 2 these cases will be sufficient to prove the theorem. It is easy to see that $|\mathscr{B}| = 2(s\lambda_q - q\lambda_s)$.

For $\mathscr{W} \subseteq \mathscr{B}$, put $\mathscr{B}(\mathscr{W}) = \bigcup_{H \in \mathscr{W}} H$, $J^{\rightarrow}(\mathscr{W}) = \{i : \mathscr{B}_i^{\rightarrow}(j) \in \mathscr{W} \text{ for some } j\}$ and $J^{\leftarrow}(\mathscr{W}) = \{i : \mathscr{B}_i^{\leftarrow}(j) \in \mathscr{W} \text{ for some } j\}$. Now $\mathscr{B}_i^{\rightarrow}(j) = \mathscr{B}_i^{\rightarrow}(1)$ so

$$|\mathscr{B}(\mathscr{W})| = \left| \left(\bigcup_{i \in J^{\rightarrow}(\mathscr{W})} \mathscr{B}_i^{\rightarrow}(1) \right) \cup \left(\bigcup_{i \in J^{\leftarrow}(\mathscr{W})} \mathscr{B}_i^{\leftarrow}(1) \right) \right|.$$

Note that $J^{\rightarrow}(\mathscr{W}) \cap J^{\leftarrow}(\mathscr{W}) = \emptyset$ implies $|\mathscr{W}| \leq |\mathscr{B}|/2$. We will divide the analysis into several cases and for each case we will show that $|\mathscr{B}(\mathscr{W})| \geq |\mathscr{W}|$; the cases are expressed in terms of $J^{\rightarrow}(\mathscr{W})$ but the corresponding cases for $J^{\leftarrow}(\mathscr{W})$ follow by symmetry and we will often use this fact without explicitly mentioning it in the arguments below.

Case 1. Suppose $J^{\rightarrow}(\mathscr{W})$ contains i_1 and i_2 satisfying $|i_1 - i_2| > 1$. Because $b \geq a$, for each i and j, at least one of $A_i^{\rightarrow}(j) \in \mathscr{B}_{i_1}^{\rightarrow}(1)$ and $A_i^{\rightarrow}(j) \in \mathscr{B}_{i_2}^{\rightarrow}(1)$ holds and at least one of $A_i^{\leftarrow}(j) \in \mathscr{B}_{i_1}^{\rightarrow}(1)$ and $A_i^{\leftarrow}(j) \in \mathscr{B}_{i_2}^{\rightarrow}(1)$ holds. Hence we have $\mathscr{B}(\mathscr{W}) = \mathscr{Y}^{\rightarrow} \cup \mathscr{Y}^{\leftarrow} = \mathscr{Y}$ and $|\mathscr{B}(\mathscr{W})| = |\mathscr{B}| \geq |\mathscr{W}|$.

Thus we can assume that $|J^{\rightarrow}(\mathscr{W})| \leq 2$. Furthermore, if $|J^{\rightarrow}(\mathscr{W})| = 2$, then $J^{\rightarrow}(\mathscr{W}) = \{k, k+1\}$ for some integer k so that $\mathscr{B}(\mathscr{W})$ contains at least $|\mathscr{Y}^{\rightarrow}| - (s - \lambda_s + \lambda_q - q)$ members of $\mathscr{Y}^{\rightarrow}$. This follows because $A_i^{\rightarrow}(j) \in \mathscr{B}(\mathscr{W})$ if $kb \notin A_i^{\rightarrow}(j)$. Similarly $\mathscr{B}(\mathscr{W})$ contains at least $|\mathscr{Y}^{\leftarrow}| - (s - \lambda_s)$ members of $\mathscr{Y}^{\leftarrow}$.

Case 2. Suppose $J^{\rightarrow}(\mathscr{W})$ contains an $i \leq q$. We then have $\mathscr{B}_i^{\rightarrow}(\mathscr{W}) \supseteq \mathscr{Y}^{\leftarrow}$ and $|\mathscr{B}(\mathscr{W})| \geq |\mathscr{B}|/2$. Hence $|\mathscr{B}(\mathscr{W})| \geq |\mathscr{W}|$ holds when $J^{\rightarrow}(\mathscr{W}) \cap J^{\leftarrow}(\mathscr{W}) = \emptyset$ or $J^{\leftarrow}(\mathscr{W})$ contains a $k \leq q$. Thus, by Case 1, we only have to consider the case $J^{\rightarrow}(\mathscr{W}) = \{q, q+1\}$, $q+1 \in J^{\leftarrow}(\mathscr{W})$ and $J^{\rightarrow}(\mathscr{W}) \cap \{1, 2, \ldots, q\} = \emptyset$.

If $\lambda_s \neq 0$, $\mathscr{B}_{q+1}^{\rightarrow}(j)$ contains at least $s - \lambda_s + \lambda_q - q$ members of $\mathscr{Y}^{\rightarrow}$ so $|\mathscr{B}(\mathscr{W})| \geq |\mathscr{B}|/2 + s - \lambda_s + \lambda_q - q \geq |\mathscr{W}|$ because $|J^{\rightarrow}(\mathscr{W}) \cap J^{\leftarrow}(\mathscr{W})| = 1$.

If $\lambda_s = 0$, $s \geq 3$ because $2b < 1$. Now $A_i^{\rightarrow}(j) \in \mathscr{B}(\mathscr{W})$ if $qb \notin \text{int } A_i^{\rightarrow}(j)$. There are at most $s - \lambda_s$ members of $\mathscr{Y}^{\rightarrow}$ having qb as an interior point so $\mathscr{B}(\mathscr{W})$ contains at least $(s\lambda_q - q\lambda_s) - (s - \lambda_s) = s\lambda_q - s$ members of $\mathscr{Y}^{\rightarrow}$. Now $s \leq q + \lambda_q$ because $(q + \lambda_q)b \geq qb + \lambda_q a \geq 1 > (s-1)b$. Thus $|\mathscr{B}(\mathscr{W})| \geq |\mathscr{B}|/2 + s\lambda_q - s \geq |\mathscr{B}|/2 + (s-1)\lambda_q - q \geq |\mathscr{B}|/2 + \lambda_q - q \geq |\mathscr{W}|$ because $J^{\rightarrow}(\mathscr{W}) \cap J^{\leftarrow}(\mathscr{W}) = \{q+1\}$.

By Cases 1 and 2 we can now assume

$$|J^{\rightarrow}(\mathscr{W})| \leq 2, \quad \text{and} \quad |J^{\leftarrow}(\mathscr{W})| \leq 2 \tag{9.7}$$

and

$$i \in J^{\rightarrow}(\mathscr{W}) \cup J^{\leftarrow}(\mathscr{W}) \text{ implies } i > q. \tag{9.8}$$

Thus

$$|\mathscr{W}| \leq (\lambda_q - q)(|J^{\rightarrow}(\mathscr{W})| + |J^{\leftarrow}(\mathscr{W})|). \tag{9.9}$$

Case 3. Suppose $J^{\rightarrow}(\mathscr{W}) = \{k, k+1\}$ where $k > q$. We have already shown that, for this case, $\mathscr{B}(\mathscr{W})$ contains at least $s\lambda_q - q\lambda_s - (s - \lambda_s)$ of the $\mathscr{Y}^{\leftarrow}$ and at least $s\lambda_q - q\lambda_s - (s - \lambda_s + \lambda_q - q)$ of the $\mathscr{Y}^{\rightarrow}$. Furthermore, if $|J^{\leftarrow}(\mathscr{W})| = 2$, $\mathscr{B}(\mathscr{W})$ contains at least $s\lambda_q - q\lambda_s - (s - \lambda_s)$ members of $\mathscr{Y}^{\rightarrow}$. Thus

$$|\mathscr{B}(\mathscr{W})| \geq 2(s\lambda_q - q\lambda_s) - 2(s - \lambda_s) - (\lambda_q - q) = 3(\lambda_q - q) + 2H(s,q) \qquad (9.10)$$

where $H(s,q) = (s - \lambda_s)(q - 1) + (\lambda_q - q)(s - 2)$.

Note that, if $|J^{\leftarrow}(\mathscr{W})| = 2$, we have the stronger inequality

$$|\mathscr{B}(\mathscr{W})| \geq 4(\lambda_q - q) + 2H(s,q) \qquad (9.11)$$

Because $s \in \Lambda^+$ and $q \in \Lambda^-$, $\lambda_q \geq q$ and $s > \lambda_s$ so $s \geq 2$ because $2b < 1$. Thus $H(s,q) \geq 0$ if $q \geq 1$. Furthermore $\lambda_0 \geq s$ because $b \geq a$ so, if $s \geq 3$, $H(s,0) \geq -s + \lambda_s + \lambda_0 \geq 0$.

Hence, by (9.7), (9.9)–(9.11), $|\mathscr{B}(\mathscr{W})| \geq |\mathscr{W}|$ if $q \geq 1$ or ($q = 0$ and $s \geq 3$). Thus we have to consider only the case $q = 0$, $s = 2$ which implies $k = 1$ and $\lambda_2 = 1$ because $2b < 1$ and $s \in \Lambda^+$. Using Remark 2 we need only consider the case $2b + a = 1$. But then $\lambda_s a = a \leq b = kb$ and so, from the definition of $\mathscr{Y}^{\rightarrow}$, (9.10) can be strengthened to

$$|\mathscr{B}(\mathscr{W})| \geq 4(\lambda_q - q) + 2H(s,q) = 4\lambda_0 - 2.$$

As $b \geq a$ and $2b < 1$, $\lambda_0 \geq 3$. Thus $|\mathscr{B}(\mathscr{W})| \geq |\mathscr{W}|$ when $|J^{\leftarrow}(\mathscr{W})| \leq 1$. However, if $1 \in J^{\leftarrow}(\mathscr{W})$, $\mathscr{B}(\mathscr{W})$ contains $\mathscr{Y}^{\rightarrow}$ and $\mathscr{Y}^{\leftarrow}$ because $B_1^{\rightarrow}(j)$ and $B_1^{\leftarrow}(j)$ are distance a apart. Thus $|\mathscr{B}(\mathscr{W})| \geq |\mathscr{W}|$.

In addition to (9.8) and (9.9), we can now assume

$$|J^{\rightarrow}(\mathscr{W})| \leq 1, \quad \text{and} \quad |J^{\leftarrow}(\mathscr{W})| \leq 1 \qquad (9.12)$$

If $\lambda_s > 0$, $|J^{\rightarrow}(\mathscr{W})| = 1$ implies $\mathscr{B}(\mathscr{W})$ contains at least $s - \lambda_s + \lambda_q - q$ members of $\mathscr{Y}^{\leftarrow}$ and, from (9.9), (9.12) and symmetry, it follows that $|\mathscr{B}(\mathscr{W})| \geq |\mathscr{W}|$. Hence we can also suppose that $\lambda_s = 0$ and so, by Remark 1, that

$$qb + \lambda_q a = 1. \qquad (9.13)$$

Case 4. Suppose $J^{\rightarrow}(\mathscr{W}) = \{k\}$ where $k > q$ by (9.8). If $qb \geq 1/2$, then, using (9.13), $\mathscr{B}(\mathscr{W}) \subseteq \mathscr{Y}^{\rightarrow}$ so that $|\mathscr{B}(\mathscr{W})| \geq |\mathscr{W}|$) Thus we may assume $qb < 1/2$ so $s > 2q$ because $sb \geq 1$. Hence $s \geq q + 2$ because $s \geq 3$.

If $s = q + 2$, then $s = 3$ and $q = 1$ because $s > 2q$ so that $b + \lambda_1 a = 1$ by (9.13). For this case $\mathscr{B}_3^{\rightarrow}(j)$ contains $\mathscr{Y}^{\rightarrow}$ while, because $\lambda_1 \geq 2$, $\mathscr{B}_2^{\rightarrow}(j)$ contains at least $s\lfloor \lambda_1/2 \rfloor = 3\lfloor \lambda_1/2 \rfloor \geq 3(\lambda_1 - 1)/2 \geq \lambda_1 - 1$ members of $\mathscr{Y}^{\rightarrow}$. Hence, using the symmetric result if $|J^{\leftarrow}(\mathscr{W})| = 1$, we have $|\mathscr{B}(\mathscr{W})| \geq |\mathscr{W}|$.

Now suppose $s \geq q + 3$ and let $\eta = \lfloor b/a \rfloor$. Then $B_k^{\rightarrow}(j)$ intersects at most $(\eta + 2)s$ members of $\mathscr{Y}^{\leftarrow}$, $[qb, (k-1)b]$ contains at least $(k - q - 1)\eta s$ members of $\mathscr{Y}^{\leftarrow}$ and the interval $[kb, 1]$ contains at least $(s - k - 1)^+ \eta s$ members of $\mathscr{Y}^{\leftarrow}$ where $(s - k - 1)^+ = \max\{0, s - k - 1\}$. Thus $\mathscr{B}_k^{\rightarrow}(j)$ contains at least

$$\frac{\rho(k)s}{s\rho(k) + (\eta + 2)s}|\mathscr{Y}^{\leftarrow}| = \frac{\rho(k)s}{\rho(k) + (\eta + 2)}\lambda_q$$

members of $\mathscr{Y}^{\leftarrow}$ where

$$\rho(k) = \begin{cases} \eta(s-q-2) & \text{if } k \neq s, \\ \eta(s-q-1) & \text{if } k = s \end{cases}$$

and so at least λ_q if $\rho(k)(s-1) \geq (\eta+2)$. Hence, using the symmetric result if $|J^{\leftarrow}(\mathscr{W})| = 1$, we have $|\mathscr{B}(\mathscr{W})| \geq |\mathscr{W}|$ if $\rho(k)(s-1) \geq (\eta+2)$.

For $s \geq q+3$, $\rho(k)(s-1) \geq \eta(s-1)$ so $|\mathscr{B}(\mathscr{W})| \geq |\mathscr{W}|$ for $\eta \geq 2$ and for $\eta = 1$ and $s \geq 4$. However $\eta = 1$ and $s < 4$ gives $s = 3$ and $q = 0$. By Remark 1 we only need to consider the case $b = 1/3$ and $a = 1/\lambda_q$. Because $\eta = 1$, $\lambda_q \leq 5$. It is easy to see that $\mathscr{B}(\mathscr{W})$ contains at least $6 > \lambda_q$ members of $\mathscr{Y}^{\leftarrow}$ so $|\mathscr{B}(\mathscr{W})| \geq |\mathscr{W}|$.

Using the Remarks, we have therefore shown that, for all $\mathscr{W} \subseteq \mathscr{B}$ which are needed for the proof of the theorem $|\mathscr{B}(\mathscr{W})| \geq |\mathscr{W}|$. Thus, for all relevant $\mathscr{B}$, $\mathscr{B}$ has a set of distinct representatives by Hall's Theorem. Thus $B_i^{\rightarrow}(j) \in \mathscr{X}^{\rightarrow}$ gives rise to a pure Defender strategy with intervals $B_i^{\rightarrow}(j)$ and A where $A \in \mathscr{Y}$ is the representative of $\mathscr{B}_i^{\rightarrow}(j)$; we say that A is the correspondent of $B_i^{\rightarrow}(j)$. Pure Defender strategies involving $B_i^{\leftarrow}(j)$ are obtained similarly. Thus we have a total of $2(s\lambda_q - q\lambda_s)$ pure strategies and every member of $\mathscr{Y} \cup \mathscr{X}^{\rightarrow} \cup \mathscr{X}^{\leftarrow}$ occurs in precisely one of them. Let S denote the Defender strategy which selects one of these pure strategies at random. We show that, for $w \in [0,1]$, S has at least $2(s - \lambda_s + \lambda_q - q)$ pure strategies which have intervals containing w. Now

$$\bigcup_{i=1}^{s} B_i^{\rightarrow}(j) \cup \bigcup_{i=1}^{\lambda_s} A_i^{\leftarrow}(j) \qquad j = 1, \ldots, \lambda_q - q$$

$$\bigcup_{i=1}^{q} B_i^{\rightarrow}(j+\lambda_q-q) \cup \bigcup_{i=1}^{\lambda_s} A_i^{\leftarrow}(j+\lambda_q-q) \cup \bigcup_{i=1+\lambda_s}^{\lambda_q} A_i^{\leftarrow}(j) \qquad j = 1, \ldots, s - \lambda_s$$

are coverings of $[0,1]$ and so are the above expressions with the arrows reversed. Denote the set of these coverings by $\mathscr{C}$. Note that every member of $\mathscr{Y} \cup \mathscr{X}^{\rightarrow} \cup \mathscr{X}^{\leftarrow}$ occurs in precisely one member of $\mathscr{C}$.

For $w \in [0,1]$, a covering $C \in \mathscr{C}$ has a (unique) first interval $I_C(w)$ containing w, by which we mean that the left-hand endpoint of $I_C(w)$ is strictly less than the left-hand endpoint of any other interval of C containing w. Note that $I_C(w)$ starts strictly to the left of w if $w \neq 0$. Now $I_C(w)$ is in precisely one of the pure strategies of S for any w so we can define a mapping of $\mathscr{C}$ into the pure strategies of S by mapping C to the pure strategy containing $I_C(w)$. We show that this mapping is an injection. Suppose two different coverings C_1 and C_2 map into the same pure strategy. This pure strategy therefore comprises the intervals $I_{C_1}(w)$ and $I_{C_2}(w)$, one of which is of length b and the other (its correspondent) is of length a. Because an interval of length b does not contain its correspondent, $w > 0$. Hence $I_{C_1}(w)$ and $I_{C_2}(w)$ have an interval $(w - \varepsilon, w]$ in common for some $\varepsilon > 0$. Thus, by the definition of the $\mathscr{B}_i(j)$, the interval of length b must be of the form $B_i^{\rightarrow}(j)$ or $B_i^{\leftarrow}(j)$ where $i \in \{q,s\}$ for some j and the interval of length a of the form $A_{\lambda_i}^{\leftarrow}(j_1)$ or $A_{\lambda_i}^{\rightarrow}(j_1)$ respectively where $i \in \{q,s\}$ for some j_1. If $B_i^{\rightarrow}(j)$ contains w so must $B_i^{\rightarrow}(j_1)$. Similarly, if $A_{\lambda_i}^{\rightarrow}(j_1)$

contains w so must $A_{\lambda_i}^{\rightarrow}(j)$. By definition, $B_i^{\rightarrow}(j_1)$ starts to the left of $A_{\lambda_i}^{\leftarrow}(j_1)$ and $A_{\lambda_i}^{\rightarrow}(j)$ starts to the left of $B_i^{\leftarrow}(j)$. As $A_{\lambda_i}^{\leftarrow}(j)$ and $B_i^{\rightarrow}(j)$ are in the same covering, as are $A_{\lambda_i}^{\rightarrow}(j_1)$ and $B_i^{\leftarrow}(j_1)$, $I_C(w)$ cannot be of the form $A_{\lambda_i}^{\leftarrow}(j_1)$ or $B_i^{\leftarrow}(j)$ for $i \in \{q,s\}$ and we have a contradiction. Thus there are at least $2(s - \lambda_s + \lambda_q - q)$ pure strategies in S which have intervals containing w and the theorem follows. □

From Theorems 1 and 3 we have

Theorem 4. *For $a \leq b < 1/2$,*

$$v(a,b) = 1 - \max_{s \in \Lambda^+, q \in \Lambda^-} \left\{ \frac{s - \lambda_s + \lambda_q - q}{s\lambda_q - q\lambda_s} \right\}.$$

Note that when $q = \lambda_q$, $(s - \lambda_s + \lambda_q - q)/(s\lambda_q - q\lambda_s) = 1/q$.

9.7 Comments on Our Results

Although we have obtained its value, we have not produced explicit optimal strategies for either player in the general game. Evidence suggests that Infiltrator optimal strategies are more difficult to find than optimal Defender strategies and Woodward [14] has produced explicit Defender strategies for all values of a and b. Indeed finding an optimal Defender strategy is fairly straightforward using arguments similar to those employed for the special example in Sect. 9.5. Once the value of the game is obtained, one can find $s \in \Lambda^+$ and $q \in \Lambda^-$ such that $v(a,b) = 1 - G(s,q)$ so that the family $\mathscr{B}$ in Sect. 9.6 can be constructed. Algorithms exist for finding systems of distinct representatives (see, for instance, Marshall Hall [9]) and then a strategy can be obtained as in the proof of Theorem 3.

Of course Woodward's result [13] that $\Gamma(a,b)$ is equivalent to a finite game means that optimal strategies for any particular values of a and b can be obtained by using linear programming. In practice matters are not that straightforward because the strategy spaces in Woodward's finite game can be comparatively large; for instance, in the example in Sect. 9.5, the Defender has 81 and the Infiltrator 9 pure strategies. Furthermore, as mentioned in Chap. 6, linear programming can give very different optimal strategies for two games with slightly different values of a and b even though they have the same value. By contrast, our approach not only enabled the example to be solved easily but tells us that games with values of a and $b \geq a$ which give rise to the same Λ^+ and Λ^- as $a = 6/21$ and $b = 7/21$ also have the value 5/12. Thus games with (a,b) in the triangle with vertices $(1/4, 1/3)$, $(1/3, 1/3)$ and $(1/4, 3/8)$ and its interior with the closed line segment joining $(1/3, 1/3)$ to $(1/4, 3/8)$ removed all have value 5/12. Defender optimal strategies for these games are easily derived from Fig. 9.2. Although optimal Infiltrator strategies for the games are not quite so immediate, they follow a similar pattern to the particular example. Noting that the values of a and b in the given region satisfy $1 - a - 2b < 0$, intervals of length a and b can contain at most two of the points $0, a+, b+, (1-b)-, (1-a)-, 1$ where $x+$ and $x-$ represent points an appropriately small distance to the left and right of x respectively. The argument used for the particular case now show that choosing these points with probabilities $10/36, 3/36, 5/36, 5/36, 3/36, 10/36$ respectively is an optimal Infiltrator strategy.

One might have hoped that the expression for $v(a,b)$ in Theorem 4 also covers the case $b \geq 1/2$. However this is not the case because Baston and Bostock [2] showed that $v(1/3, 1/2) = 1/4$ whereas, for $a = 1/3$ and $b = 1/2$, Theorem 3 gives $v(1/3, 1/2) = 1 - \max\{(1+\lambda_1)/(2\lambda_1), (2+\lambda_0)/(2\lambda_0)\} = 1/6$.

Theorem 4 enables us to express the value of $\Gamma(a,b)$ when $1/3 \leq b < 1/2$ in a simpler form than Lee [8].

Theorem 5. *Let $a \leq b$, $1/3 \leq b < 1/2$ and λ_i defined by (9.2).*

(i) *If $\lambda_2 \geq 2$, $v(a,b) = (2m-6)/3m$ where $m = \min\{2\lambda_0, 3\lambda_1, 6\lambda_2\}$.*
(ii) *If $\lambda_2 = 1$, $v(a,b) = (m-1)/2m$ where $m = \min\{\lambda_0, 2\lambda_1 - 1\}$.*

Proof. (i) Let $\lambda_2 \geq 2$, then $\Lambda^+ = \{3\}$ and $\lambda_3 = 0$ so, by Theorem 4,

$$v(a,b) = 1 - \max\{\frac{3+\lambda_0}{3\lambda_0}, \frac{2+\lambda_1}{3\lambda_1}, \frac{1+\lambda_2}{3\lambda_2}\} = \frac{2}{3} - \max\{\frac{2}{2\lambda_0}, \frac{2}{3\lambda_1}, \frac{2}{6\lambda_2}\}$$

and (i) follows.
(ii) Let $\lambda_2 = 1$, then $\Lambda^+ = \{2,3\}$ so, by Theorem 4,

$$v(a,b) = 1 - \max\{\frac{3+\lambda_0}{3\lambda_0}, \frac{2+\lambda_1}{3\lambda_1}, \frac{1+\lambda_0}{2\lambda_0}, \frac{\lambda_1}{2\lambda_1 - 1}\}$$

Now $(3+\lambda_0)/(3\lambda_0) \leq (1+\lambda_0)/(2\lambda_0)$ because $\lambda_0 \geq 3$ and $(\lambda_1)/(2\lambda_1 - 1) \geq (2+\lambda_1)/(3\lambda_1)$ because $\lambda_1 \geq 2$. Thus

$$v(a,b) = 1/2 - \max\{\frac{1}{2\lambda_0}, \frac{1}{2(2\lambda_1 - 1)}\}$$

and (ii) follows. $\square$

We now give examples to show that every case in the theorem arises:
$a = 1/8$, $b = 3/8$ gives $\lambda_2 \geq 2$ and $6\lambda_2 < 3\lambda_1 < 2\lambda_0$,
$a = 3/39$, $b = 13/39$ gives $\lambda_2 \geq 2$ and $2\lambda_0 < 3\lambda_1 < 6\lambda_2$,
$a = 2/15$, $b = 5/15$ gives $\lambda_2 \geq 2$ and $3\lambda_1 < 2\lambda_0 < 6\lambda_2$,
$a = 3/10$, $b = 4/10$ gives $\lambda_2 = 1$ and $2\lambda_1 - 1 < \lambda_0$,
$a = 2/20$, $b = 9/20$ gives $\lambda_2 = 1$ and $\lambda_0 < 2\lambda_1 - 1$.

Acknowledgements The first author was supported, in part, by NATO Grant PST.CLG.976391.

References

1. M. Aigner: *Combinatorial Theory*, Springer-Verlag, Berlin (1979).
2. V. J. Baston and F. A. Bostock: A Continuous Game of Ambush, *Naval Research Logistics* 34 645–654 (1987).
3. V. J. Baston and K. Kikuta: K. An Ambush Game with an Unknown Number of Infiltrators, *Operations Research* 52 597–605 (2004).
4. V. J. Baston and K. Kikuta: An Ambush Game with a Fat Infiltrator, *Operations Research* 57 514–519 (2009).
5. P. J. Cameron: *Combinatorics: Topics, Techniques, Algorithms*, Cambridge University Press, Cambridge (1994).
6. A. Y. Garnaev: On a Ruckle Problem in Discrete Games of Ambush, *Naval Research Logistics* 44 353–364 (1997).
7. A. Y. Garnaev: *Search Games and Other Applications of Game Theory*, Springer, London (2000).
8. K. T. Lee: On Ruckle's Game of Ambush, *Naval Research Logistics* 37 355–363 (1990).
9. Marshall Hall Jr: *Combinatorial Theory*, Blaisdell, Waltham, MA (1967).
10. W. H. Ruckle: *Geometric Games and Their Applications*, Pitman, Boston, MA (1983).
11. W. H. Ruckle: Ambushing Random Walks II; Continuous Models, *Operations Research* 29 108–120 (1981).
12. H. Weyl: Elementary Proof of a Minimax Theorem due to Von Neumann, Contributions to the Theory of Games, 1 19–25 (1950).
13. I. D. Woodward: Discretization of the Continuous Ambush Game, *Naval Research Logistics* 50 515–529 (2003).
14. I. D. Woodward: Cable Laying Ambush Games, Ph.D. thesis, University of Southampton 2002.
15. N. Zoroa, P. Zoroa and M. J. Fernández-Sáez: A Generalisation of Ruckle's Results for an Ambush Game, *European Journal of Operational Research* 119 353–364 (1999).

Chapter 10
How to Poison Your Mother-in-Law, and Other Caching Problems

Robbert Fokkink, Joram op den Kelder, and Christos Pelekis

Abstract We discuss an open problem on caching games, which in its full generality is due to Alpern, Kikuta and Ruckle. The problem is related to current work in the theory of hypergraphs.

R. Fokkink (✉) • J. op den Kelder • C. Pelekis
Delft Institute of Applied Mathematics, TU Delft, P.O.Box 5031, 2600GA Delft, Netherlands
e-mail: r.j.fokkink@tudelft.nl; J.M.opdenKelder@student.tudelft.nl;
c.pelekis@tudelft.nl

S. Alpern et al. (eds.), *Search Theory: A Game Theoretic Perspective*,   155
DOI 10.1007/978-1-4614-6825-7_10, © Springer Science+Business Media New York 2013

10.1 The Kikuta-Ruckle Conjecture

The Kikuta-Ruckle conjecture arose from a series of papers on search games and optimal allocation that were written by Ken Kikuta and William Ruckle [9–11] over a number of years. In fact, there is not one, but there are several conjectures that can be found dispersed over these papers. In this chapter we discuss one of these conjectures, as well as its ramifications into search games, which have led to some fascinating new results on expanding search that have been reported by Tom Lidbetter in Chap. 2. The interplay of game theory and combinatorics is a hallmark of Ruckle's work. It is manifest in the two cable ambush game, as described in the previous chapter by Vic Baston and Ian Woodward. The Kikuta-Ruckle conjecture is no different.

Suppose your mother-in-law comes over for tea. You know that she is going to eat k biscuits from a tray that contains n biscuits in total, but you do not know which biscuits she is going to take. Each biscuit is equally likely. You possess h grammes of arsenic, where $h > 1$ is a real number. The lethal dose of arsenic is one gramme. Unfortunately, you cannot put the poison in her tea, you have to put it in the biscuits. How should you distribute the poison to maximize the probability that your mother-in-law gets the lethal dose?

> **Kikuta-Ruckle Conjecture** ([11]) It is optimal to put a dose of $1/j$ in as many biscuits as possible, for a natural number j that depends on h, k, n.

Here, optimal means that the distribution maximizes the number of k-element subset such that the cumulative amount is lethal. To illustrate this conjecture we first present a solution for the particular case of $k = 3$ and $n = 5$, which involves the Petersen graph, see Fig. 10.1. Each vertex represents a possible selection of three biscuits, and two vertices are neighbors if and only if they have exactly one biscuit in common. The solution of $k = 3$ and $n = 5$ breaks up into two parts, depending on the value of h:

$\frac{3}{2} \leq \mathbf{h} < \frac{5}{3}$: Any circuit of length five contains contains each biscuit exactly thrice. So the amount of poison in the circuit adds up to $3h < 5$. Therefore, any circuit contains at least one non-lethal vertex. Any pair of vertices can be avoided by a circuit of length five, so there are at least three non-lethal vertices. Now it is not hard to see that a poison distribution of $\frac{1}{2}, \frac{1}{2}, \frac{1}{2}, 0, 0$ is optimal. So the conjecture holds with $j = 2$ in this case.

$1 \leq \mathbf{h} < \frac{3}{2}$: There must be adjacent vertices that are lethal, otherwise there would hardly be any. Adjacent vertices share one common biscuit. If both vertices are lethal, the common biscuit contains a dose of $> \frac{1}{2}$ since $h < 3/2$. It follows that any other lethal vertex contains this particular biscuit. So we may just as well put a unit dose in it. The conjecture holds with $j = 1$ in this case. □

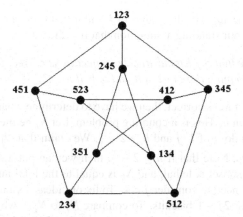

Fig. 10.1 The nodes in the Petersen graph represent all selections of your mother-in-law if $k = 3$ and $n = 5$

The example displays the combinatorial flavor of the conjecture and it is not surprising that there exists related work in combinatorics. Recently Alon et al. [1, Conjecture 1.4] have proposed a conjecture that is equivalent to the Kikuta-Ruckle conjecture. This work of Alon et al. is motivated by an old but still unsolved conjecture of Erdös on the matching number of hypergraphs [6].

The solution of the conjecture for $k = 3$ and $n = 5$ breaks up into two parts, depending on the value of h. If h is small, then it is optimal to put a unit dose. If h is large, then it is optimal to put smaller doses. This applies to all known solutions of the Kikuta-Ruckle conjecture, such as the case of the odd graph that we settle in the next section, or the case of cyclic graphs that has been solved in [4], also see [2]. The optimal j increases with h. Such monotonicity has also been encountered in probabilistic allocation problems, see e.g. [7].

10.2 The Kikuta-Ruckle Conjecture for Odd Graphs

The odd graph $\mathcal{O}_k$ has one vertex for each of the k-element subsets of a $(2k-1)$-element set. Two vertices are connected by an edge if and only if the corresponding subsets have one common element.[1] The Petersen graph is equal to the odd graph for $k = 3$. Norman Biggs [5] already remarked that if one wants to understand a graph theory problem, the odd graph is a good place to start. So we consider the

---

[1] The original definition of the odd graph takes $(k-1)$-element subsets as its vertices. They are connected by an edge if and only if they are disjoint. So for each edge there is one element that is not contained in the two vertices: the odd one out. This is where the graph gets its name from. Our definition is equivalent and more convenient for the poisoning problem. An edge represents the odd one in.

Kikuta-Ruckle conjecture for the values of k and n that correspond to odd graphs: in this section, it is our standing assumption that $\mathbf{n} = \mathbf{2k - 1}$.

Lemma 1. *Suppose that the Kikuta-Ruckle conjecture is correct for the odd graph. Then it is optimal to put a dose of $1/j$ if and only if $h \in [2 - \frac{1}{j}, 2 - \frac{1}{j+1})$.*

Proof. If the conjecture is correct, then we have to determine what the optimal dose $1/j$ is, depending on h. This is a counting problem. Let N_j be the number of lethal subsets if we put a dose of $1/j$ and $h \geq 2 - \frac{1}{j}$. We claim that $N_1 < \cdots < N_k$ is an increasing sequence. Note that if $h \geq 2 - \frac{1}{k}$, then we can put a dose $1/k$ in every biscuit, so every k-subset is lethal, and N_k is equal to the total number of biscuits. Therefore, we only need to consider $j < k$. To fix our ideas, assume that we put the dose $1/j$ in the first $2j - 1$ biscuits. To compare N_j to N_{j+1} we need to consider the effect of reducing the amount of poison in the first $2j - 1$ biscuits from $1/j$ to $1/(j+1)$, while putting a dose of $1/(j+1)$ in the next two biscuits that previously did not contain any poison. Such a redistribution can only change the lethality of a k-subset if it contains either j or $j+1$ elements from $\{1, \ldots, 2j+1\}$. A lethal subset becomes non-lethal if it contains j elements from $\{1, \ldots, 2j - 1\}$ and none from $\{2j, 2j+1\}$. There are exactly

$$\binom{2j-1}{j}\binom{2k-2j-2}{k-j}$$

such subsets. Conversely, a non-lethal subset becomes lethal if it contains $j - 1$ elements from $\{1, \ldots, 2j - 1\}$ and both $2j$ and $2j+1$. There are exactly

$$\binom{2j-1}{j-1}\binom{2k-2j-2}{k-j-1}$$

such subsets. Dividing the first binomial product by the second gives $\frac{k-j-1}{k-j} < 1$, so the number of k-subsets that become lethal exceeds those that become non-lethal. Which proves that $N_j < N_{j+1}$.

If we put a dose of $1/j$ while $h < 2 - \frac{1}{j}$, then there are at most $2j - 2$ poisonous biscuits. Let M_j be the number of lethal k-subsets in this case. We claim that $M_1 < \cdots < M_k$ is again an increasing sequence. To compare M_j to M_{j+1} we need to consider the effect of reducing the amount of poison in the first $2j - 2$ biscuits, while putting a dose of $1/(j+1)$ in biscuit $2j - 1$ and $2j$. The number of lethal subsets that become non-lethal now is

$$\binom{2j-2}{j}\binom{2k-2j-1}{k-j}$$

while the number of subsets that become lethal is

$$\binom{2j-2}{j-1}\binom{2k-2j-1}{k-j-1}.$$

The quotient of these two binomial products is $\frac{i-1}{j} < 1$, so the number of subsets that become lethal upon redistribution again exceeds the number of those that become non-lethal. Now we claim that $M_k < N_1$, so it is better to put a single unit dose. Indeed $M_k = \binom{2k-2}{k}$ while $N_1 = \binom{2k-2}{k-1}$. So putting a dose $1/j$ for $j > 1$ is only optimal once $h \geq 2 - \frac{1}{j}$. □

We have an amount of poison h that we distribute over the biscuits, putting a dose h_i in the i-th biscuit. A k-subset V is lethal if and only if $h(V) = \sum_{i \in V} h_i \geq 1$. We number the biscuits in decreasing order of their doses, putting the most poisonous biscuit first, i.e., $h_1 \geq \cdots \geq h_{2k-1}$. Let $\mathscr{P}$ be the family of poisonous k-subsets. We want to distribute the poison in such a way that $\mathscr{P}$ has maximum cardinality. We adopt hypergraph terminology. We say that $V \in \mathscr{P}$ is an edge, and $\deg_{\mathscr{P}}(i)$ is equal to the number of edges that contains i.

Lemma 2. *If* $h < 2 - \frac{1}{j+1}$ *then* $\deg_{\mathscr{P}}(2j+1) \leq \frac{1}{2}\binom{2k-2}{k-1}$.

Proof. By the decreasing dosage of poison

$$(2j+1)h_{2j+1} \leq h_1 + \cdots + h_{2j+1} \leq h < \frac{2j+1}{j+1},$$

and so $h + h_{2j+1} < 2$. If V is any k-subset that contains $2j+1$ then let $\bar{V} = V^c \cup \{2j+1\}$. In other words, $\bar{V}$ is the neighbor of V in the odd graph $\mathscr{O}_k$ that is connected by the edge that has $2j+1$ as the odd one in. Then $h(V) + h(\bar{V}) = h + h_{2j+1} < 2$. So if V is poisonous then $\bar{V}$ is not, and we conclude that $\deg_{\mathscr{P}}(2j+1)$ is at most half of the degree of $2j+1$ in the complete hypergraph on all k subsets. The degree of the complete hypergraph is $\binom{2k-2}{k-1}$. □

Lemma 3. *If* $h < 2 - \frac{1}{j+1}$ *then the number of lethal edges is at most*

$$\frac{1}{2}\binom{2j}{j}\binom{2k-2j-1}{k-j} + \sum_{i=j+1}^{k}\binom{2j}{i}\binom{2k-2j-1}{k-i}.$$

Proof. We maximize the number of edges V under the constraint that the hypergraph has maximal $\sum_{i \geq 2j+1} \deg(i)$, which by the previous lemma is bounded by

$$\frac{n-2j}{2}\binom{2k-2}{k-1}.$$

The greedy solution is to first take all k-subsets that have no elements in $\{2j+1,\ldots,2k-1\}$, then to take all k-subsets that have one element in $\{2j+1,\ldots,2k-1\}$, etc, until the sum of the degrees exceed the given bound. We need to show that this happens exactly when we have taken all k-subsets that contain $> j$ elements from $\{1,\ldots,2j\}$ and half of the k-subsets that contain exactly j elements from this set. In other words, we need to show that

$$\frac{1}{2}\binom{2j}{j}\binom{2k-2j-1}{k-j}(k-j)+\sum_{i=j+1}^{\min\{2j,k-1\}}\binom{2j}{i}\binom{2k-2j-1}{k-i}(k-i)$$

is equal to $\frac{n-2j}{2}\binom{2k-2}{k-1}$. This equality can be rewritten to

$$\frac{\binom{2j}{j}\binom{2k-2j-2}{k-j-1}}{2\binom{2k-2}{k-1}}+\sum_{i=j+1}^{\min\{2j,k-1\}}\frac{\binom{2j}{i}\binom{2k-2j-2}{k-i-1}}{\binom{2k-2}{k-1}}=\frac{1}{2}.$$

Let X be a hypergeometric random variable that describes the number of successes in $k-1$ draws from a population of $N=2k-2$ with $2j$ successes. Then this equation is equal to

$$\frac{1}{2}\mathbf{P}(X=j)+\mathbf{P}(X>j)=\frac{1}{2}.$$

In other words, the median of X is at j. To see why this is true, notice that drawing $k-1$ from $2k-2$ is equivalent to leaving $k-1$ from $2k-2$. Since the number of successes is $2j$, this implies that $\mathbf{P}(X>j)=\mathbf{P}(X<j)$. □

Theorem 1. *The Kikuta-Ruckle conjecture is true for odd graphs, i.e., if $n=2k-1$.*

Proof. If we put $2j-1$ doses of $1/j$ then an edge is lethal if and only if it contains at least j out of the first $2j-1$ biscuits. So the number of lethal edges is equal to

$$\sum_{i=j}^{k}\binom{2j-1}{i}\binom{2k-2j}{k-i}.$$

By the previous lemma, it suffices to show that this is equal to

$$\frac{1}{2}\binom{2j}{j}\binom{2k-2j-1}{k-j}+\sum_{i=j+1}^{k}\binom{2j}{i}\binom{2k-2j-1}{k-i}.$$

If we divide both sums by $\binom{2k-1}{k}$ then the first quotient is $\mathbf{P}(X_1\geq j)$ for a hypergeometric random variable that counts the number of successes if we draw k times with $2j-1$ successes. The second quotient is $\frac{1}{2}\mathbf{P}(X_2=j)+\mathbf{P}(X_2\geq j+1)$ if the number of successes is $2j$. To see why these probabilities are the same, start with the population that has $2j$ successes and call one of them a failure, which transforms X_2 into X_1. Let $\mathbf{U}$ be the event that the draw does not contain the success which turns into a failure. Then $X_1\geq j$ is equal to

$$(X_2\geq j+1)\cup\{\mathbf{U}\cap X_2=j\}.$$

Now observe that

$$\mathbf{P}(\mathbf{U}\cap X_2=j)=\mathbf{P}(\mathbf{U}\mid X_2=j)\mathbf{P}(X_2=j)=\frac{1}{2}\mathbf{P}(X_2=j).$$

The computation in our proof, which is essentially a double counting argument, seems to work more or less by coincidence. To check the validity of the Kikuta-Ruckle conjecture, other sets of parameters need to be tested. The case $n < 2k$ seems to be a logical next step.

10.3 Alpern's Caching Games

The Kikuta-Ruckle conjecture presents a combinatorial problem, it is not a game. It be turned into a game by adding some geometry. Suppose that a hider places n objects on a graph and that the searcher wins if and only if he retrieves k of these objects, but he is not allowed to search the entire graph. This is a difficult game to solve, even for relatively simple graphs. Elementary examples suggest that if the searcher is allowed to search a large part of the graph, then the hider wants to put all the objects in a single place, hoping that the searcher won't find them. If the searcher can only search a small part of the graph, then the hider spreads out the objects, so that they become out of the searcher's reach. If the game parameter changes to his advantage, the hider spreads out. This should be compared to the poisoning problem, in which the poisoner distributes poison over ever more biscuits if the parameter h changes to his advantage.

Consider the following search game. The hider can place two objects on the line, at positions x_1 and x_2. The hider starts from the origin and has to dig a tunnel, so it takes an effort to place these objects. The hider cannot carry on forever. He can only dig a tunnel of unit length. This could be a tunnel that goes one way only, from 0 to 1, or from 0 to -1. It can also be a tunnel that goes both ways, from $-x$ to $1 - x$ for some $0 < x < 1$. The hider places the two objects somewhere in the tunnel, and after he is done, he fills up the tunnels again. One should imagine that the hider is a squirrel that buries nuts, caching them for later use. Steve Alpern, who first thought of these games, has coined the term *caching game*.

Fig. 10.2 The searcher's dilemma in the caching game

After the hider is done, the searcher looks for these two objects, looks for these two objects, facing the dilemma that is illustrated in Fig. 10.2. He is at least as powerful as the hider and can dig a tunnel of total length $h \geq 1$. The searcher wins if he retrieves both objects, otherwise the hider wins. If $h \geq 2$ then the searcher always wins, by digging a tunnel that runs from -1 to 1. If $h < 2$ then the hider ensures a 50 % probability win by putting both nuts either at $+1$ or -1. Conversely, if $h \geq \frac{3}{2}$ then the searcher ensures a 50 % probability win by digging either to $+1$ or -1 all the way, and digging $h - 1$ in the opposite direction. To see why this guarantees a 50% probability, observe that the hider wins if he digs in the direction that contains the hidden object that is farthest away from the origin. If $h < 3/2$ then the hider places the objects either at $\{-1, -\frac{1}{2}\}$ or $\{-\frac{1}{2}, +\frac{1}{2}\}$ or $\{+\frac{1}{2}, +1\}$, equiprobably. The distance between these three positions is such that the searcher can only reach one of these three placements. So the hider wins with probability two thirds, at least. The searcher, on the other hand, has a strategy that guarantees that he wins with probability one third, as follows. He digs into one directions, and if he finds an object, then he continues digging in that direction with probability $\frac{2}{3}$, or he starts digging in the other direction with probability $\frac{1}{3}$. This guarantees that the searcher finds both objects with probability one third. So the value of the game, which we define as the probability of a searcher win, is equal to one third if $1 \leq h < 3/2$.

In the caching game with two objects and two directions the hider either places both objects in the same location, or he places them in such a way that if the hider finds one object, then the remaining object is optimally placed. Indeed, if the searcher finds the first object at say $+\frac{1}{2}$, then the remaining object is either at $+1$ or at $-\frac{1}{2}$, equiprobably. In the remaining game, the searcher is looking for a single object, for which he has to dig a distance $\frac{1}{2}$, either to the left or to the right. The hider has made sure that the remaining object is optimally placed. This seems to be a general principle that applies to all versions of the caching game that we can solve.

One can increase the number of directions in which the hider can dig, or the number of objects that he hides. The game with three tunnels and two objects has been solved in [3]. The game with four tunnels and two objects appears to be difficult, and has not been fully solved yet. The value of the game and the optimal strategies for the players have been determined for a substantial range of the parameter h and can be found in [8]. The following table for the game value has been taken from that report:

h	value
$[0, 1)$	0
$[1, \frac{3}{2})$	$\frac{1}{10}$
$[\frac{3}{2}, \frac{5}{3})$	$\frac{3}{20}$
$[\frac{5}{3}, \frac{7}{4})$	$\frac{1}{5}$
$[\frac{7}{4}, 2)$	$?$
$[2, \frac{11}{5})$	$\frac{2}{3}$
$[\frac{11}{5}, \frac{7}{3})$	$?$
$[\frac{7}{3}, 3)$	$\frac{1}{2}$
$[3, 4)$	$\frac{3}{4}$
$[4, \infty)$	1

The solution of the game for h in $[\frac{7}{4}, 2) \cup [\frac{11}{5}, \frac{7}{3})$ remains open. For all instances of the game in which the solution has been found, the hider places the two objects in such a way that once the searcher finds one object, the remaining object is placed equiprobably in one of the four directions, as far away as possible. Even though the amount of evidence is not overwhelming, there may be an underlying principle:

> **A Kikuta-Ruckle Conjecture for Caching Games** Let $\Gamma(j, k, n, h)$ denote the caching game in which the hider can dig in n directions for a total length of one unit, hiding k objects, of which the searcher has to retrieve j and he can dig a total length h. Then the hider places the objects in such a way that once the searcher finds a single object at distance x, then the remaining objects are optimally placed in the remaining game $\Gamma(j-1, k-1, n, \frac{h-x}{1-x})$.

If such a recursive principle exists, it should also apply to the Kikuta-Ruckle conjecture that we exhibited in the first section, and other versions of that conjecture which can be found in the papers by Kikuta and Ruckle on accumulation games.

References

1. N. Alon, P. Frankl, H. Huang, V. Rödl, A. Ruciński, B. Sudakov, Large matchings in uniform hypergraphs and the conjectures of Erdös and Samuels, J. Combin.Theory, Series A, **119**, 1200–1215, (2012)
2. S. Alpern, R. Fokkink, K. Kikuta, On Ruckle's conjecture on accumulation games, SIAM J. Control Optim. **48** no 8, 5073–5083, (2010)
3. S. Alpern, R. Fokkink, J. op den Kelder, T. Lidbetter, Disperse or unite: a mathematical model for coordinated attack, Decision and Game Theory for Security (GAMESEC2010), LNCS 6442, 221–233, (2010)
4. S. Alpern, R. Fokkink, C. Pelekis, A solution of the Kikuta-Ruckle conjecture on cyclic caching of resources, J. Optim. Theory Appl. **153** no 3, 650–661, (2012)
5. N. Biggs, Some odd graph theory, Annals New York Academy of Sciences **319** no 1, 71–81 (1979)
6. P. Erdös, A problem on independent r-tuples, Ann. Univ. Sci. Budapest Eötvös Sect. Math. **8**, 93–95, (1965)
7. K. Jogdeo, S.M. Samuels, Monotone convergence of binomial probabilities and a generalization of Ramanujan's equation, Ann. Math. Statistics **39** no 4, 1191–1195, (1968)
8. J. op den Kelder, Disperse or unite: a mathematical model for coordinated attack, TU Delft technical report, http://repository.tudelft.nl/, (2012)
9. K. Kikuta, W. Ruckle, Accumulation games, Part 1: noisy search, J. Optim. Theory Appl. **94** no 2, 395–408, (1997)
10. K. Kikuta, W. Ruckle, Continuous accumulation games in continuous regions, J. Optim. Theory Appl. **106** no 3, 581–601, (2000)
11. K. Kikuta, W. Ruckle, Continuous accumulation games on discrete locations, Naval Res. Logistics, **49** no 1, 60–77, (2002)

Part III
Rendezvous

Chapter 11
Rendezvous Problem

Leszek Gąsieniec

Abstract The rendezvous problem refers to the algorithmic challenge in which two or more *mobile entities* (depending on the context) called *players*, *agents* or *robots*, are expected to meet at the same time and point in space. The meeting challenge can be a task on its own or it may form a part of a more complex communication or coordination process in which the agents are involved. The space can be either a network of discrete nodes between which the agents can move along existing edge connections, or a geometric environment in which movement of agents is only restricted by topological properties of the space. In order to meet, the agents must agree in advance on a rendezvous mechanism. The feasibility and efficiency of the adopted rendezvous solution depends on agents' ability to move, observe and communicate. In this chapter we give a short introduction to the rendezvous problem including motivation, models of considered networks and participating agents. We also provide some examples and discuss instances of the considered problem.

11.1 Introduction

With the recent advent of ad-hoc, not well-structured, large, and (very often) dynamic network environments there is a strong need for more robust, universal, and inexpensive distributed network protocols. The purpose of these protocols is to support basic network integrity mechanisms as well as more dedicated tasks such as information dissemination, network search and discovery, frequent monitoring including handling emergencies.

L. Gąsieniec (✉)
Department of Computer Science, University of Liverpool, Ashton Street,
Liverpool, L69 3BX, UK
e-mail: L.A.Gasieniec@liverpool.ac.uk

S. Alpern et al. (eds.), *Search Theory: A Game Theoretic Perspective*,
DOI 10.1007/978-1-4614-6825-7_11, © Springer Science+Business Media New York 2013

One of the novel, promising, and perhaps challenging alternatives in supporting such network protocols are dedicated teams of mobile entities that can work independently on top of basic network processes. Mobile entities may, e.g., represent software agents [22] residing in nodes or traversing through network connections, autonomous mobile robots [27] located in a (real) geometric environment, or a group of people that have to meet in a city whose streets form a road network [2]. The structure of the network environment can be stable or it can change in time due to accidental failures, mobility or instability of objects including malicious performance of nodes, unwanted visits of intruders, etc.

Apart from populating network environments, teams of mobile entities can be also seen as more complex systems on their own. For example, a traditional communication network can be replaced by a more arbitrary environment in which a collection of networked or free-standing agents representing groups of humans, animals, vehicles or specialised robots are asked to perform a dedicated computational task. This could be done in the form of a fully-coordinated effort or as a collection of (semi-)independent individual (possibly greedy) performances.

Rendezvous of mobile entities (agents, players) is often a goal on its own. Alternatively, it can be used as a subroutine in a range of basic network integrity and coordination mechanisms. The agent's ability to act autonomously including observation, communication and relocation impels the design and further implementation of efficient communication and navigation mechanisms.

11.2 Models

In this section we briefly survey basic properties of network environments and agents populating them. A specific choice of network and agents attributes results in a certain type of the rendezvous problem. This type refers to the difficulty of the problem and in turn to the efficiency of possible algorithmic solutions.

11.2.1 Networks

Recall that we consider two types of network environments: graph based and geometric. In the graph based representation the nodes of the network may not have distinct identities. In such case we say that the network is *anonymous*. In anonymous networks two nodes of the same degree are virtually impossible to distinguish. In order to enable navigation in such networks all edges incident to a given node are either explicitly or implicitly arranged in a periodic order.

A network can be either *finite* or of *unbounded in size*. In the latter case one needs to design search protocols that preserve locality of the solution. Otherwise the complexity of the solution could be unbounded. Another important network attributes refer to *localisation mechanism* (knowledge of the current location) and *sense of direction*. For example, in the geometric setting these two refer to the system of coordinates accompanied by geographic directions. Sense of direction turned out to greatly effect the solvability and efficiency of solution of a number of problems in distributed computing [19] and has been shown to be important in rendezvous as well [8].

Another crucial property refers to global clock availability. In particular, in a *synchronous* network one assumes access to the global clock allowing agents to coordinate their actions, including moves, using time frames. In contrast, in asynchronous networks the speed with which agent compute and move cannot be determined. In this case rendezvous is obtained either by adoption of predefined trajectories [29] or through analysis of the current configuration of the network [25].

The network can be reliable or it can report to its users imprecise information. In such error prone network rendezvous time can be largely elongated or meeting may prove to be impossible [18].

11.2.2 Agents

One of the major attributes of agents is their identity (e.g., a distinct label) that for some reason may be missing. Agents without identities are referred to as *anonymous* agents. Anonymous agents must execute the same procedure while agents with unique identities have the potential to behave differently. Another important attribute of agents is their initial *knowledge*. This may refer to the network size and topology as well as to the number, identities and location of available agents. In this context it is important whether agents can learn (*adapt*) throughout the rendezvous process or whether their control mechanism remains unchanged. In the latter case we say that the agents are oblivious. The process of learning, *adaptivity* of agents depends on their memory as well as on observation and communication abilities. For example, in some models it is assumed that agents are memoryless, where the agents rely on the use of random walk procedure [12]. The random walk is an example of a randomised procedure requiring access to random bits. As discussed later in this chapter in some instances of the rendezvous problem feasibility of the solution relies on symmetry breaking that cannot be implemented without a random number generator.

Agents may also have zero visibility without being able to communicate remotely [14]. In such cases the only way in which agents can learn about presence of one another is via spatial rendezvous. In some other extreme cases agents can constantly monitor movement of the others as it is assumed in the *Look-Compute-Move* model [25].

An interesting aspect of movement coordination of agents equipped with *different maximal speeds* has been recently studied in the context of network patrolling [13]. The authors proposed a number of algorithms that allow agents to patrol linear spaces efficiently. In [5], an alternative performance measure based on power consumption was used to assess efficiency of proposed rendezvous strategies.

Finally, somewhere between the model of networks and agents is agents ability to interact with the environment. This includes ability to release special marks in the form of stationary or mobile *tokens* [21] or longer messages stored in special message repositories such as *whiteboards* [15].

11.3 Rendezvous

Recall that the rendezvous problem refers to the algorithmic challenge in which two or more agents are expected to meet at the same time and point is space. The first reference to rendezvous problem is very often attributed to political science monograph [26] by Schelling, which initiated the discussion of coordination problems. Schelling considered approach in which each of the two players have only one attempt to choose the meeting location, and if they miss each other at the first attempt they fail.

The rendezvous problem as we now know it was first informally introduced by Steve Alpern in mid 1970s. He posed several problems including famous two:

Astronaut Problem Two astronauts land on a spherical body that is much larger than the detection radius (within which they can see each other). The body does not have a fixed orientation in space, nor does it have an axis of rotation, so that no common notion of position or direction is available to the astronauts for coordination. Given unit walking speeds for both astronauts, how should they move about so as to minimize the expected meeting time T (before they come within the detection radius)?

Telephone Problem In each of two rooms, there are n telephones randomly strewn about. They are connected in a pairwise fashion by n wires. At discrete times $t = 0, 1, 2, \ldots$ players in each room pick up a phone and say *"hello"*. They wish to minimize the time T when they first pick up paired phones and can communicate. What common randomization procedure should they adopt for choosing the order in which they pick up the phones?

The field was later popularised by Anderson and Weber in their seminal paper [7] on discrete location rendezvous. The continuous formalisation of the problem was later given by Alpern in [1]. Since then, the field grew substantially and attracted interest of researchers from a number of fields including Mathematics, Operations Research and Computer Science. A comprehensive collection of rendezvous problems including their rigorous classification can be found in the second part (Rendezvous Theory) of the book [4] from Alpern and Gal.

In general, Computer Science (CS) and Operations Research (OR) communities tend to study different models and aspects of the rendezvous problem. While OR research focuses mainly on minimisation of the expected time to meet, and sometimes maximisation of the probability of meeting within a given time, CS community tends to study efficiency trade-offs based on the use of resources. Despite differences, both communities expressed strong interest in rendezvous on a (possibly infinite) line. A number of randomised as well as deterministic rendezvous strategies have been proposed and analysed in this environment.

Alpern in [1] introduced the symmetric rendezvous search problem on the line and proposed a strategy with the expected meeting time of $5d$, where d is known and it refers to the original distance between the agents. Alpern's idea was to iterate for as long as it is needed the following procedure. Pick a random direction and move distance d in this direction and later distance $2d$ in the opposite direction, all at speed one. In addition, Alpern and Gal in [3] gave the proof that all symmetric strategies have expected time of rendezvous at least $3.25 \cdot d$. Also in 1995, Anderson and Essegaier in [6] improved the upper bound to $4.5678 \cdot d$ using a novel idea of mixed movements. Baston in [9] further improved the upper bound to $4.4182 \cdot d$ by accumulating and using all information before rendezvous takes place. More recently Uthaisombut in [30] presented tuned up mixed strategy imposing a better upper bound of $4.3931 \cdot d$. He also provided argument for the lower bound $3.9546 \cdot d$. These two bounds were further improved by Han et al. in [20] to $4.2574 \cdot d$ and $4.1520 \cdot d$ respectively. These two results required strong reference to Markov chains, fractional quadratic programming and semidefinite programming. The authors also conjectured that the rendezvous value is asymptotically equal to $4.25 \cdot d$.

The case when the distance d is not known in advance have been discussed in [10] where the competitive ratios (that compares efficiency of the proposed strategy to the best possible solution) 17.686 for the total distance traveled and 24.843 for the total time are established.

In case of deterministic rendezvous one needs to break symmetry between agents. For example, if the agents were anonymous (indistinguishable) they would perform the same moves and be always separated by distance d. The symmetry between the agents can be broken in many different ways including further information about network size and topology, labels given to agents, or through awareness of their own location.

In the context of the line most of the deterministic rendezvous strategies refer to asynchronous models, where the cost of the solution corresponds to the cumulative distance walked by any agent before rendezvous takes place. In [17] the authors discuss efficient rendezvous strategies for trees which impose rendezvous on the path of length n with the cost $O(n)$. In [16] agents are labelled and two algorithms for the infinite line are considered. If d is known to both agents the cost of rendezvous is $O(d|L_{min}|^2)$, where L_{min} refers to the size of the smaller label. If d is not known in advance the rendezvous cost rises to $O(d^3 + |L_{max}|^3)$, where L_{max} represents the size of the larger label. Performance of the latter algorithm is improved in [28], where we find an algorithm with cost $O(d \cdot \log^2 d + d \cdot d \log d |L_{max}| + d |L_{min}|2 + |L_{max}||L_{min}| \log |L_{min}|)$.

A different approach to rendezvous on the line was adopted by Collins *et al.* [11], where they assumed that the agents are not aware of d but they know their own location on the line. In particular, they proved that two agents in the synchronised model can always meet in time at most $6d$. Further, they also showed that their approach can be adopted in the asynchronous model with the rendezvous cost $O(d)$.

11.4 Also in This Volume

In this short introduction to the rendezvous problem the emphasis is mainly on major features of considered models of networks and mobile agents. Two more comprehensive survey type documents can be found in Chaps. 12 and 13. The list of ten open problems from the perspective of Operational Research, including the Astronaut Problem and the Telephone Problem, can be found in Chap. 14.

Chapter 12 surveys rendezvous in several models of distributed networks where the emphasis is on deterministic algorithms for networks with unknown topology. The authors consider several types of networks including those containing 'malicious' nodes (known in literature as *black holes*) and networks in which no consistent ordering on the edges at each node is imposed. The chapter provides a selection of algorithmic solutions (upper bounds), non trivial complexity analysis as well as it introduces more general techniques developed for the worst case scenarios in the rendezvous problem. This work is a nice complement of the recent survey on the topic written by Pelc [24] that adopts stronger assumptions about the network environment including distinct node identifiers, synchronicity, or restricted network topologies such as the ring, mesh or tree topologies.

Chapter 13 refers to the rendezvous problem in asynchronous networks in which the oblivious memory *Look-Compute-Move* model is assumed [25]. The mobile entities cannot leave any marks at visited nodes, nor send messages to other robots. The movement of entities depends solely on snapshots of the network configuration (position of all entities) taken independently by each entity. This chapter surveys recent results obtained in important network topologies such as rings, grids, and trees. This work include impossibility results on rings including the case where global-strong multiplicity detection (ability to detect whether to entities occupy the same node) is assumed. Further, the global-weak multiplicity detection model is considered in which all possible gatherable configurations have been determined. Finally, this survey provides also partial results for the case of local-weak multiplicity detection.

Finally, in Chap. 14 one can find a list of open rendezvous problems asked from Operations Research perspective, where optimization of the search process is interpreted as minimising the expected time to meet, or possibly maximising the probability of meeting within a given time.

11.5 Further Comments

The readers are strongly encouraged to advance their knowledge in the field. The book by Alpern and Gal [4] is a jewel on the shelf of any researcher thinking seriously about working on searching games and rendezvous problems. Other recommended survey type sources include a monograph on rendezvous in the ring co-authored by Kranakis et al. [21] and a more recent survey by Pelc [24] that focuses on deterministic mechanisms used in efficient rendezvous. The rendezvous problem has been also discussed in the context of consensus problems in networked dynamic systems, flocking, fast consensus in small-world networks, Markov processes and gossip-based algorithms, load balancing in networks, distributed sensor fusion in sensor networks, and belief propagation [23].

References

1. S. Alpern, The rendezvous search problem, *SIAM J. Control Optimisation*, 33, 673–678, 1995.
2. S. Alpern, Bilateral street searching in Manhattan (line-of-sight rendezvous on a planar lattice), *CDAM Research Report Series*, LSE-CDAM-2004-09.
3. S. Alpern and S. Gal, Rendezvous Search on the Line with Distinguishable Players, *SIAM J. Control and Optimisation* 33, 1270–1276, 1995.
4. S. Alpern and S. Gal, Theory of Search Games and Rendezvous, *Kluwer Academic Publishers*, 2003.
5. J. Anaya, J. Chalopin, J. Czyzowicz, A. Labourel, A. Pelc, and Y. Vaxes, Collecting Information by Power-Aware Mobile Agents, *DISC 2012*, 46–60.
6. E.J. Anderson and S. Essegaier, Rendezvous search on the line with indistinguishable players, *SIAM J. Control Optimisation* 33, 1637–1642, 1995.
7. E.J. Anderson and R.R. Weber, The rendezvous problem on discrete locations, *J. Applied Probability* 27, 839–851, 1990.

8. L. Barriere, P. Flocchini, P. Fraigniaud, and N. Santoro, Rendezvous and Election of Mobile Agents: Impact of Sense of Direction, *Theory Computer Systems* 40(2), 143–162 (2007).
9. V.J. Baston, Two rendezvous search problems on the line, *Naval Research Logistics* 46, 335–340, 1999.
10. A. Beveridge, D. Ozsoyeller, and V. Isler, Symmetric Linear Rendezvous with an Unknown Initial Distance, *Technical Report*, University of Minnesota - Computer Science and Engineering, March 2011.
11. A. Collins, J. Czyzowicz, and L. Gasieniec, A. Kosowski, and R. Martin, Synchronous rendezvous for location-aware agents *DISC 2011*, 447–459.
12. C. Cooper, A.M. Frieze, and T. Radzik, Multiple Random Walks and Interacting Particle Systems, *ICALP 2009*, 399–410.
13. J. Czyzowicz, L. Gąsieniec, A. Kosowski, and E. Kranakis, Boundary Patrolling by Mobile Agents with Distinct Maximal Speeds, *ESA 2011*, 701–712.
14. J. Czyzowicz, A. Pelc, and A. Labourel, How to meet asynchronously (almost) everywhere, *ACM Transactions on Algorithms* 8(4), paper 37, 2012.
15. S. Das, Distributed computing with mobile agents: solving rendezvous and related problems, *PhD Dissertation*, University of Ottawa, 2007.
16. G. De Marco, L. Gargano, E. Kranakis, D. Krizanc, A. Pelc, and U. Vaccaro, Asynchronous deterministic rendezvous in graphs, *Theoretical Computer Science* 355, 315–326, 2006.
17. A. Dessmark, P. Fraigniaud, and A. Pelc, Deterministic rendezvous in graphs, *ESA 2003*, 184–195.
18. Y. Dieudonné, A. Pelc, and D. Peleg, Gathering despite mischief, *SODA 2012*, 527–540.
19. P. Flocchini, B. Mans, and N. Santoro, Sense of direction in distributed computing, *Theoretical Computer Science* 291(1), 29–53, 2003.
20. Q. Han, D. Du, J. Vera, and L.F. Zuluaga, Improved Bounds for the Symmetric Rendezvous Value on the Line *Operations Research* 56(3), 772–782, 2008.
21. E. Kranakis, D. Krizanc and E. Markou, The Mobile Agent Rendezvous Problem in the Ring, *Synthesis Lectures on Distributed Computing Theory*, 2010.
22. D.B. Lange and M. Oshima, Seven Good Reasons for Mobile Agents, *Communications of the ACM* 42(3), 88–89, 1999.
23. R. Olfati-Saber, J.A. Fax, and R.M. Murray, Consensus and Cooperation in Networked Multi-Agent Systems, *Proc. IEEE* 95(1), 215–233, 2007.
24. A. Pelc, Deterministic rendezvous in networks: A comprehensive survey, *Networks* 59(3), PP. 331–347, 2012.
25. G. Prencipe, Impossibility of gathering by a set of autonomous mobile robots, *Theoretical Computer Science* 384, 222–231, 2007.
26. T. Schelling, The strategy of conflict, *Oxford University Press*, Oxford, 1960.
27. R. Siegwart, I.R. Nourbakhsh, and D. Scaramuzza, Introduction to Autonomous Mobile Robots (2nd Ed.), *Intelligent Robotics & Autonomous Agents Series*, MIT Press, 2011.
28. G. Stachowiak, Asynchronous Deterministic Rendezvous on the Line. *SOFSEM 2009*, 497–508.
29. A. Ta-Shma and U. Zwick, Deterministic rendezvous, treasure hunts and strongly universal exploration sequences, *SODA 2007*, 599–608.
30. P. Uthaisombut, Symmetric rendezvous search on the line using moving patterns with different lengths, *Working paper*, Department of Computer Science, University of Pittsburgh, 2006.

Chapter 12
Deterministic Symmetric Rendezvous in Arbitrary Graphs: Overcoming Anonymity, Failures and Uncertainty

Jérémie Chalopin, Shantanu Das, and Peter Widmayer

Abstract We consider the rendezvous problem of gathering two or more identical agents that are initially scattered among the nodes of an unknown graph. We discuss some of the recent results for this problem focusing only on deterministic algorithms for the general case when the graph topology is unknown, the nodes of the graph may not be uniquely labeled and the agents may not be synchronized with each other. In this scenario, the objective is to solve rendezvous whenever deterministically feasible, while optimizing on the amount of movement by the agents or the memory required (for the nodes or the agents) in the worst case. Further we also investigate some special scenarios such as (i) when the graph contains some *dangerous* nodes or, (ii) when there is no consistent ordering on the edges of a node. We present positive results, complexity analysis and some general techniques for dealing with such worst case scenarios for the symmetric rendezvous problem.

12.1 Introduction

The problem of rendezvous requires two or more entities (called *agents*) located in distinct vertices of a graph, to meet at one vertex of the graph. This problem occurs in many natural contexts [3] and requires different strategies depending on the scenario and the particular objective. In the original definition of the problem [2],

J. Chalopin (✉)
LIF, CNRS & Aix-Marseille University, Marseille, France
e-mail: jeremie.chalopin@lif.univ-mrs.fr

S. Das
BGU & Technion-Israel Institute of Technology, Haifa, Israel
e-mail: shantanu@tx.technion.ac.il

P. Widmayer
Institute of Theoretical Computer Science, ETH Zürich, Zürich, Switzerland
e-mail: widmayer@inf.ethz.ch

S. Alpern et al. (eds.), *Search Theory: A Game Theoretic Perspective*,   175
DOI 10.1007/978-1-4614-6825-7__12, © Springer Science+Business Media New York 2013

the objective was to minimize the expected time to meet. If we are restricted to deterministic strategies, the objective may be to minimize the worst-case time to meet, over all possible starting configurations. Moreover if there is no common notion of time, we may wish to minimize the total distance traveled by the agents until rendezvous. In some cases, there are other parameters to consider, for example the size of memory used by the agents or the number of additional resources (e.g. flags for marking) used by the agents.

This chapter considers the deterministic rendezvous of two or more agents in a finite connected graph, placed initially at locations chosen by an adversary. The agents are assumed to be identical and they execute the same algorithm, without global knowledge about the graph. Other than finiteness and connectivity, we make no other assumptions about the topology of the graph. In an arbitrary connected graph, it is not always possible to solve rendezvous using deterministic means. For instance consider a ring of size n, where two agents are placed at a distance of $n/2$ from each other; if each agent follows the same strategy (any combination of moving clockwise, counterclockwise or remaining stationary) the agents may forever be the same distance apart from each-other. In most cases, the ability to distinguish vertices in some way allows the agents to rendezvous even if they are using identical strategies. Given any graph and the starting locations of the agents in the graph, it is possible to determine whether rendezvous is possible for the particular instance. Thus, it is possible to characterize the instances where deterministic rendezvous is feasible. The prior knowledge of certain graph parameters (such as the size or diameter) or the ability to mark vertices of the graph also influences the feasibility of rendezvous.

The model considered here is very generic in the sense that we do not assume any global clock (the agents act asynchronously), nor do we assume unique identifiers for the nodes of the graph or for the agents (the graph and the agents may be anonymous); and the graph topology is not known to the agents (i.e. the topology could be any arbitrary connected graph). In stronger models, e.g. when the agents have distinct identifiers [11, 16] or, when they are synchronous [12, 13], or if the environment is restricted to specific topologies such as the ring [19, 22], grid [4] or tree [13] topologies, then it becomes easier to solve rendezvous and the set of solvable instances may become relatively larger. For results on rendezvous in such models, please see the recent survey [23]. Another significant difference between the results in this chapter and those of [23] is that we allow the agents to meet only at a node, whereas most results from the above paper also allow meeting on an edge when two agents are traversing it from opposite sides.[1] The rendezvous problem has also been studied in a completely different model where the agents move in a continuous terrain [18] or in a graph [21] but have global visibility. Finally there exist many results on solving rendezvous using randomized algorithms (see [3] for a survey).

---

[1] This difference implies that in our model it is not possible to rendezvous even on the trivial graph consisting of two nodes and a single edge.

This chapter is organized as follows. The next section defines the model and the problem and describes some basic properties of graphs which we use in solving rendezvous. Section 12.3 provides a characterization of those instances where deterministic rendezvous is possible. In the rest of the chapter, we focus on the problem of Rendezvous-with-Detect where agents solve rendezvous whenever possible and otherwise detect the fact that rendezvous is not possible. Section 12.4 provides some minimum conditions required for solving the problem. We present algorithms for solving Rendezvous-with-Detect both for the model where agents are not allowed to mark nodes (Sect. 12.5) and the model where marking is allowed (Sect. 12.6). In Sect. 12.7, we consider the model where each agent is provided with a pebble, allowing it to mark at most one node of the graph. We also discuss how to tolerate failures and uncertainties in this model. Finally, in Sect. 12.8, we study rendezvous in dangerous environments where some of the agents may disappear (e.g. they are devoured by some faulty node), and show how to rendezvous the surviving agents. Section 12.9 concludes this chapter with a discussion of some open directions.

12.2 The Model and Basic Properties

The Environment. The environment is represented by a simple undirected connected graph $G = (V(G), E(G))$ and a set $\mathscr{Q}$ of mobile agents that are located in the nodes of G. The initial placement of the agents is denoted by the function $p : \mathscr{Q} \to V(G)$. We denote such a distributed mobile environment by $(G, \mathscr{Q}, p)$ or by (G, χ_p) where χ_p is a vertex-labeling of G such that $\chi_p(v) = 1$ if there exists an agent a such that $p(a) = v$, and $\chi_p(v) = 0$ otherwise. For simplicity, we assume the agents to be initially located in distinct nodes, but the algorithms can be generalized to the case when two or more agents start from the same location (e.g. if two agents happen to be initially co-located, they will move together as a single merged agent).

For the rest of this chapter, $n = |V(G)|$ and $m = |E(G)|$ denotes the numbers of vertices and of edges of G, while $k = |\mathscr{Q}|$ denotes the number of agents. We shall use the words vertex and node interchangeably.

In order to enable navigation of the agents in the graph, at each node $v \in V(G)$, the edges incident to v are distinguishable to any agent a at node v. In other words, there is a bijective function

$$\delta_{a,v} : \{(v,u) \in E(G) : u \in V(G)\} \to \{1, 2, \ldots d(v)\}$$

which assigns unique labels to the edges incident at node v (where $d(v)$ is the degree of v). The function $\delta_a = \{\delta_{a,v} : v \in V(G)\}$ is called the local orientation or port-numbering[2] and it is usually assumed that all agents have the same consistent port-numbering (i.e. $\delta_a = \delta$, $\forall a \in \mathscr{Q}$). In Sect. 12.7.2, we shall consider the special case when this is not true. For the rest of the paper, we assume a common port-numbering δ (and thus remove the subscript a).

---

[2] The labels on the edges may correspond to port numbers on a network

The vertices of G are labeled over the set of symbols L by $\lambda : V(G) \to L$ which is the labeling function. However, note that this labeling is not necessarily injective, i.e. two vertices may have the same label. This means that we must design algorithms that work for any such labeling, and in particular for the constant labeling which labels all nodes with the same label $c \in L$ (in this case, the nodes of the graph are said to be *anonymous*).

The environment is thus represented by the tuple $(G, \lambda, \mathscr{Q}, p, \delta)$ or equivalently by $(G, \lambda, \chi_p, \delta)$. In case the nodes of the graph are anonymous, we shall omit λ.

The Agents. Each agent a starts from the node $p(a)$, called the *homebase* of a, and executes a sequence of steps. The agents start from the same initial state but may not necessarily start at the same time, and every action they perform (computing, moving, etc.) takes a finite but otherwise unpredictable amount of time (i.e. the actions of the agents are not synchronized). The actions that an agent a located at a node v can perform depend on the state of the agent and the state of the node v (including the degree of v, the label of v, and the presence of other agents or marks left by other agents). An agent can see another agent only when they are both located at the same node. However, an agent may not even detect the presence of another agent if both are traversing the same edge. Two agents may traverse the same edge at different speeds; thus if agents a and b start traversing the same edge (u, v) one after the other, it is possible that agent b arrives at the other end-point earlier than agent a.

Communication model: Whiteboards and Tokens. As mentioned before, two agents may communicate (i.e. exchange information) directly only when they are at the same node. To facilitate the task of rendezvous, sometimes the agents may be allowed to leave marks on a node as a signal for other agents. In the *white-board* model, the agents communicate by reading and writing information on public whiteboards locally available at the nodes of the network. Each node $v \in G$ has a whiteboard (which is a shared region of its memory) and any agent visiting node v can read or write to the whiteboard. Access to the whiteboard is restricted by fair mutual exclusion, so that at most one agent can access the whiteboard of a node at the same time, and any requesting agent will be granted access within finite time.

A more restrictive model is the *token* model, where no whiteboards are available but each agent has one or more identical *tokens* (sometimes called pebbles) to mark nodes. An agent that contains a token can place it on a node v before leaving the node; this token will be visible to any agent visiting node v, i.e. the visiting agent can determine whether or not there is a token at that node. Similar to the whiteboard model, we assume mutually exclusive access to node; Thus two agents may not place their tokens at the same node simultaneously, they must do so sequentially. The tokens are moveable, i.e. an agent can pick up a token, carry the token and place it on a different node that the agent visits.

Cost Measures. The cost of an algorithm can be the time taken until rendezvous is achieved. Since we consider the actions of the agents to be asynchronous, a more useful measure for the efficiency of the algorithm is the amount of movement made by the agents, called the move complexity. In other words, whenever an agent traverses an edge of the graph, this incurs a unit cost and the total cost of the algorithm is the total number of edge-traversals made by all the agents together, in the worst case execution of the algorithm.

Other than optimizing on the time or the movement cost, one can consider the space cost of an algorithm e.g. the memory that needs to be allocated at each node of the graph, or the memory used by an agent during the algorithm. Thus, we associate three different cost measures for a rendezvous algorithm: (i) Movement Cost, (ii) Node Memory, and (iii) Agent Memory.

Problem Definition

Definition 1 (Rendezvous). Given a distributed mobile environment $(G, \lambda, \mathcal{Q}, p, \delta)$, an algorithm $\mathcal{A}$ is said to solve *rendezvous* if for any distributed execution of the algorithm by the agents, there exists a node $v \in G$ such that all agents in $\mathcal{Q}$ eventually reach node v and do not move thereafter.

In the definition above, we do not require the agents to terminate explicitly (i.e. an agent may not be aware when rendezvous has been achieved). Note that even though we consider only deterministic algorithms, the outcome of the algorithm may depend on the particular sequence of events and actions during the (asynchronous) execution of the algorithm by the individual agents. We define a *distributed execution* of an algorithm as one possible sequence of actions and events that is consistent with the environment and the algorithm.

We say that rendezvous is feasible in $(G, \lambda, \mathcal{Q}, p, \delta)$, if and only if there exists a deterministic algorithm that solves *rendezvous* in $(G, \lambda, \mathcal{Q}, p, \delta)$.

Definition 2 (Rendezvous-with-Detect). Given a distributed mobile environment $(G, \lambda, \mathcal{Q}, p, \delta)$, an algorithm is said to solve *Rendezvous-with-Detect* if the following holds for any distributed execution of the algorithm. If rendezvous is feasible in $(G, \lambda, \mathcal{Q}, p, \delta)$, then all agents in $\mathcal{Q}$ must eventually terminate at one unique node v of G and if not, then each agent must terminate in its homebase and output "Rendezvous is not solvable".

When there are more than 2 agents, i.e. $|\mathcal{Q}| > 2$, we can define the concept of *partial rendezvous* where at least $w < |\mathcal{Q}|$ agents are required to gather at a node of the graph. This will be discussed further in Sect. 12.8.

Properties of Graphs: Coverings and Universal Exploration Sequences

The notions presented in this section were introduced in [6]. Any connected (undirected) graph G can be represented by a strongly connected symmetric digraph $D = Dir(G)$, where each edge of G is represented by a pair of symmetric arcs in D, one in each direction. In this section, we present some definitions and results related to directed graphs and their coverings, which we use to characterize the solvable instances for rendezvous. A directed graph (digraph) $D = (V(D), A(D), s_D, t_D)$

possibly having parallel arcs and self-loops, is defined by a set $V(D)$ of vertices, a set $A(D)$ of arcs and by two maps s_D and t_D that assign to each arc two elements of $V(D)$: a source and a target (in general, the subscripts will be omitted). A *symmetric* digraph D is a digraph endowed with a symmetry, that is, an involution $Sym : A(D) \to A(D)$ such that for every $a \in A(D), s(a) = t(Sym(a))$ and $Sym(Sym(a)) = a$. Given a simple connected graph G, a vertex labeling function λ, and a port-numbering δ, we will denote by $(Dir(G), \lambda, \delta)$ the labeled digraph constructed in the following way. The vertices of $Dir(G)$ are the vertices of G and they have the same labels as in (G, λ). Each edge $\{u, v\}$ is replaced by two arcs $a_{(u,v)}, a_{(v,u)} \in A(Dir(G))$ such that $s(a_{(u,v)}) = t(a_{(v,u)}) = u$, $t(a_{(u,v)}) = s(a_{(v,u)}) = v$, $\delta(a_{(u,v)}) = (\delta_u(v), \delta_v(u))$, $\delta(a_{(v,u)}) = (\delta_v(u), \delta_u(v))$ and $Sym(a_{(u,v)}) = a_{(v,u)}$ (See Fig. 12.1).

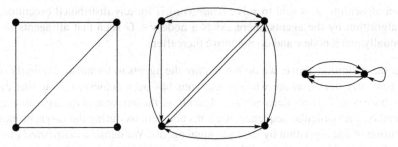

Fig. 12.1 A graph G, the corresponding digraph Dir(G), and its minimum-base H

A *covering projection* is a homomorphism φ from D to D' satisfying the following: (i) For each arc a' of $A(D')$ and for each vertex v of $V(D)$ such that $\varphi(v) = v' = t(a')$ there exists a unique arc a in $A(D)$ such that $t(a) = v$ and $\varphi(a) = a'$. (ii) For each arc a' of $A(D')$ and for each vertex v of $V(D)$ such that $\varphi(v) = v' = s(a')$ there exists a unique arc a in $A(D)$ such that $s(a) = v$ and $\varphi(a) = a'$. A symmetric digraph D is a *symmetric covering* of a symmetric digraph D' via a homomorphism φ if φ is a covering projection from D to D' such that for each arc $a \in A(D)$, $\varphi(Sym(a)) = Sym(\varphi(a))$.

A digraph D is *symmetric-covering-minimal* if there does not exist any graph D' not isomorphic to D such that D is a symmetric covering of D'. The notions of coverings extend to labeled digraphs in an obvious way: the homomorphisms must preserve the labeling. Given a simple labeled graph (G, λ) with a port-numbering δ, we say that (G, λ, δ) is symmetric-covering-minimal if $(Dir(G), \lambda, \delta)$ is symmetric-covering-minimal. For any simple labeled graph (G, λ) with a port-numbering δ, there exists a unique digraph (D, μ_D) such that (i) $(Dir(G), \lambda, \delta)$ is a symmetric covering of (D, μ_D) and (ii) (D, μ_D) is symmetric-covering-minimal. This labeled digraph (D, μ_D) is called the *minimum base* of (G, λ, δ) (See Fig. 12.1).

The main result that we will use from the theory of graph coverings is the following. Given an environment $(G, \lambda, \chi_p, \delta)$, if the corresponding labeled digraph $(Dir(G), \mu_G)$ is not symmetric-covering-minimal, i.e. $(Dir(G), \mu_G)$ covers a smaller digraph (H, μ_H), then the vertices of G can be partitioned into equivalence classes, each of size $q = |V(G)|/|V(H)|$ such that the vertices in the same class are symmetric and indistinguishable from each other. This is also related to the concept of views introduced in [24]. Nodes having the same view belong to the same equivalence class.

Definition 3. Given a labeled graph (G, λ) with a port numbering δ, the *view* of a node v is the infinite rooted tree denoted by $T_G(v)$ defined as follows. The root of $T_G(v)$ represents the node v and for each neighbor u_i of v, there is a vertex x_i in $T_G(v)$ (labeled by $\lambda(u_i)$) and an edge from the root to x_i with the same labels as the edge from v to u_i in (G, δ). The subtree of $T_G(v)$ rooted at x_i is again the view $T_G(u_i)$ of node u_i.

For traversal of an unknown graph we will use the notion of a *Universal Exploration Sequence* (UXS) [20]. For any node $u \in G$, we define the ith successor of u, denoted by $succ(u, i)$ as the node v reached by taking port number i from node u (where $0 \le i < d(u)$). Let $(a_1, a_2, \ldots, a_k)$ be a sequence of integers. An *application* of this sequence to a graph G at node u is the sequence of nodes $(u_0, \ldots, u_{k+1})$ obtained as follows: $u_0 = u$, $u_1 = succ(u_0, 0)$; for any $1 \le i \le k$, $u_{i+1} = succ(u_i, (p + a_i) \bmod d(u_i))$, where p is the port-number at u_i corresponding to the edge $\{u_{i-1}, u_i\}$. A sequence $(a_1, a_2, \ldots, a_k)$ whose application to a graph G at any node u contains all nodes of this graph is called a UXS for this graph. A UXS for a class of graphs is a UXS for all graphs in this class. For any positive integers $n, d, d < n$, there exists a UXS of length $O(n^3 d^2 \log n)$ for the family of all graphs with at most n nodes and maximum degree at most d [1].

12.3 Feasibility of Deterministic Rendezvous

Deterministic rendezvous is not always possible in arbitrary graphs, as we have seen before (recall the example of the two agents symmetrically placed in a ring). Given an environment $(G, \lambda, \mathcal{Q}, p, \delta)$, the feasibility of rendezvous may depend on the structure of G, the labeling λ, the port numbering δ as well as the initial placement of the agents. When the agents do not have the capability of marking nodes, the feasibility of rendezvous depends on the labeled graph (G, λ, δ) and not on the starting locations. This is equivalent to the feasibility of electing a leader among the nodes of a graph, a well-studied problem for which there exists a known characterization of solvable instances. The following properties are based on the results from [5, 24, 25].

Theorem 1. *Rendezvous is solvable in (G, λ, δ) irrespective of the number of agents and their starting locations if and only if (G, λ, δ) is symmetric-covering-minimal with respect to any covering projection that preserves the edge-labeling δ and the node-labeling λ.*

On the other hand, if the agents are allowed to mark the nodes of the graph then the placement p of the agents in G influences the solvability of rendezvous. In this case, we can assume that the starting locations of the agents are distinctly labeled by the function χ_p and thus consider the node-labeling $\lambda' = \lambda \times \chi_p$.

Theorem 2. *Rendezvous is solvable in the environment* $(G, \lambda, \mathcal{Q}, p, \delta)$ *if and only if* (G, λ', δ) *is symmetric-covering-minimal with respect to any label-preserving covering projection, where* $\lambda' = \lambda \times \chi_p$.

We can assume that the edge-labeling and node labeling of the graph is given by an adversary. Thus, it makes sense to characterize the family of graphs where rendezvous is possible for any labeling (assuming that the labeling provides a local orientation at each node).

Theorem 3. *Given any connected graph G, the following statements are equivalent:*

1. *For any port-numbering* δ, *and any placement* χ_p *of agents in G, rendezvous can be solved in* (G, χ_p, δ);
2. *For any port-numbering function* δ, *each vertex of* (G, δ) *has a distinct view;*
3. *There is no partition* $V_1, V_2, \ldots V_k$ *of* $V(G)$ *with* $k \in [1, |V(G)| - 1]$ *such that for any distinct* $i, j \in [1, k]$, *the following conditions hold:*

    (i) *$G[V_i]$ is d-regular for some d, and if d is odd, it contains a perfect matching,*
    (ii) *$G[V_i, V_j]$ is regular.*

4. *Dir(G) is symmetric-covering-minimal.*

12.4 Impossibility Results

From the results of the previous section, we know rendezvous can be solved only in an environment $(G, \lambda, \mathcal{Q}, p, \delta)$ where the corresponding labelled graph (G, λ', δ) is symmetric-covering minimal. Given such an instance, it is possible to construct another instance $(H, \lambda'_H, \delta_H)$ such that $|V(H)| = 2|V(G)|$ and $(H, \lambda'_H, \delta_H)$ covers (G, λ', δ), and thus, rendezvous is not possible in $(H, \lambda'_H, \delta_H)$. Any algorithm that solves *Rendezvous-with-Detect* must be able to distinguish between these two instances. It is not possible to distinguish between these two instances unless the agents are provided with some prior knowledge which allows them to deduce the size of the graph with some accuracy. In fact, if the agents know an upper bound B such that $B < 2n$ this is already sufficient to solve *Rendezvous-with-Detect*. (Recall that for any graph H that covers G and is not isomorphic to G, the size of H must be strictly a multiple of the size of G and thus $|V(H)|$ is at least twice of $|V(G)|$.)

Theorem 4. *The knowledge of only an arbitrary upper bound on n is not sufficient for solving Rendezvous-with-Detect in an environment* $(G, \lambda, \mathcal{Q}, p, \delta)$.

We now consider the move complexity of Rendezvous-with-Detect. It is easy to see that each edge of the graph must be traversed by at least one agent. Moreover, in a symmetric environment (e.g. a ring with agents placed equidistant apart) each agent may need to make $O(n)$ moves before it can detect the impossibility of rendezvous. This gives us the following lower bound.

Theorem 5. *Solving Rendezvous-with-Detect with k agents in an arbitrary graph of n nodes and m edges requires* $\Omega(m + nk)$ *moves in the worst case.*

12.5 Rendezvous Without Marking

In this section we assume that the agents have no means of marking the nodes of the graph (i.e. no whiteboards or tokens are available). The knowledge of n (or, at least some upper bound on it) is required to even explore the graph unless the graph happens to be a tree. In asymmetric trees, rendezvous is possible without marking and without knowledge of n. It is possible to traverse an anonymous tree and find the central edge or central node in the tree (every tree has either a central node or a central edge). The usual technique for rendezvous is to gather at the central node or at one of the endpoints of the central edge. In the latter case, the agent needs to do a comparison of the subtrees at either end of the central edge e, in order to choose among the two end-points of e. This problem has been studied for agents having small memory (see Sect. 12.5.2).

12.5.1 Agents Having Unbounded Memory

When an upper bound on n is known a priori, and the agents have sufficient memory, it is possible to solve rendezvous in an arbitrary graph (G, λ, δ) by constructing the minimum-base of the labeled graph and then moving to a unique node of the minimum-base. Note that according to Theorem 1, rendezvous is solvable in this case only if (G, λ, δ) is covering minimal. If that condition is satisfied then the constructed minimum-base is isomorphic to G and thus all the agents will reach the same node, hence solving rendezvous.

In case the exact value of n is provided, it is possible to use the same algorithm to check for symmetric-covering-minimality and thus, solve Rendezvous-with-Detect. We now discuss the algorithm (see [8] for more details). The first part of the algorithm is a traversal of the graph visiting every vertex of G at least once. The second part is the classification of the visited vertices into equivalence classes. Initially all vertices are put in the same class and in subsequent rounds, the algorithm refines the classes until each class corresponds to one vertex of the minimum-base. For the traversal we use a UXS $U(N, d)$ where $N \geq n$ is an upper bound on n and d is some upper bound on the maximum degree of the graph G. We now describe the class refinement process.

Given a graph G and node u of G and a sequence of edge-labels

$$Y = ((p_1, q_1), (p_2, q_2), \ldots, (p_j, q_j)),$$

we say that Y is *accepted* from u if there exists a path $P = (u = u_0, u_1, \ldots, u_j)$ in G such that $\delta(P) = Y$, i.e. for each i, $1 \leq i \leq j$, $(p_i, q_i) = \delta(u_{i-1}, u_i)$. For any $k > 0$, two vertices u, v that have the same view up to depth k are said to be k-equivalent; we denote it by $u \sim_k v$. The k-class of u is the set of all vertices that are k-equivalent to u and this set is denoted by $[u]_k$. Given any two k-classes C, C', a (C, C')-*distinguishing path* is a sequence of edge-labels $Y_{C,C'} = ((p_1, q_1), (p_2, q_2), \ldots, (p_j, q_j))$ such that

---

Algorithm 1: Class-Refinement(N)

Let $v_1, v_2, \ldots v_t$ be the sequence of nodes visited by $U(N,d)$, possibly containing duplicate nodes ;

Follow $U(N,d)$ and **for** *each node v_i* **do**

$\quad\lfloor$ Store the labels of each edge incident to v_i;

Compute the number of 1-classes and store a distinguishing path for each pair of distinct classes ;

$k := 2$;

repeat

$\quad\mid$ Follow $U(N,d)$ and **for** *each node v_i* **do**

$\quad\mid\quad$ **for** *each edge (v_i, w) incident to v_i* **do**

$\quad\mid\quad\quad\lfloor$ Compute the $(k-1)$-class of w (using the distinguishing paths);

$\quad\mid\quad\quad\quad$ Store the label of (v_i, w) and the index of the $(k-1)$-class of w ;

$\quad\mid$ Compute the number of k-classes and store a distinguishing path for each pair of distinct k-classes ;

$\quad\mid$ Increment k;

until *the number of k-classes is equal to the number of $(k-1)$-classes*;

Move to a vertex of class one;

$Y_{C,C'}$ is accepted from each node $u \in C$ and it is not accepted from any node $v \in C'$. Given any two distinct k-classes C, C', either there exists a (C, C')-*distinguishing path* of length at most k, or there exists a (C', C)-*distinguishing path* of length at most k.

For $k = 1$, it is easy to determine the k-class of any node v by traversing each edge incident to v and noting the labels. From this information, one can find the distinguishing paths for any pair of 1-classes. For $k \geq 2$, it is possible to identify the k-classes and the corresponding distinguishing paths (from knowledge of the $k-1$ classes) using the properties below.

Property 1. For $k \geq 2$, two nodes u and v belong to the same k-class, i.e. $[u]_k = [v]_k$, if and only if (i) $[u]_1 = [v]_1$ and (ii) for each i, $0 \leq i \leq deg_G(u) = deg_G(v)$, the ith neighbor u_i of u and the ith neighbor v_i of v belong to the same $(k-1)$-class and $\delta(u, u_i) = \delta(v, v_i) = (i, j)$, for some $j \geq 0$.

Theorem 6 ([8]). *Algorithm 1 builds the quotient graph of any graph of size $n \leq N$ in $O(|U(N,d)| \cdot n^3 d)$ moves and requires $O(n^3 \log n + |U(N,d)| \log d)$ memory for each agent.*

There exists a UXS for graphs of size at most N and maximum degree at most d, that is of length $O(N^3 d^2 \log N)$ [1]. Using such a sequence for the traversal gives us an algorithm of move complexity $O(N^3 n^3 d^3 \log N)$ for solving rendezvous.

12.5.2 Agents Having Little Memory

In the algorithms discussed above, the agent needs to have enough memory to construct and to remember a map of the graph or a part of it. In this section we consider the effect of limiting the memory of the agent. The task of rendezvous in tree networks has been studied in synchronous environments for agents with small memory and it was shown that logarithmic memory is required for rendezvous even on the line [13]. Note that this lower bound does not apply directly in our setting since the set of solvable instances of rendezvous is strictly larger in a synchronous environment than in an asynchronous one when the agents cannot mark the vertices. However, it is well known that $o(\log n)$ memory is not sufficient for exploration of an arbitrary graph with termination, if marking is not allowed. Consequently, if the agents cannot mark the nodes, one can show that $\Omega(\log n)$ bits of memory are necessary to solve rendezvous of two agents in an arbitrary graph in an asynchronous environment.

In a synchronous environment without the ability to mark nodes, it is known [12] that $O(\log n)$ memory is sufficient for rendezvous of two agents starting on asymmetric positions in an arbitrary graph (even if the agents do not necessarily start at the same time). The idea of the algorithm in [12] is to obtain a unique ordering on distinct equivalence classes of nodes (without having to construct the views of the nodes). Each agent can then use the index given to its initial position as its unique identifier and rendezvous can be achieved using the standard algorithm for agents having distinct identifiers in synchronous environments.

In the asynchronous setting, when the agent cannot mark nodes, we know that the starting positions of the agents cannot be used to break symmetry. Thus, rendezvous is solvable only if (G, λ, δ) is symmetric-covering-minimal. If each agent initially knows an upper bound on the size of the graph, the agent can execute the first part of the algorithm of [12] to distinguish between equivalence classes of nodes and to order them. Thereafter, each agent could move to a node that belongs to the class appearing first in this ordering. If (G, λ, δ) is symmetric-covering-minimal, this node is unique and the agents would have achie

Theorem 7. *Agents with $O(\log n)$ memory can solve rendezvous in any environment (G, λ, δ) where deterministic rendezvous is feasible without marking.*

12.6 Rendezvous with Marking

In this section, we assume that the agents can write on whiteboards present in the nodes of G. If we assume no bounds on the memory available to the agent or at a node then there is an optimal algorithm to solve rendezvous (or Rendezvous-with-Detect) for two agents using $\Theta(m)$ moves. For the general case of $k \geq 2$ agents, this generalizes to an algorithm that requires $O(mk)$ moves to solve Rendezvous-with-Detect and $O(m \log k)$ to solve rendezvous.

The algorithm proceeds in two phases. In the first phase, the agents construct a spanning forest of the graph using a distributed DFS-type algorithm (described below as procedure DDFS). At the end of this procedure there is exactly one agent in each tree in the forest and each agent a has a map of the tree that it belongs to (we call this the agent's territory T_a). The second phase of the algorithm is a competition between neighboring agents, during which each losing agent merges its territory

with the corresponding winning agent. This process is repeated with the objective
of eventually forming a single tree spanning the graph G so that all the agents gather
at the root of this spanning tree. We show that this is always possible whenever the
condition of Theorem 2 is satisfied.

Procedure DDFS: An agent A starts from its homebase a depth-first search traversal marking the nodes that it visits (unless they are already marked) and labeling
them with numbers $1, 2, 3, \ldots$ etc. Each node marked by the agent and the edge used
to reach it are added to its tree. If the agent reaches an already marked node, it back-
tracks to the previous node and tries the other edges incident to the node. The agent
stops when there are no unexplored edges incident to the nodes of its tree. This tree
is the territory T_A of the agent.

Partial-view (PV): Based on the territory of an agent, we define the *Partial-View*
PV_A of an agent A having territory $\mathscr{T}_A$, as the finite rooted tree (see Fig. 12.2), such
that: (i) The root corresponds to the homebase v_0 of agent A. (ii) For every other
node v_i in $\mathscr{T}_A$, there is a vertex x_i in PV_A. (iii) For each edge (v_i, v_j) in $\mathscr{T}_A$, there is
an edge (x_i, x_j) in PV_A. (iv) For each outgoing edge $e = (v_i, u_i)$ such that $v_i \in \mathscr{T}_A$
but $e \notin \mathscr{T}_A$, PV_A contains an extra vertex y_i (called an external vertex) and an edge
$\hat{e} = (x_i, y_i)$ that joins x_i to it. (v) Each edge in PV_A is marked with two labels, which
are same as those of the corresponding edge in G. (vi) Each vertex x_i in PV_A is
labeled with $\lambda(v_i)$ and $\chi_p(v_i)$, where v_i is the node in G corresponding to x_i. (vii)
Each vertex is also labeled with the numeric identifier assigned to the corresponding
node during procedure DDFS.

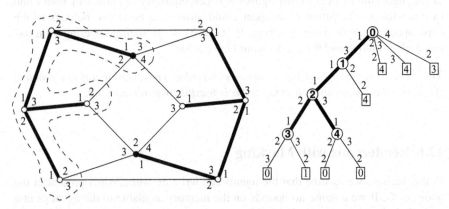

Fig. 12.2 Territories and Partial-Views: (**a**) A graph G with ten nodes and two agents A and B
(whose territories are marked by **bold** edges). (**b**) The *Partial-View* PV_A for agent A

The algorithm proceeds by comparing the partial-views of neighboring agents
(we use a fixed ordering on the partial-views). We say that an agent A is a neighbor
to agent B, if there exists an edge (u, v) such that $u \in T_A$, $(u, v) \notin T_A$ and $v \in T_B$.
By this definition, an agent may be its own neighbor. The communication between
neighbors works as follows. To send any information w, the agent writes w on each

whiteboard of its territory (function "WRITE-ALL"). To read the partial-view of neighboring agents, an agent visits each external node x and reads the contents of the whiteboard at x (function "READ-PV"). In any round i, if agent A reads a partial-view $PV_{i,x}$ greater than its own partial-view $PV_{i,A}$ in this round, then agent A is defeated (i.e. it becomes passive and does not participate in the algorithm anymore) and the edge connecting node x to the tree T_A is used to merge the two trees. This process is repeated for k iterations or until the territory of an agent spans the whole graph.

The algorithm assumes the prior knowledge of $k = |\mathscr{Q}|$. Alternately if the value of n is known (but not k) then the algorithm may be modified accordingly to use this information. In this case, the main loop of the algorithm will be executed for at most n iterations and the agent will terminate the algorithm successfully if its territory contains n nodes.

Algorithm 2: Make-Tree(k)

Execute procedure DDFS to construct the territory T_A;
$PV_{1,A} \leftarrow$ COMPUTE-PV(T_A) ;
for *phase* $i = 1$ *to* k **do**
    if *Number of Agents in T_A is k* **then**
        Collect all agents to root;
        Return("Success");
    WRITE-ALL($PV_{i,A}, i$);
    $S \leftarrow$ READ-PV (i);
    State $\leftarrow$ COMPARE-PV(PV_{iA}, S);
    if *State = Passive* **then**
        SEND-MERGE(i);
        WRITE-ALL("Defeated", i);
        Return to homebase and execute WAIT();
    else
        RECEIVE-MERGE(i);
        execute UPDATE-PV() and continue;
WRITE-ALL("Failure");

Lemma 1 ([14]). *Algorithm* Make-Tree *solves Rendezvous-with-Detect using* $O(mk)$ *moves in total and requires* $O(m \log n)$ *whiteboard memory for each node.*

A modified version of the algorithm solves rendezvous for all solvable instances in at most $O(m \log k)$ moves. The only modification is during the comparison of the partial-views; if the agent A finds that all neighboring agents have the same

Partial-view, then agent A returns to its homebase and waits (instead of continuing with the competition rounds for k iterations). This algorithm would not have an explicit termination.

Another possible modification to the algorithm is to reduce the memory required for the whiteboards (see [14]). If the whiteboards at each node are limited to $O(\log n)$ bits, then a modified version of Algorithm *Make-Tree* can be used to solve Rendezvous-with-Detect using $O(\log n)$ bit whiteboards and $O(m^2 k)$ moves in total. The idea of this algorithm is to perform the comparisons of partial-views by traversing the territories of the neighbors. Thus the agents have to perform additional moves, but the only information that we need to write on the whiteboards is the label assigned to the node and a link to its parent in the tree.

12.7 Rendezvous Using Tokens

In this section we consider the rendezvous problem in the token model where each agent has a token which they can place on any node they visit. Note that tokens used by all agents are identical (and thus indistinguishable). As opposed to the previous section, there are no public whiteboards on the nodes. If every agent puts its token on its starting location, we have a bicoloring on the nodes of the graph representing the function χ_p on $V(G)$. The agent can now execute Algorithm 1 with the following modification. The initial classification partitions the nodes into two classes–those that contain a token and those that do not! The algorithm will succeed in solving rendezvous whenever the conditions of Theorem 2 are satisfied. Moreover, the algorithm can solve Rendezvous-with-Detect, if the exact value of n is known. However this algorithm is not efficient in terms of the moves complexity as we have seen. We present below a different algorithm which is more efficient [8].

12.7.1 Rendezvous with a Single Unmovable Token

The algorithm for rendezvous presented in this section is for two agents, though the same idea may be used to rendezvous any $k \geq 2$ agents using a more involved algorithm. We assume that an agent always places the token at its starting location. First, suppose there is a single agent exploring a graph G. The fact that the starting node r of the agent is marked and can be distinguished from other nodes, makes it easier to perform an exploration of G. The agent can perform a breadth-first traversal building a BFS-tree T rooted at r. During the traversal, whenever the agent explores a new edge and reaches a node v, it checks whether v is same as some node u in its tree. This can be done by successively applying the label-sequences for the back-paths from each node $u \in T$ to the root r, and checking if one of these hits the marked node. Based on this idea, we have an algorithm for building a map of G with a single agent starting from a unique marked homebase in G (see Algorithm 3). The algorithm maintain·a BFS-tree T containing the visited nodes and a data structure called ROOT_PATHS that stores the edge-labeled path P in T from any node v to the homebase r. For such a stored path P, Start(P) refers to the node v. For any path $P = (u_0, u_1, \ldots u_t)$ in the tree T, the label sequence of path P is $\Lambda(P) = (\delta(u_0, u_1), \ldots \delta(u_{k-1}, u_t))$. Other than the tree T, the algorithm also maintains the cross-edges which together with T, give the complete map of G.

Algorithm 3: BFS-Tree-Construction

$Map := T := \{r\}$;
Add r to Queue;
ROOT_PATHS $:= \{\phi\}$;
while *Queue is not empty* **do**
    Get next node v from Queue and go to v using Map;
    while *node v has unexplored edges* **do**
        Traverse the next unexplored edge $e = (v, u)$;
        for *each path $P \in ROOT\_PATHS$* **do**
            Apply sequence $\Lambda(P)$ at node u ;
            if *successfully reached a marked node* **then**
                Add to Map a cross-edge from v to Start(P);
                Update the number of explored edges at the node Start(P);
                Return to node v using T and exit Loop;
            else
                Backtrack to node u ;
        if *All path sequences failed to reach a marked node* **then**
            Add a new node u to T and Map ;
            Add edge (v, u) to T and Map ;
            Insert u to Queue ;
            ROOT_PATHS $:=$ ROOT_PATHS $\cup$ Path$_T(u, r)$;
            Backtrack to node v ;

When two identical agents execute the Algorithm 3 from marked homebases, it is clear that the agents will not have a map of the complete graph. However, the following properties are satisfied.

Lemma 2 ([8]). *During algorithm BFS-Tree-Construction: (i) The graph T constructed by each agent will be an acyclic connected subgraph of G, and (ii) if the maps constructed by the two agents are identical then the views from the two homebases are identical.*

The tree constructed by an agent in the above algorithm, is similar to the territory of an agent as in Sect. 12.6. Due to the above properties, we know that when the maps obtained by the two agents are identical, then rendezvous is not solvable deterministically. So, we only need to consider the case when the maps are distinct. In this case if we could compare the maps of the agents, we can elect one of the agents and the agents could rendezvous at the homebase of the elected agent. This algorithm (called Algorithm RDVwithToken) was presented in [8] and we have the following result.

Theorem 8 ([8]). *Algorithm RDVwithToken solves Rendezvous-with-Detect for two agents on a graph of size n and maximum degree d, and requires $O(n^4 d^2)$ moves by each agent. Each agent requires a private memory of size $O(nd \log n)$.*

12.7.2 Tolerating Failures and Uncertainty

In this section we consider two special cases. The first scenario is when tokens placed by the agent are subject to failures (i.e. they may disappear during the execution). This problem has been studied for the ring [22] and a solution is provided for $f < k$ failures, assuming certain conditions on the parameters n and k. A more general solution for arbitrary graphs is provided in [15] which works if at least one token does not fail (irrespective of the values of n and k). The idea of the algorithm is the following. If there are no failures then any standard algorithm (e.g. the one at the beginning of this section) can be used to determine a unique rendezvous location, whenever (G, χ_p, δ) is symmetric-covering-minimal. On the other hand if there are $1 \le f < k$ failures, then the agents whose tokens failed are distinguished from the agents whose tokens are still in their homebase. The former agents (called *Runners*) traverse the graph and carry information to each marked homebase, while the latter agents (called *Owners*) wait at their homebase to receive information from each *Runner* agent. Using this information, each *Owner* agent can determine the location of the missing tokens and thus reconstruct a map of the original environment, and solve rendezvous. The challenging part of the algorithm is to switch from the procedure for the fault-free scenario to the procedure for the faulty scenario, in case faults do occur at arbitrary times during the execution of the algorithm.

Another scenario that has been studied recently is the rendezvous of agents in graphs having no common port-numbering [7]. Note that all the algorithms considered so far are based on the fact that any two agents have the same *view* from any given vertex $v \in G$. If we consider the situation where two agents a and b may have distinct port-numbering functions δ_a and δ_b, then this assumption is no longer true. In this case, any agent a may navigate in the graph using its own port-numbering δ_a and build a map representing the minimum-base of the environment $(G, \lambda, \chi_p, \delta_a)$ but the maps built by the two agents may not necessarily be identical (though they will be isomorphic). Thus, the agents may not agree on a unique ordering of the vertices and rendezvous is not possible without any additional assumptions. Surprisingly, if the agents are provided with an additional token (i.e. each agent now has 2 identical tokens) then it is possible to solve Rendezvous-with-Detect in all environments that satisfy the conditions of Theorem 3. The algorithm that achieves this, works as follows. Once an agent a has built a map of G, the vertices of G are partitioned into automorphism classes (ignoring the port-numbering δ_a). The agents then iteratively refine this partitioning by a process of selective marking of the nodes, eventually obtaining a total order on the set of nodes. During each phase of this iterative process, an agent uses one token to mark the selected node and the other token to synchronize with other agents. The full details of the algorithm can be found in [7] and we only state the main result here.

Theorem 9 ([7]). *Two tokens per agent are necessary and sufficient to solve Rendezvous-with-Detect in the absence of common port-numbering in any environment $(G, \lambda, \mathcal{Q}, p)$ such that $(Dir(G), \lambda')$ is symmetric-covering minimal where $\lambda' = \lambda \times \chi_p$.*

12.8 Rendezvous in Dangerous Graphs

We now consider the scenario where some of the nodes of the graph may be dangerous and inaccessible to the agents. This is inspired by communication networks where some nodes or links between nodes may develop faults. If an agent attempts to traverse a faulty link or to move to a faulty node it simply disappears (i.e. the agent is destroyed without leaving a trace). Given a graph with multiple faulty nodes, we can merge them all into one dangerous node x, called the *black hole*. The question of whether rendezvous can be solved in a graph containing a black hole was first studied in [17] where an algorithm was provided for ring graphs containing a single black hole. We study the problem for arbitrary graphs with both faulty nodes and faulty edges, where the agents do not have prior knowledge of the graph topology or the possible location of faults. Throughout this section we assume the whiteboard model of communication, i.e. the agents may write any information on the nodes they visit.

Since the location of a black hole is unknown and cannot be determined unless an agent falls into it and is destroyed, this means that rendezvous of all agents is not possible. Thus, the objective is to achieve rendezvous of as many agents as possible while avoiding the black hole. It can be shown that in an unknown arbitrary graph with τ links that lead to a black hole, it is not possible to solve rendezvous of more than $k - \tau$ agents [9]. Note that no agent starts from the black hole (i.e. the homebases are distinct from the black hole node). For agents starting from arbitrary locations, rendezvous of any two agents is possible only if the graph obtained after removing the black hole is connected. We now present some other conditions that must be satisfied for feasibility of deterministic rendezvous in an environment $(G, \lambda, \mathcal{Q}, p, \delta, \eta)$ where the function $\eta : E(G) \to \{0, 1\}$ denotes which edges are safe ($\eta(e) = 1$) and which are faulty ($\eta(e) = 0$). Given a graph with multiple faulty edges and faulty nodes, we can replace each faulty edge (u, v) with two edges (u, x) and (v, y) leading to two distinct (dangerous) nodes x and y respectively. We denote by τ, the number of faulty links (each faulty edge accounts for two faulty links).

We define the *extended-map* of the environment $(G, \lambda, \mathcal{Q}, p, \delta, \eta)$ as the labeled digraph (H, μ_H) such that, H consists of two disjoint vertex sets V_1 and V_2 and a set of arcs $\mathscr{A}$ as defined below:

- $V_1 = V(G)$;
- $\mu_H(v) = (\lambda(v), \chi_p(v)), \forall v \in V_1$;
- For every safe edge $e = (u, v) \in E(G)$, there are two arcs $a_1, a_2 \in \mathscr{A}$ such that $s(a_1) = t(a_2) = u$, $s(a_2) = t(a_1) = v$, and $\mu_H(a_1) = (\delta_u(e), \delta_v(e))$, $\mu_H(a_2) = (\delta_v(e), \delta_u(e))$.

- For every faulty edge $e = (u, v)$, there are vertices u' and $v' \in V_2$ with $\mu_H(u') = \mu_H(v') = -1$ and arcs $(u, u'), (u', u), (v, v')$ and $(v', v) \in \mathscr{A}$ with labels $(\delta_e(u), 0)$, $(0, \delta_e(u))$, $(\delta_e(v), 0)$, and $(0, \delta_e(u))$ respectively;

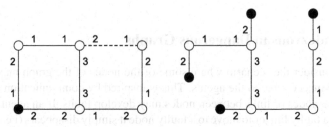

Fig. 12.3 The extended-map of an environment containing faulty edges (marked in *dashed lines*) and black-holes (colored *black*)

The *extended-map* can be thought of as a canonical representation of the environment (see Fig. 12.3). It can be shown that any execution of a deterministic algorithm on the environment $(G, \lambda, \mathscr{Q}, p, \delta, \eta)$ can be simulated on the extended-map (H, μ_H). Based on this we have the following result from [9].

Theorem 10. *It is not possible to rendezvous $k - \tau$ agents in an environment whose extended-map is not symmetric-covering minimal.*

In the following we will briefly discuss an algorithm that solves the rendezvous of $k - \tau$ agents in an environment that satisfies the conditions above. We present below a lower bound on the moves complexity of any algorithm solving rendezvous of $k - \tau$ agents (see [9] for a proof).

Theorem 11. *For solving rendezvous of $(k - \tau)$ agents in an environment $(G, \lambda, \mathscr{Q}, p, \delta, \eta)$ without any knowledge other than the size of G, the agents need to make at least $\Omega(m(m + k))$ moves in total.*

We can ensure that no more than one agent dies while traversing the same link, using the *cautious walk* technique as in [17]. At each node, all the incident edges are considered to be unexplored in the beginning. Whenever an agent A at a node u has to traverse an unexplored edge $e = (u, v)$, agent A first marks link $\delta_u(e)$ as "Being Explored" and if it is able to reach the other end v successfully, it immediately returns to node u and re-marks the link $\delta_u(e)$ as "safe". During the algorithm we follow the rule that no agent ever traverses a link that is marked "Being Explored". This ensures no more than τ agents may die during the algorithm.

The algorithm is based on similar ideas as in Algorithm 2. Recall that the algorithm proceeded with each agent exploring a part of the graph and marking its territory, followed by comparison between the territories of agents over multiple rounds. In the present algorithm, several improvisations are required to account for the fact that some agents may die during the execution of the algorithm. First of all, rounds of exploration are alternated with rounds of competition between agents. During an exploration round, an agent may fall into a black hole and die, without completing the process of marking its territory. Thus, during the competition an agent can win over another agent based on either comparison of territories or comparison of round number (a dead agent would not be able to increment its round number). When an agent A wins the territory of another agent (that may have died) there may be unexplored edges incident to this territory and the agent A needs to expand the territory during the next exploration round. Recall that it is not possible to distinguish between a dead agent and an agent that is slow in moving along an edge. This means that an agent cannot wait for another agent to complete its exploration. Thus, there could be multiple agents expanding a given territory simultaneously. Those agents which are in the same territory need to coordinate with each other in the exploration and competition tasks. This is done by communicating using messages written on the whiteboard of the root node. The algorithm ensures that there is always a unique root node in every territory during the execution of the algorithm.

At any stage of the algorithm, there are teams of agents, each team possessing a *territory* which is a connected acyclic subgraph of G (disjoint from other territories). Each team of agents tries to expand its territory until it spans a majority of the nodes. Once a team is able to acquire more than half the nodes of the network, it wins and agents from all other teams join the winning team to achieve rendezvous. We call this algorithm as Algorithm *RDV_BH*.

Theorem 12 ([9]). *Algorithm* RDV_BH *correctly solves rendezvous for* $k - \tau$ *agents in any network whose extended-map is symmetric-covering minimal provided the agents initially knows a bound B such that $n \leq B < 2n$. The moves complexity of algorithm* RDV_BH *is $O(m(m + k))$.*

The above result implies that the algorithm described in this section is optimal in terms of the moves complexity.

12.9 Conclusion

We considered the problem of symmetric asynchronous rendezvous in graphs whose nodes are anonymous and whose edges are locally ordered. Since the agents are identical and follow the same deterministic algorithm, solving the problem requires breaking the symmetry and finding a unique location to meet. This is possible only if there is some asymmetry in the structure of the graph (for agents that cannot mark the graph) or if the agents start from asymmetric locations within the graph (in the case when agents are allowed to mark their starting location). It is possible determine for exactly which instances rendezvous is feasible and then solve rendezvous in those cases. We presented solutions for rendezvous with detection and also discussed techniques for dealing with exceptional situations involving faulty nodes, token failures and inconsistencies in local labelling of the edges. While all solutions studied here assume that the edges of the graph are bidirectional, some of the techniques could be extended to work for (strongly connected) directed graphs (e.g. see [10]). In general, solving rendezvous in directed graphs is more difficult due to the inability of agents to backtrack and this is one of directions for future research. Another open problem is solving rendezvous in dangerous graphs assuming the weaker model of communication when there are no whiteboards and the agents are only provided with a few pebbles.

References

1. R. Aleliunas, R.M. Karp, R.J. Lipton, L. Lovász, and C. Rackoff. Random walks, universal traversal sequences, and the complexity of maze problems. In Proc. of 20th Annual Symposium on Foundations of Computer Science (FOCS), pp. 218–223, 1979.
2. S. Alpern. Hide and seek games. Seminar, Institut fur Hohere Studien, Wien, July 1976.
3. S. Alpern and S. Gal. *The Theory of Search Games and Rendezvous*. Kluwer, 2003.
4. E. Bampas, J. Czyzowicz, L. Gasieniec, D. Ilcinkas, A. Labourel. Almost Optimal Asynchronous Rendezvous in Infinite Multidimensional Grids. In Proc. 24th International Symposium on Distributed Computing (DISC), pp. 297–311, 2010.
5. P. Boldi, B. Codenotti, P. Gemmell, S. Shammah, J. Simon, and S. Vigna. Symmetry breaking in anonymous networks: Characterizations. In Proc. 4th Israeli Symposium on Theory of Computing and Systems, pp. 16–26, 1996.
6. P. Boldi and S. Vigna. Fibrations of graphs. Discrete Mathematics, 243(1–3):21–66, 2002.
7. J. Chalopin and S. Das. Rendezvous of Mobile Agents without Agreement on Local Orientation. In Proc. 37th Int. Coll. on Automata, Languages and Programming (ICALP), pp. 515–526, 2010.
8. J. Chalopin, S. Das, and A. Kosowski. Constructing a Map of an Anonymous Graph: Applications of Universal Sequences. In Proc. 14th International Conference on Principles of Distributed Systems (OPODIS), pp. 119–134, 2010.
9. J. Chalopin, S. Das, and N. Santoro. Rendezvous of Mobile Agents in Unknown Graphs with Faulty Links. In Proc. 21st International Symposium on Distributed Computing (DISC), pp. 108–122, 2007.
10. J. Chalopin, S. Das, and P. Widmayer. Rendezvous of Mobile Agents in Directed Graphs. In Proc. 24th International Symposium on Distributed Computing (DISC), pp. 282–296, 2010.
11. J. Czyzowicz, A. Labourel, and A. Pelc, How to meet asynchronously (almost) everywhere, In Proc. 21st Annual ACM-SIAM Symposium on Discrete Algorithms (SODA), pp. 22–30, 2010.
12. J. Czyzowicz, A. Kosowski, A. Pelc. How to meet when you forget: log-space rendezvous in arbitrary graphs. Distributed Computing 25(2): 165–178, 2012.
13. J. Czyzowicz, A. Kosowski, A. Pelc, Time vs. space trade-offs for rendezvous in trees, In Proc. 24th ACM Symposium on Parallelism in Algorithms and Architectures (SPAA), pp. 1–10, 2012.
14. S. Das, P. Flocchini, A. Nayak, and N. Santoro. Effective elections for anonymous mobile agents. In Proc. 17th Symp. on Algorithms and Computation (ISAAC), pp. 732–743, 2006.
15. S. Das, M. Mihalak, R. Sramek, E. Vicari, P. Widmayer, Rendezvous of mobile agents when tokens fail anytime, In Proc. 12th Int. Conf. on Principles of Distributed Systems (OPODIS), pp. 463–480, 2008.
16. A. Dessmark, P. Fraigniaud, D. Kowalski, A. Pelc. Deterministic rendezvous in graphs, Algorithmica, 46: 69–96, 2006.
17. S. Dobrev, P. Flocchini, G. Prencipe, and N. Santoro. Multiple agents rendezvous in a ring in spite of a black hole. In Proc. 7th Int. Conf. on Principles of Distributed Systems (OPODIS), pp. 34–46, 2003.

18. Paola Flocchini, Giuseppe Prencipe, Nicola Santoro, Peter Widmayer: Gathering of asynchronous robots with limited visibility. Theor. Comput. Sci. (TCS) 337(1–3):147–168, 2005.
19. L. Gasieniec, E. Kranakis, D. Krizanc, X. Zhang. Optimal Memory Rendezvous of Anonymous Mobile Agents in a Unidirectional Ring. In Proc. 32nd Conference on Current Trends in Theory and Practice of Computer Science (SOFSEM), pp. 282–292, 2006.
20. M. Kouckỳ, Universal traversal sequences with backtracking, J. Comput. Syst. Sci. 65:717–726, 2002.
21. R. Klasing, E. Markou, and A. Pelc, Gathering asynchronous oblivious mobile robots in a ring, Theor. Comput. Sci. 390(1):27–39, 2008.
22. E. Kranakis, D. Krizanc, and E. Markou, The mobile agent rendezvous problem in the ring, Morgan and Claypool Publishers, 2010.
23. A. Pelc. Deterministic rendezvous in networks: A comprehensive survey, Networks, 59: 331–347, 2012.
24. M. Yamashita and T. Kameda. Computing on anonymous networks: Part I–Characterizing the solvable cases. IEEE Transactions on Parallel and Distributed Systems, 7(1):69–89, 1996.
25. M. Yamashita and T. Kameda. Leader election problem on networks in which processor identity numbers are not distinct. IEEE Transactions on parallel and distributed systems, 10(9):878–887, 1999.

Chapter 13
Gathering Asynchronous and Oblivious Robots on Basic Graph Topologies Under the Look-Compute-Move Model*

Gianlorenzo D'Angelo, Gabriele Di Stefano, and Alfredo Navarra

Abstract Recent and challenging models of robot-based computing systems consider identical, oblivious and mobile robots placed on the nodes of anonymous graphs. Robots operate asynchronously in order to reach a common node and remain with it. This task is known in the literature as the *gathering* or *rendezvous* problem. The target node is neither chosen in advance nor marked differently compared to the other nodes. In fact, the graph is anonymous and robots have minimal capabilities. In the context of robot-based computing systems, resources are always limited and precious. Then, the research of the minimal set of assumptions and capabilities required to accomplish the gathering task as well as for other achievements is of main interest. Moreover, the minimality of the assumptions stimulates the investigation of new and challenging techniques that might reveal crucial peculiarities even for other tasks. The model considered in this chapter is known in the literature as the *Look-Compute-Move* model. Identical robots initially placed at different nodes of an anonymous input graph operate in asynchronous Look-Compute-Move cycles. In each cycle, a robot takes a snapshot of the current global configuration (Look), then, based on the perceived configuration, takes a decision to stay idle or to move to one of its adjacent nodes (Compute), and in the latter case it makes an instantaneous

G. D'Angelo (✉)
MASCOTTE Project, INRIA/I3S(CNRS/UNSA), Sophia-Antipolis Cedex, France
e-mail: gianlorenzo.d_angelo@inria.fr

G. Di Stefano
Dipartimento di Ingegneria e Scienze dell'Informazione e Matematica,
Università degli Studi dell'Aquila, L'Aquila, Italy
e-mail: gabriele.distefano@univaq.it

A. Navarra
Department of Mathematics and Computer Science, University of Perugia, Perugia, Italy
e-mail: alfredo.navarra@unipg.it

* This work has been partially supported by the Fondazione Cassa di Risparmio della Provincia dell'Aquila (Italy) within project "ARISE" (Arising Robust Internetworked System for Emergency contexts).

S. Alpern et al. (eds.), *Search Theory: A Game Theoretic Perspective*,
DOI 10.1007/978-1-4614-6825-7_13, © Springer Science+Business Media New York 2013
197

move to this neighbor (Move). Cycles are performed asynchronously for each robot. This means that the time between Look, Compute, and Move operations is finite but unbounded, and it is decided by the adversary for each robot. Hence, robots may move based on significantly outdated perceptions. The only constraint is that moves are instantaneous, and hence any robot performing a Look operation perceives all other robots at nodes of the ring and not on edges. Robots are all identical, anonymous, and execute the same deterministic algorithm. They cannot leave any marks at visited nodes, nor can they send messages to other robots. In this chapter, we aim to survey on recent results obtained for the gathering task over basic graph topologies, that are rings, grids, and trees. Recent achievements to this matter have attracted many researchers, and have provided interesting approaches that might be of main interest to the community that studies robot-based computing systems.

13.1 Introduction

The chapter surveys on recent results in robot-based computing systems. Two or more robots, starting from distinct initial positions, have to meet at some place and remain there. The problem is known in the literature as the *gathering* problem while sometimes it is referred to as the *rendezvous* problem.

Different assumptions on the capabilities of the robots as well as on the environment where they move, lead to very different scenarios. To have an idea of the work done during the recent years, it is enough to mention that already five different surveys deal with such a problem from different perspectives. The first distinction considers the way the robots may take their decisions in order to move towards some directions. In fact, randomized algorithms can be applied for this purpose or full determinism might be required. For the former case, there is a comprehensive survey book [3] which also includes results contained in an older survey paper [2]. The latter case has captured more attention in recent studies. In particular, for the case where robots are considered to move along the nodes and edges of an input graph, the survey paper [25] and in a more extended form [26] present various scenarios and techniques for different graph topologies. Whereas, the survey book [24] focuses on the gathering over ring networks. In the literature, many results also concern the gathering of robots moving on a continuous two-dimensional Euclidean space have been devised. The interested reader may refer to [8, 15, 20, 27, 29] for the continuous case. However, a recent trend is to study discrete models like the case where robots move over graphs rather than the continuous case.

In this chapter, the aim is to provide in more details the strategies applied to accomplish the gathering task on basic graph topologies like rings, grids, and trees, under a very specific model that has attracted many researchers during the last years. Very few of such results are already contained in the aforementioned surveys. In fact, most of the results come from very recent papers and the last section contains original results for tree topologies.

The model considered in this chapter (sometimes also referred to as CORDA [27]) is known in the literature as the *Look-Compute-Move* model. Robots asynchronously run an operative cycle where first they perceive the global configuration of the robots over the graph (Look phase). That is, during the Look phase a robot is able to perceive the relative locations of the other robots with respect to its own position on the graph. The only cases where a robot might be misled concern the so called *multiplicities*, i.e., when more than one robot occupy the same node. In this case, different assumptions might be considered. Based on the perceived configuration which might reveal the exact disposal of all the robots or just which nodes are occupied, a robot evaluates whether to stay idle or to move towards one of its neighboring nodes (Compute phase). Note that, since robots are asynchronous, the Compute phase might be accomplished by a robot based on outdated configurations perceived while other robots are performing their movements. Finally, the robot enters to the Move phase, where it simply applies the computed movement. Hence, it either remains on its current position or it moves towards the computed neighboring node. The only assumption in this phase is that the movements are instantaneous and hence robots are always perceived over nodes, and never over edges. Robots are all identical, anonymous, and execute the same deterministic algorithm. They cannot leave any marks at visited nodes, nor send messages to other robots. The scheduler that wakes the robots up is assumed to be fair, i.e., all the robots will wake up, eventually, and perform their Look-Compute-Move cycles infinitely many times.

Another assumption that can be considered concerns the ability for the robots to perceive information about the number of robots occupying the same node, during the Look operation. This ability is called the *multiplicity detection* capability and it has been sometimes exploited in various forms. In any case, a robot perceives whether a node is empty or not, but in the *global-strong* version, a robot is able to perceive the exact number of robots that occupy each node. In the *global-weak* version, a robot perceives only whether a node is occupied by one robot or if a multiplicity occurs, i.e., a node is occupied by an undefined number of robots greater than one. In the *local-strong* version, a robot can perceive only whether a node is occupied or not, but it is able to perceive the exact number of robots occupying the node where it resides. Finally, in the *local-weak* version, a robot can perceive the multiplicity only on the node where it resides but not the exact number of robots composing it.

In the context of robot-based computing systems, resources are always limited and precious. Then, the research of the minimal set of assumptions and capabilities required to accomplish the gathering task as well as for other achievements is of main interest. Moreover, the minimality of the assumptions stimulates the investigation of new and challenging techniques that might reveal crucial peculiarities even for other tasks.

Depending on the multiplicity detection capability version chosen for the robots, some scenarios may be unsolvable while some others are solvable. Intuitive concepts like symmetry or periodicity might be involved and sometimes are fundamental to the feasibility of the studied problems. Depending on the assumptions made, the definition of such concepts may vary and require different approaches. This is why in what follows, the same concept might be re-defined according to the current scope. Moreover, the considered scenarios lead to very interesting and different strategies that can be considered also for other areas of applications.

Besides the gathering problem, the Look-Compute-Move model has been studied also for the problem of *graph exploration with stop* and *exclusive perpetual graph exploration* [4–6, 12–14]. In the first problem [12–14], it is required that each node (or each edge) of the input graph is visited for a finite number of times by at least one robot and, eventually, all the robots have to stop. This implies that after performing the exploration step, the algorithms need some mean to empower the robots by the capability of recording the part of the graph that has been already explored. Since the robots are oblivious, this task is performed by identifying particular configurations of the robots indicating that the exploration task has been accomplished. The exclusive perpetual graph exploration [4–6] requires that each robot visits each node of the graph infinitely many times. Moreover, it adds the constraint that no two robots should concurrently be on the same node or cross the same edge.

13.1.1 Outline

The chapter is organized in three main sections, dictated by the graph topologies considered. The next section provides techniques and results for the gathering on ring networks. In particular, the section is divided in three parts. First, impossibility results concerning the gathering on rings are summarized. Those hold even though the global-strong multiplicity detection is assumed. Then, results for the case of global-weak multiplicity detection are shown. Under such assumptions, all possible initial gatherable configurations have been addressed. Finally, partial results for the case of local-weak multiplicity detection are described. In Sect. 13.3, the problem for grids is fully characterized even when no multiplicity detection is assumed. Similarly, in Sect. 13.4 a full characterization without any multiplicity detection capability is provided for tree topologies. This is indeed an original contribution of the chapter. Finally, Sect. 13.5 concludes the chapter and outlines some possible research directions for robot-based computing systems.

13.2 Gathering on Rings

In this section, the gathering over ring networks is presented. After providing some necessary definitions, impossibility results are summarized when the global-strong multiplicity detection is assumed. Then, differences between the case of global-weak and local-weak multiplicity detection assumptions are presented. In particular, when the global-weak multiplicity detection is assumed, a full characterization of the gatherable configurations is provided. Whereas, when the local-weak multiplicity detection is assumed, only some sub-cases are solved. However, the different techniques used to accomplish the gathering task among the approached scenarios are very interesting for further investigations in robot-based computing systems.

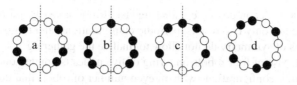

Fig. 13.1 Symmetric and periodic initial configurations on a ring. *White nodes* are empty while each *black node* is occupied by one robot

The model assumes that k robots are placed over the n nodes of a ring, and in the *initial configurations*, nodes are occupied by at most one robot. Depending on the movements imposed by the running algorithms, multiplicities may occur. A configuration is called *symmetric* if the ring admits a geometrical *axis of symmetry*, that defines a bijective function among the robots residing in the two halves of the ring cut by the axis. When the global-weak multiplicity is considered, a configuration is called symmetric if the ring admits a geometrical axis of symmetry that reflects single robots into single robots, multiplicities into multiplicities, and empty nodes into empty nodes. In this case, a configuration might be considered symmetric even though the two halves of the ring cut by the axis do not contain the same number of robots. This can happen if two symmetric multiplicities at the two halves are composed of a different number of robots. If the local-strong (or the local-weak) multiplicity detection is assumed, then a configuration might result symmetric for some robots while asymmetric for others. For instance, if robots are part of a multiplicity and the configuration does not admit an axis of symmetry passing through such a node, then the configuration would result asymmetric for all the robots composing the multiplicity, while it might be symmetric with respect to the perception of all the other ones. However, symmetric peculiarities of initial configurations are invariant with respect to the assumed multiplicity detection, as multiplicities are not allowed at the beginning.

As shown in Fig. 13.1, a symmetric configuration with an axis of symmetry has an *edge-edge symmetry* if the axis goes through two edges (Fig. 13.1a); it has a *node-edge symmetry* if the axis goes through one node and one edge (Fig. 13.1a); it has a *node-node symmetry* if the axis goes through two nodes (Fig. 13.1c); it has a *robot-on-axis symmetry* if there is at least one node on the axis of symmetry occupied by a robot (both Fig. 13.1b, c).

A configuration is called *periodic* if it is invariable under non-trivial (i.e., non-complete) rotations (Fig. 13.1d).

13.2.1 Impossibility Results

In [22], it is proved that the gathering is unsolvable if the multiplicity detection capability is completely removed in either of its forms. When the multiplicity detection is assumed, even in its strong and global form, still there are configurations for which it is impossible to accomplish the gathering task. More precisely, initial configurations composed of only two robots, periodic configurations, and those admitting an edge-edge axis of symmetry do not allow to finalize the gathering.

In [21], the case of four robots on a ring of five nodes is pointed out as a case of symmetric initial configurations with an even number of robots that does not allow any gathering algorithm. This has been also studied in [16] along with other solvable cases. In general, symmetric configurations of type node-edge with four robots and the odd interval cut by the axis bigger than the even one are ungatherable. In the rest of the chapter these configurations are denoted with the set $SP4$. The case of four robots on a five nodes ring belongs to $SP4$. Actually, some configurations in $SP4$ could be gatherable but they require strategies that are difficult to generalize.

For all the remaining initial configurations, various gathering algorithms have been provided, depending also on the assumptions concerning the multiplicity detection capability. Whenever clear by the context, we refer to initial configurations simply as configurations.

13.2.2 Global-Weak Multiplicity Detection

In this section, a description of the techniques taken from the specific literature are described. Based on the global-weak multiplicity detection capability, the next algorithms cope with all the cases left from the impossibility results previously shown.

13.2.2.1 Asymmetric Configurations

The asymmetric initial configurations have been firstly handled in [22]. When such configurations are aperiodic, they were referred to as *rigid* configurations. The gathering is performed by exploiting the perception of the robots. Perception allows robots to agree and move exactly one robot at time although the model does not allow communication. More precisely, each robot detects which one must perform the next move based on the configuration perceived during the Look phase. This is done until the first (and only) multiplicity occurs. Since the scheduler that wakes the robots up is assumed to be fair, the robot that is allowed to move will eventually wake up and perform all its Look-Compute-Move cycle. This will ensure the robots perform all required moves until the desired multiplicity is created. Once the multiplicity has been created, the robots with only free nodes between themselves and the multiplicity are allowed to move towards the multiplicity, and joining it, until all the robots gather at the same node.

At each step of the proposed strategy, and before creating the multiplicity, the robot allowed to move will be chosen in such a way that the configuration will never lose its original "rigidity". Once captured the current configuration during the Look phase, a robot looks for the pair of robots that are at the maximum distance (in terms of empty nodes in between) from each other. If only one pair of robots is detected, the one allowed to move is the robot with the closest neighbor on the other side of the maximum interval. Possible ties are easily broken by considering the next intervals and so forth, until a difference occurs. Since the configuration is asymmetric, there must be a difference somewhere, and one robot is elected to move. If more than one pair provides the same maximum distance, ties are broken by considering the global view, hence ordering lexicographically the views (in terms of sequences of distances) and choosing the interval that appears as first in the largest view. Once the single robot has been elected to move, it performs the movement that enlarges the maximum interval. This ensures that in the next step there is exactly one maximum distance, with one interval at its side smaller than the one at the other side. Hence, from now on, only the same robot will be allowed to move until creating a multiplicity.

13.2.2.2 Odd Number of Robots

Another type of initial configurations addressed and solved in [22] concerns all the configurations with an odd number of robots. In this case, the configuration can be either asymmetric or symmetric. In the former case, the gathering is solved as described in the previous section. In the latter case, it can be observed that one robot resides on the axis of symmetry, necessarily. Then the following property is exploited:

Property 1 ([22]). Let C be a symmetric configuration with an odd number of robots, without multiplicities. Let C' be the configuration resulting from C by moving the unique robot on the axis to any of its adjacent nodes. Then C' is either asymmetric or still symmetric but aperiodic. Moreover, by repeating this procedure a finite number of times, eventually the configuration becomes asymmetric (with possibly one multiplicity).

When Property 1 holds, symmetric configurations with an odd number of robots will allow only one robot to move until either a multiplicity occurs or the configuration becomes asymmetric and the gathering algorithm changes to that described above.

13.2.2.3 Even Number of Robots

The cases left open by the techniques described above are all the symmetric initial configurations with an even number of robots. Note that, configurations with only two robots are ungatherable as well as configurations with four robots in *SP4*.

A first study that addresses the case of an even number of robots comes from [21]. In that chapter, the authors solved all the symmetric cases with an even number of robots grater than 18. When robots are on the axis of symmetry it may be possible to design algorithms which break the symmetry by moving one of the robots located on the axis, as in the case of an odd number of robots described by Property 1.

When no robots reside on the axis, the algorithm works in four phases. During the first phase, since the configuration is symmetric, two robots are always allowed to move. In order to detect the two symmetric robots that must perform their moves, the robots have to elaborate the perceived configuration during their Compute phase. Based on the sequence of distances between robots along the ring, two symmetric minimal interval are detected and reduced concurrently, until two multiplicities are created. The number of robots grater than 18 comes from this computational step. In fact, the need to guarantee to break possible ties among minimal intervals, and the fact that some robots are needed between the detected intervals and the two poles defined by the axis of symmetry on the ring, gives a minimal number of required robots equal to 20.

It is very important to remark that the proposed technique is the first one that forces robots to maintain the original symmetry rather than breaking it. In fact, based on the perceived configurations, robots are always able to detect whether the current configuration is at one step from a reachable symmetry or not. In the latter case, the algorithm from [22] for asymmetric configurations can be applied. In the former case, the robot that can re-establish the symmetry will be the only one allowed to move. Note that, such a robot could have been already started its Look-Compute-Move cycle concurrently with its symmetric one, or it simply starts later. In any case, the algorithm guarantees to recover the original symmetry with two steps less towards the desired configuration with two multiplicities where the second phase starts.

When two multiplicities have been created, the idea is to move all the remaining single robots but few of them towards the two multiplicities. During the second phase, it is necessary to decide on one of the two poles of the axis of symmetry as the gathering point (the North pole). The poles are chosen so that the northern arc between multiplicities contains more robots than the southern arc; in the case of a tie, the side on which the nearest robots are closer to the multiplicities is the northern one. The robots are moved in symmetrical pairs towards their respective multiplicities, starting from the robots on the northern arc. In this way, North and South are consistently preserved throughout the phase. The phase ends with two multiplicities, two symmetric robots located at the southern part far from the multiplicities of at least one node, and two symmetric robots located at the northern part neighbors of the multiplicities. The two robots on the south are called *guards*.

The third phase is based on the position of the guards that maintain the direction to the gathering node. During this phase, the remaining single robots and those belonging to the multiplicities can move towards the North pole. The movement is performed always maintaining the robots associated to each multiplicity either as part or as neighbors of it. In this way, the configuration pattern is maintained throughout the process, until all robots except for the guards gather at the North pole in a single multiplicity.

The fourth phase simply moves the guards towards the multiplicity, and the gathering will be eventually finalized.

The algorithm has been also integrated with the one from [22], hence obtaining a full characterization of the gatherable configurations with an odd number of robots,

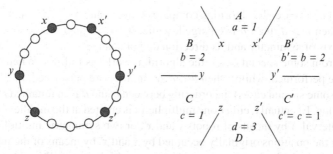

Fig. 13.2 A symmetric configuration and its representation

with an even number of robots but asymmetric, with an even number of robots admitting a robot-robot axis of symmetry, and with more than 18 robots admitting a node on the axis of symmetry. This has left open the cases of an even number of robots between 4 and 18 admitting a node on the axis of symmetry. Note that, the cases left for few robots might require more effort and different techniques for the resolution. In fact, lesser the robots lesser the information encoded by their disposal. This encouraged further investigation on configurations with few robots.

Gatherable configurations with four robots have been addressed in [16, 23]. The main idea is still to define a North and a South pole on the axis of symmetry (of type node-node). Then similarly to [21], the two northern nodes are moved while preserving the symmetry until creating a multiplicity on the North pole. After that, the other two robots join the multiplicity, hence finalizing the gathering.

The case of six robots is more intriguing as it requires different techniques from the older ones in order to fully characterize the gatherable configurations. It has been addressed in [9]. A symmetric configuration can be represented as shown in Fig. 13.2. In detail, without multiplicities, the ring is divided by the robots into 6 intervals: A, B, C, B', C', and D with a, b, c, b, c, and d free nodes, respectively. In the case of node-edge symmetry, A is the interval where the axis passes through a node and D is the interval where the axis passes through an edge; in the case of node-node symmetry, A and D are the intervals such that either $a < d$ or $a = d$ and $b < c$; the case where $a = d$ and $b = c$ cannot occur as it generates two axis of symmetry. Note that, in the case of node-node symmetry, a and d are both odd, while, in the case of node-edge symmetry, a is odd and d is even. Robots between A and B (B', respectively) are denoted by x (x', respectively); those between B and C (B' and C', respectively) are y (y', respectively); those between C (C', respectively) and D are z (z', respectively), see Fig. 13.2.

A robot $r \in \{x,y,z,x',y',z'\}$ can perform only two moves: it moves *up* ($r\uparrow$) if it goes towards A; it moves *down* ($r\downarrow$) if it goes towards D.

The main idea of the algorithm is to perform moves $x\uparrow$, $x'\uparrow$, $y\uparrow$ and $y'\uparrow$, with the aim of preserving the symmetry and gathering in the middle node of interval A, where the axis is directed. In some special cases, it may happen that the axis of symmetry changes at run time. Before multiplicities are created, the algorithm in a symmetric configuration allows only two robots to move in order to create a new symmetric configuration.

In the general case, the algorithm compares b and d, and performs a pair of moves such as when $b > d$, then b is enlarged, while, if $b < d$, then b is reduced. In this way, the axis of symmetry and its direction do not change.

Apart from some special cases, the algorithm works as follows. When $b > d$, $x\uparrow$ and $x'\uparrow$ are performed, while, when $b < d$, $y\uparrow$ and $y'\uparrow$ are performed. In both cases, (apart for some special cases) the ordering between b and d is maintained in the new configuration. Eventually, either one multiplicity is created at the middle node of the original interval A by means of robots x and x', or two symmetric multiplicities are created on the positions originally occupied by x and x' by means of the moves of y and y', respectively. In the second case, the two multiplicities will move up again to the middle node of the original interval A by allowing at most four robots to move all together. Once such a multiplicity has been created, the remaining robots join it, and conclude the gathering. In the special case of $b = d$, which can only happen in the initial configuration, the algorithm tries to break this equality by enlarging or reducing d by means of either $z\uparrow$ and $z'\uparrow$ (when $C > 0$) or $z\downarrow$ and $z'\downarrow$ (when $C = 0$ and $D > 0$). The special cases when $C = D = 0$ require specific arguments that can be found in [9].

13.2.2.4 Unifying Algorithm

Recently in [11], a new technique has been proposed for addressing all the gatherable initial configurations by means of a single algorithm that exploits some of the described strategies while also solving the remaining cases left open. In particular, existing algorithms are used as subroutines for solving the basic gatherable cases with four or six robots from [23] and [9], respectively. Also, Property 1 is exploited in some cases. Then, the main strategy is based on the definition of a particular read of the configurations perceived by the robots during their Look phase.

The current configuration of the system can be described in terms of the view of a robot r that is performing the Look operation. A configuration seen by r is represented as a tuple $Q(r) = (q_0, q_1, \ldots, q_j)$, $j \leq k - 1$, that represents the sequence of the numbers of free consecutive nodes broken up by robots when traversing the ring in one direction, starting from r. Unless differently specified, $Q(r)$ represents the configuration providing the lexicographical minimum among the two possible views. For instance, in the configuration of Fig. 13.2a, robot x can read either $Q = (1, 2, 1, 3, 1, 2)$ or $Q' = (2, 1, 3, 1, 2, 1)$, hence $Q(x) = Q$. A multiplicity is represented as $q_i = -1$ for some $0 \leq i \leq j$, regardless the number of robots composing it.

Given a generic configuration $C = (q_0, q_1, \ldots, q_j)$, let $\overline{C} = (q_0, q_j, q_{j-1}, \ldots, q_1)$, and let C_i be the configuration obtained by reading C starting from q_i as first interval, that is $C_i = (q_i, q_{(i+1) \bmod j+1}, \ldots, q_{(i+j) \bmod j+1})$. The above definitions imply:

Property 2. Given a configuration C,

(i) There exists $0 < i \leq j$ such that $C = C_i$ iff C is periodic;
(ii) There exists $0 \leq i \leq j$ such that $C = \overline{(C_i)}$ iff C is symmetric;
(iii) C is aperiodic and symmetric iff there exists only one axis of symmetry.

The next definition represents the key feature for the gathering algorithm.

Definition 1. Given a configuration $C = (q_0, q_1, \ldots, q_j)$ such that $q_i \geq 0$, for each $0 \leq i \leq j$, the view defined as $C^{SM} = \min\{C_i, \overline{(C_i)}, \mid 0 \leq i \leq j\}$ is called the *supermin configuration view*. An interval is called *supermin* if it belongs to the set $I_C = \{q_i \mid C_i = C^{SM} \text{ or } \overline{(C_i)} = C^{SM}, 0 \leq i \leq j\}$.

Once a robot is able to distinguish where a supermin is located, the next lemma provides a useful mean for computing whether the current configuration is gatherable or not.

Lemma 1 ([11]). *Given a configuration $C = (q_0, q_1, \ldots, q_j)$ with $q_i \geq 0$, $0 \leq i \leq j$:*
1. *$|I_C| = 1$ if and only if C is either asymmetric and aperiodic or it admits only one axis of symmetry passing through the supermin;*
2. *$|I_C| = 2$ if and only if C is either aperiodic and symmetric with the axis not passing through any supermin or it is periodic with period $\frac{n}{2}$;*
3. *$|I_C| > 2$ if and only if C is periodic, with period at most $\frac{n}{3}$.*

The above lemma already provides useful information for a robot when it wakes up. In fact, during the Look operation, it can easily recognize if the configuration contains only two robots, or if it belongs to the set *SP4*, or if $|I_C| > 2$ (i.e., the configuration is periodic), or in case $|I_C| = 2$, if the configuration admits an edge-edge axis of symmetry or it is again periodic. After this check, a robot knows if the configuration is gatherable, and proceeds with its computations. Indeed, all the remaining configurations are shown to be gatherable.

The main strategy allows only the movements which affect the supermin. In fact, if there is only one supermin, and the configuration allows its reduction, the subsequent configuration would still have only one supermin (the same as before but reduced), or a multiplicity is created. In general, such a strategy would lead asymmetric configurations or also symmetric ones with the axis passing through the supermin to create one multiplicity where the gathering will be easily finalized by collecting at turn the closest robots to the multiplicity. This strategy reminds the one used in [22] but with the difference to deal with the minimum rather than with the maximum.

For gatherable configurations with $|I_C| = 2$, the algorithm requires more phases before creating the final multiplicity where the gathering ends. In this case, there are two supermins that can be reduced. If both are reduced simultaneously, then the configuration is still symmetric and gatherable. Possibly, it contains two symmetric multiplicities. In fact, this is the status that one wants to reach even when only one of the two supermins is reduced. In general, the algorithm tries to preserve the original symmetry or to create a gatherable symmetric configuration from an asymmetric one. It is worth to remark that in all symmetric configurations with an even number of robots, the algorithm always allows the movement of two symmetric robots. Then after the initial movement, it is possible to obtain a symmetric configuration or an asymmetric one with a possible *pending* move. In fact, if only one robot (among the two allowed to move) performs its movement, it is possible that its symmetric one

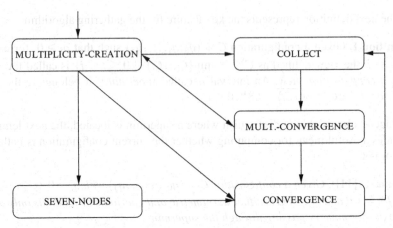

Fig. 13.3 Phases interchanges

either has not yet started its Look phase, or it is taking more time. If there might be a pending move, then the algorithm forces it before any other decision.

In contrast, asymmetric configurations cannot produce pending moves as the algorithm allows the movement of only one robot. In fact, it reduces the unique supermin by deterministically distinguish among the two adjacent robots, until one multiplicity is created. Finally, all the other robots will join the multiplicity one-by-one. In some special cases, from asymmetric configurations at one "allowed" move from symmetry (i.e., with a possible pending move), robots must guess which move would have been realized from the symmetric configuration, and force it in order to avoid unexpected behaviors. By doing this correctly, the algorithm eventually brings the configuration to have two symmetric multiplicities as above. From here, a new phase that collects all the other robots but two into the multiplicities starts. Still the configuration may move from symmetric configurations to asymmetric ones at one move from symmetry. Once the desired symmetric configuration with two multiplicities and two single robots is reached, a new phase starts and moves the two multiplicities to join each other. The node where the multiplicities join represents the final gathering location. This strategy reminds the one used in [21] as it tries to preserve the symmetry until the guards can join all the other robots in the gathering node.

Actually, sometimes the strategy that affects only the supermin cannot be applied, as a move may produce some undesired "side-effects", i.e., leading the configuration to ungatherable cases. In order to cope with such cases, two other moves have been defined. However, it can be shown that a robot is always able to understand the correct move to be performed.

An alternative move is to try to reduce the second supermin, i.e., the supermin of the configuration is evaluated after the real one. Another move, called XN, is applied when specific configurations occur. The definition of XN and the description of the cases where it must be applied are not provided in this chapter.

The algorithm works in five phases and depends on the configuration perceived by the robots, see Fig. 13.3. First, it starts from a configuration without multiplicities and performs phase MULTIPLICITY-CREATION whose aim is to create one multiplicity where all the robots will eventually gather, or a symmetric configuration with two multiplicities. In the former case, phase CONVERGENCE is performed to gather all the robots into the multiplicity. In the latter case, phases COLLECT and then MULTIPLICITY-CONVERGENCE are performed in order to first collect all the robots but two into the two multiplicities and then to join the two multiplicities into a single one. After that, phase CONVERGENCE is performed. Special cases of six robots on a seven nodes ring are considered separately in phase SEVEN-NODES.

In each phase, the robots can distinguish the type of configuration and apply the suitable strategy/move. The way a robot can identify the type of configuration is based on basic and simple calculations. Given a configuration $C = (q_0, q_1, \ldots, q_j)$, a robot compute the following parameters:

1. Number of nodes in the ring, $n(C)$;
2. Number of multiplicities, $m(C)$;
3. Number of nodes occupied or number of robots in the case without multiplicities, OCCUPIED(C);
4. Distance between single robots and multiplicities;
5. If C is symmetric;
6. If C is at one move from one of the symmetries allowed by the algorithm.

Parameters 1–3 can be computed by formulas $n(C) = \sum_{q_i \geq 0}(q_i + 1)$, $m(C) = |\{q_i = -1, 0 \leq i \leq j\}|$, and OCCUPIED$(C) = j + 1 - m(C)$, respectively. The distance between single robots and multiplicities is easily computed by summing the intervals between a single robot and a multiplicity. The symmetry of a configuration is computed by checking whether $C = \overline{C}_i$ for some $0 \leq i \leq j$.

To understand when C is at one move from a symmetry allowed by the algorithm, it is sufficient to simulate such a move backwards and checking whether the obtained configuration is symmetric.

Based on the perceived configuration, and once calculated the above parameters, a robot is able to answer to basic questions that check the accomplishment of the gathering task. In particular, a robot can distinguish if the current configuration is gatherable, which type of configuration it perceived, which strategy/move should be applied, if it is allowed to move and towards which direction.

The algorithm solves all the gatherable cases, hence closing also the ones left open by other strategies. However, different assumptions on the model may constitute very interesting directions for further investigations.

13.2.3 Local-Weak Multiplicity Detection

Using the local-weak multiplicity detection capability, not all the cases has been addressed so far. In [17], it has been proposed an algorithm for the case of rigid initial configurations where the number of robots k is strictly smaller than $\lfloor \frac{n}{2} \rfloor$. In [18], the case where k is odd and strictly smaller than $n - 3$ has been solved.

In [19], the authors provide and algorithm for the case where n is odd, k is even, and $10 \leq k \leq n-5$. The remaining cases are still open and a unified algorithm like that for the case where the global-weak multiplicity detection capability is allowed is still not known. In the following, the mentioned algorithms are summarized.

13.2.3.1 Asymmetric Configurations with $k < \lfloor \frac{n}{2} \rfloor$

This algorithm assumes, without loss of generality, that a configuration view seen by a robot is the lexicographically maximal that the robot can read, instead of the lexicographically minimal as it was in the case of global-weak multiplicity detection. These two assumptions are equivalent thus in the rest of the chapter we keep on using the one in [17]. As the configuration is asymmetric, by Property 2, the views seen by the robots are all different. Therefore, let $C = (q_0, q_1, \ldots, q_j)$ be the lexicographically maximal configuration view, $j \leq k$, and r_i be the robot (or the set of robots in the case of a multiplicity) before the interval q_i of empty nodes. First, an algorithm to achieve the gathering for the case where $q_0 \geq 3$ and $q_1 \geq 2$ is given. Then, a strategy to create a configuration of the above type starting from a configuration where $q_0 \geq 3$ and $q_1 < 2$ is devised. Finally, the case to increase q_0 from 2 to 3 is addressed. As it is assumed that $k < \lfloor \frac{n}{2} \rfloor$ and q_0 is the maximal interval, then q_0 cannot be smaller than 2. All the three algorithms keep the configuration asymmetric and aperiodic. Here, the algorithm for the case where $q_0 \geq 3$ and $q_1 \geq 2$ is described, while the details for the other cases can be found in [17].

The idea is to generate a configuration with only two occupied nodes where $k-1$ robots are gathered on the same node and the other occupied node contains a single robot. From this configuration the robots can distinguish which is the node occupied by a single robot by using the local-weak multiplicity detection. Therefore, the single robot moves towards the multiplicity, eventually achieving the gathering.

The algorithm when at least three nodes are occupied ($j \geq 2$) is as follows.

R1: If $q_j \geq 1$ move r_0 towards q_j;
R2: If $j \neq 2, q_j = 0$, and $q_{j-1} \geq 1$

    R2-1: If q_0 is the only maximum interval of empty nodes, move r_j towards q_{j-1};
    R2-2: Otherwise move r_1 towards q_1;

R3: If $j \neq 2, q_j = 0$, and $q_{j-1} = 0$ move r_0 towards q_j;
R4: If $j = 2$ and $q_2 = 0$ move r_0 towards q_2;

First, it is assumed that q_0 is the only maximum interval of empty nodes. The algorithm allows moves where q_0 is increased, q_1 is not changed, q_j is kept shorter than q_1, and the other intervals are decreased. This ensures that q_0 remains the only maximum interval of empty nodes. The algorithm starts by moving the robots r_0 towards q_j until they become neighbors of robots in r_j (see rules **R1** and **R3**). Then robots in r_j move towards q_{j-1} until they become neighbors of robots in r_{j-1} (see rule **R2-1**). At this point the robots in r_0 join those in r_j. By applying these rules,

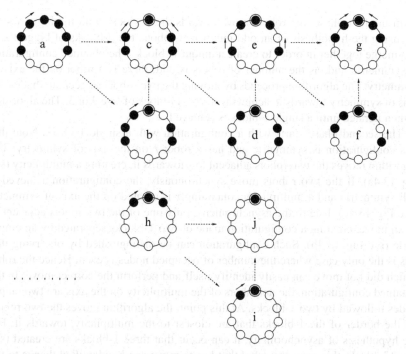

Fig. 13.4 Second phase of the gathering algorithm for an odd number of robots with local-weak multiplicity detection capability. A multiplicity is denotes as a *circle* around an occupied node

eventually a configuration with only three nodes occupied is achieved where $j = 2$ and $q_2 = 0$. In this case, all the robots in r_0 join those in r_2 (see rule **R4**) achieving a configuration where $k - 1$ robots are gathered on the same node. Finally, the single robot joins the other ones. In the case that q_0 is not the only maximum interval of empty nodes, the algorithm moves r_1 towards q_1, enlarging q_0 (see rule **R2-2**). After this moment, q_0 is the only maximum interval of empty nodes. It can occur that q_1 becomes smaller than 2 or the maximal configuration view is reversed. For instance this can happen when $q_1 = q_j$. In the first case, the algorithm for $q_0 \geq 3$ and $q_1 < 2$ is applied and in the second case the algorithm is applied on the new maximal configuration view.

13.2.3.2 Configurations with an Odd Number of Robots

The description of the algorithm requires some definitions and terminology. Let d be the size of the minimum interval of empty nodes plus one, a *d-block* is a maximal path where there is exactly one occupied node every d edges. The size of a d-block is the number of robots that it contains.

The algorithm works in two phases. The first phase builds a configuration made of a single 1-block, and the second phase achieves gathering.

In the first phase, the robots move towards the d-blocks with the biggest size. By using the hypothesis of an odd number of robots, the d-blocks of biggest size can merge together in order to create a unique d-block. The obtained configuration is symmetric and, as the number of robots is odd, there is a robot on the axis of symmetry. The algorithm proceeds by moving the two robots adjacent to that on the axis of symmetry towards it, achieving a $(d-1)$-block of size 2 or 3. The algorithm is then iterated until a single 1-block is achieved.

The second phase starts with a configuration with a single 1-block. Note that this configuration is symmetric and has a robot r on the axis of symmetry. The algorithm moves the two robots adjacent to r towards it, creating a multiplicity (see Fig. 13.4a). If the two robots move synchronously, the configuration achieved is still symmetric and a multiplicity containing r is created on the axis of symmetry (see Fig. 13.4c). Due to the asynchronicity, only one of the two robots adjacent to r can move, creating a configuration made of two 1-blocks separated by an empty node (see Fig. 13.4b). Such configuration can be distinguished by observing that this is the only case where the number of occupied nodes is even. Hence the robot which did not move can easily identify itself and perform the correct move. In the obtained configuration, the neighbors of the multiplicity on the axis are two empty nodes followed by two 1-blocks. At this point, the algorithm moves the two robots on the border of the 1-blocks that are closest to the multiplicity, towards it. For the hypothesis of asynchornicity, it can occur that three 1-blocks are created (see Fig. 13.4d). In this case, the robot that has to move can be identified thanks to the hypothesis that the number of occupied nodes is always smaller that $n-3$. In fact, in these configurations, the 1-blocks are interleaved by two single empty nodes and by a path of empty nodes of size at least three. By iterating this process, a new 1-block

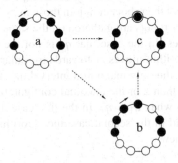

Fig. 13.5 Third phase of the gathering algorithm for an even number of robots in an odd ring with local-weak multiplicity detection capability. A multiplicity is denotes as a *circle* around an occupied node

is created with the size reduced by 2 with respect to the original 1-block and where there is a multiplicity on the axis of symmetry (see Fig. 13.4g). The algorithm is then iterated until the gathering is achieved. In the final step of the algorithm a single 1-block of size 2 can be created as a consequence of asynchronicity (see Fig. 13.4i).

At this point the algorithm exploits the local-weak multiplicity detection capability and moves the robot which is not on the multiplicity towards the other occupied node.

13.2.3.3 Configurations with an Even Number of Robots on an Odd Size Ring

In this case, the algorithm is divided into three phases. The first two phases aim at creating a *terminal* configuration, i.e., a configuration made of only two 1-blocks of size $\frac{k}{2}$ which are separated by exactly one empty node. Finally, the third phase finalizes the gathering.

The first phase starts from any allowed configuration and creates a configuration with either a single 1-block or two 1-blocks of size $\frac{k}{2}$. The idea is similar to that of the case of odd robots. First, only d-blocks are created by moving all the isolated blocks towards the d-blocks until joining them. Then, the robots move with the aim of creating a unique d-block. When all the robots belong to the same d-block, some robots move in order to decrease the d and repeat the algorithm until $d = 1$. The correctness of the algorithm relies on the fact that the number of nodes in the ring is odd and the number of robots is even. This implies that if the configuration is symmetric, then the axis of symmetry passes through exactly one empty node and one edge. In the second phase, the algorithm moves the two 1-blocks towards the empty node crossed by the axis of symmetry until a terminal configuration is created. In the case that the first phase ends with a single 1-block, this is split into two 1-blocks. All these movements are done by preserving the symmetry of the configuration. The third phase achieves the gathering from terminal configuration by moving the two robots that are on the border of the two 1-blocks and that are neighbors of a single empty node. These two robots move towards the single empty node. For an example, see Fig. 13.5. When these two robots are moved one of the following cases (depending on the activation schedule) can occur:

- The two robots move synchronously and create a symmetric configuration with a multiplicity crossed by the axis of symmetry (see Fig. 13.5c);
- Only one robot moves and creates a configuration with two 1-blocks separated by a single empty node and whose sizes differ by 2 (see Fig. 13.5b). This configuration is easy to recognize and hence the pending move can be performed, achieving again a symmetric configuration with a multiplicity crossed by the axis of symmetry (see Fig. 13.5c).

At this point, the robots on the multiplicity are not allowed to move. The other robots see an odd number of robots and perform the last phase of the algorithm for an odd number of robots (see Fig. 13.4c–j), so that the gathering is eventually achieved. This implies that the maximum number k of allowed robots has to satisfy $k - 1 < n - 3$, that is $k \leq n - 5$.

13.3 Gathering on Grids

In this section, results achieved in [10] are reported. The authors consider the gathering problem on an anonymous and undirected grid of $n \times m$ nodes, with $m \geq n$. The main assumption that distinguish these results from those obtained on rings is the lack of any multiplicity detection capability: if a node is occupied by more than one robot, it is not perceived by the robots, even if they reside on such a node.

Initially, each node is occupied by at most one robot. During a Look operation, a robot perceives the relative locations on the grid of occupied nodes, regardless of the number of robots at a node.

The current configuration of the system can be described in terms of the view of a robot r which is performing the Look operation at the current moment. A configuration seen by r is denoted as an $n \times m$ matrix M that has elements belonging ot the set $\{0, 1\}$. Value 0 represents an empty node, and 1 represents an occupied node.

Since the grid is anonymous and undirected, each robot can perceive the current configuration with respect to different rotations and reflections leading to any view of the grid satisfying the $n \times m$ dimension. In particular, when $n = m$ each of the four rotations and four reflections provides a feasible view.

Definition 2. A configuration is *periodic* if it is invariant with respect to rotations of 90° or 180°, where the rotation point coincides with the geometric center of the grid.

Definition 3. A configuration is *symmetric* if it is invariant after a reflection with respect to a vertical, horizontal, or diagonal (in case of square grids) axis passing through the geometric center of the grid.

13.3.1 Odd × Odd Grids

This case is trivially solvable, in fact in odd × odd grids, a robot can always detect, during its Look operation, the central node of the grid $M\left[\left\lceil \frac{n}{2} \right\rceil, \left\lceil \frac{m}{2} \right\rceil\right]$, regardless of its possible view. This means that all the robots can move toward the center, concurrently.

13.3.2 Odd × Even Grids

In this case, the gathering is not always feasible. In fact, similarly to the ring case on periodic or symmetric configurations of type edge-edge [22], if a configuration C is periodic, or symmetric with respect to an axis passing through the edges (i.e., dividing the grid into two halves from the even side), then C is ungatherable.

When the starting configuration does not belong to the above ungatherable configurations, it always possible to devise an algorithm achieving the gathering without multiplicity detection.

The idea is to distinguish between the two nodes that are the central nodes of the odd borders of the grid. If m (n, respectively) is odd, then the two mentioned nodes are given by positions $M[1, \lceil \frac{m}{2} \rceil]$ and $M[n, \lceil \frac{m}{2} \rceil]$ ($M[\lceil \frac{n}{2} \rceil, 1]$ and $M[\lceil \frac{n}{2} \rceil, m]$, respectively). The line connecting those two nodes will be denoted as the NS line. One of the two extreme nodes on the NS line will be the place where the gathering is finalized. In order to select the gathering node, a robot considers the line passing through the central edges of the even sides of the grid (denoted as the EW line) dividing the grid into two halves. The idea is to distinguish a north and a south part among the two halves and the gathering node will be the one in the north half. See Fig. 13.6a for a visualization. The north is the half with more nodes occupied by robots, if any. If the number of occupied nodes in the two halves is the same, then some more computations are required. In both cases, the robots move from the south to the north until all the robots will be in the north part. Note that, during such a stage, if multiplicities are created in the south, then the number of occupied nodes decreases with respect to the north part. If multiplicities are created in the north, it means that a robot has moved from the south to the north part, still preserving the required distinction.

In order to distinguish the north from the south in the case of configurations with the same number of robots among the two halves obtained by the EW line, a robot associates to each configuration C a binary string as follows. Starting from each corner of the grid, and proceeding in the direction parallel to the NS line, a robot records the elements of M row by row, or column by column (according to the direction specified by the NS line). Once it has computed the four strings, it associates to C the lexicographically largest one. For instance, starting from corner $M[1,1]$, and assuming m odd, the corresponding binary string would be composed by the sequence $M[1,1]$, $M[2,1]$, ..., $M[n,1]$, $M[1,2]$, ..., $M[n,2]$, $M[1,m]$, ..., $M[n,m]$. See Fig. 13.6a for an example.

It is possible to show that if C is a gatherable configuration, then, among the four possible strings coming from a robot view of the input grid, at most two strings can be the lexicographically largest ones. If there are two largest strings, then they represent the views of C starting from two symmetric corners with respect to the NS line, see Fig. 13.6d. Note that, instances of Fig. 13.6b, c are ungatherable as they admit an edge-edge symmetry or a periodicity.

Then, the *gathering node* is defined as the one residing on the same odd side where the corner(s) providing the lexicographically largest string resides. The gathering node will determine also the directions along the NS line: the gathering node is called the north pole.

Configurations on odd × even grids that are aperiodic and do not admit an axis of symmetry passing through edges are always gatherable, and the algorithm is the following.

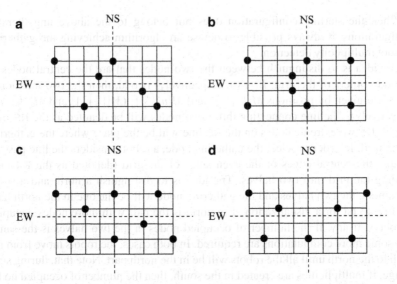

Fig. 13.6 Case of a 6×9 grid. (**a**) The two halves have the same number of nodes. The maximal binary string is that read from the *upper left* corner which starts with $(0, 1, 0, 0, 0, 0, \ldots)$. (**b**)–(**d**) Examples of 6×9 grids with two lexicographically largest strings: (**b**) The two lexicographically largest strings correspond to the views starting from two symmetric corners with respect to the EW line; (**c**) The two lexicographically largest strings correspond to the views starting from two corners residing on one of the two diagonals of the grid; (**d**) The two lexicographically largest strings correspond to the views starting from two symmetric corners with respect to the NS line

Once the gathering node has been unambiguously identified by a robot during its Compute operation, if the robot resides on the half grid where the south pole is, then it moves towards the north pole. Note that, each time a robot in the southern half of the grid performs such a movement, the gathering node cannot change. In fact, two following two cases can occur: (1) the number of occupied nodes decreases in the southern part of the grid, either because a robot moves to the northern part or because a multiplicity is created; (2) the string associated to the corners in the south are decreasing due to the robots' movements. In this case, the corresponding strings defining the current configuration starting from the northern corners are increasing. This clearly leaves unchanged the direction on the NS line. Note that the corner to which the lexicographically largest string was associated might change during the described process, but the only option is the other corner on the same odd side of the original one. This preserves the direction on the NS line. By keeping on moving in the described way, all the robots will reach the northern part. The case in which a subset of robots from a multiplicity move, increasing the number of occupied nodes, does not require any special argument. More precisely, since the initial configuration does not contain multiplicities, either the minimality of the number of robots in one half of the grid is preserved, or case (2) ensures that the lexicographically largest string is associated to a corner in the north.

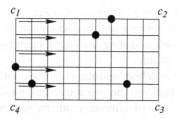

Fig. 13.7 Case of a 6×10 grid. The *arrows* indicate the horizontal direction of the reading from corner c_1, it gives $(6, 8, 14, 10, 5, 12)$. The other seven sequences read by the robots are: $(3, 6, 20, 4, 9, 13)$ from c_1 vertically, $(3, 10, 24, 2, 5, 11)$ and $(16, 1, 6, 26, 4, 2)$ from c_2 horizontally and vertically, respectively, $(12, 5, 10, 14, 8, 6)$ and $(13, 9, 4, 20, 6, 3)$ from c_3, $(11, 5, 2, 24, 10, 3)$ and $(2, 4, 26, 6, 1, 16)$ from c_4. The *minimal* sequence is $(2, 4, 26, 6, 1, 16)$ and $c = c_4$

Once all the robots belong to one half of the grid, then they are allowed to move, during their Move operation, towards the gathering node. In fact, such a node is well-defined and cannot change as the robots are not allowed to move to the other half of the grid.

13.3.3 Even × Even Grids

In this section, the case of grids whose sides are both even is studied. Also in this case, there are some configurations which are ungatherable, namely the periodic configurations and those configurations having a vertical or a horizontal axis of symmetry. In [10], it is shown that all the other cases are gatherable without any multiplicity detection, but for the case of 2×2 grids.

On 2×2 grids, configurations with two or four nodes occupied are ungatherable due to periodicity and edge-edge symmetries. If three nodes are occupied with robots having the local-weak multiplicity detection, the configuration is gatherable by moving the robot in between the other two occupied nodes arbitrarily, and then moving the robot not in the multiplicity towards the other occupied node. Hence, the remaining gatherable configurations are the aperiodic, asymmetric, and those with only one axis of symmetry passing through the diagonal of a square grid of dimensions larger than 2×2. All such configurations are referred to as the set EG (Even-Gatherable) and it is proved that all the configurations in EG are indeed gatherable without any multiplicity detection. In order to achieve this results, it is first assumed that at least one node on the border of the grid is occupied. Then, the gathering node is identified among the eight sequences of distances (number of empty nodes) between occupied nodes obtained by traversing the grid starting from the four corners and proceeding towards the two possible directions (see, e.g. Fig. 13.7).

The lexicographically smallest sequence between the two readings from any corner is associated to the corner itself. In rectangular grids, these two sequences can be equal but it is possible to distinguish one of them by assuming the reading in the direction of the smallest side.

The *minimal* sequence is defined as follows. If the configuration is symmetric, it is the smallest sequence between the two sequences associated to the two corners through which passes the axis of symmetry, otherwise it is the smallest among the four sequences associated to the four corners. In any case there exists a minimal sequence $C = (q_0, q_1, \ldots, q_j)$ which identifies a single corner c, unambiguously.

An important property of the gathering strategy is that a robot is not allowed to move to a corner different from c.

First it is proven that for any EG configuration with no corners occupied and at least one robot on the border there exists a strategy that leads to a configuration with exactly one corner occupied. The idea is to reduce q_0 by moving the robot towards c (or the two robots, when the configuration is symmetric) on the border which is (are) closest to c. The authors show that no symmetric configuration, other than the possible original one, can be created.

Then, it is shown that for any configuration in EG with more than three nodes occupied and at least two corners occupied there exists a strategy that leads to a configuration with either exactly one corner occupied or exactly three corners occupied.

Finally the main contribution is proven: Aperiodic configurations on even × even grids larger than 2×2, that do not admit an axis of symmetry passing through edges, are gatherable, without assuming any multiplicity detection.

To achieve this result, it has been first observed that the set of possible grids can be restricted by considering the minimal even × even sub-grid which is centered in the geometrical middle of the original grid and includes all the occupied nodes of it. Such minimal *wrapping* grid is still of type even × even and preserves the possible symmetry of the original one. Moreover, it always has at least an occupied node on the border. Then it is possible to apply the first partial result mentioned above.

The proposed algorithm only uses such sub-grid without changing its size, i.e., it neither enlarges nor reduces the sub-grid by moving robots outside the border or from the border to the inside.

If no corners are occupied, by reducing q_0, a configuration with one corner occupied can be reached. In this case, all the robots move towards c by reducing the Manhattan distance to c and then achieving the gathering.

When two corners are occupied, as said above, it is possible to reach a configuration with one or three corners occupied. In the former case the gathering can be easily finalized, in the latter case all the robots, but those in the corners, are moved towards the corner that does not share any coordinate with the empty corner. This process finishes with a symmetric configuration with exactly three corners occupied. In this configuration, c is the corner on the axis of symmetry, and the other two robots move one step towards c either concurrently or alternately, until creating a configuration with only one corner occupied.

If four corners are occupied, the robot which occupies the corner farthest from c is moved in an arbitrary direction, generating a configuration where only three corners are occupied.

It remains the case where the minimal wrapping even × even sub-grid which includes all the occupied nodes of the original grid has dimension 2×2. The configuration is ungatherable on this sub-grid without multiplicity detection. However, in the case of exactly three nodes occupied, it is possible to exploit the larger dimensions of the original grid in order to avoid the multiplicity detection. The cases of two or four nodes occupied clearly remain ungatherable. The strategy is then to move the robot on the corner of the 2×2 grid which is in between the other two occupied corners towards the external row or column, arbitrarily. The case where the minimal wrapping grid has dimension 4×4 is obtained and no corners are occupied.

13.4 Gathering on Trees

In this section, gathering results on trees are presented. To the best of our knowledge, these are original contributions as trees were never treated before under the considered model. Given a tree, a node at minimal distance from all the other ones is called *center*. Based on well-known results [28] about the tree topology, within a tree there is either one center or there are two neighboring centers. In the former case, no matter the initial distribution of the robots, each of them can move towards the center, concurrently. The gathering will be eventually finalized, even without any multiplicity detection assumption. In the latter case, some more specific arguments are required. In fact, some impossibility results hold.

Lemma 2. *If the two subtrees rooted at the centers along with the disposal of the robots are isomorphic, then the gathering is impossible.*

Proof. Any algorithm designed to accomplish the gathering on the tree must work regardless the delays on the decisions made by robots. In particular, also the synchronous case must be solved. Since the two considered subtrees are isomorphic, with the same disposal of robots, if one robot is allowed to move within one subtree, there must exist another robot that is allowed to accomplish the same specular movement. If both robots perform such movements, again a configuration with two isomorphic subtrees is obtained. In proceeding so, there will not exist any move that can break such a situation. Hence, the gathering cannot be finalized as it requires to distinguish one single node belonging to one of the two subtrees.

In the case the isomorphism among the two subtrees along with the disposal of the robots does not hold, the following strategy can be applied. Let c_1 and c_2 be the two centers of the input tree T, and let T_1 and T_2 be the two subtrees rooted at c_1 and c_2, respectively, when the edge connecting c_1 and c_2 is removed. If the number of nodes occupied in T_1 is smaller than that in T_2, then all robots in T_1 are moved towards c_2. Once T_1 gets empty, all robots in T_2 should be moved towards

c_2 in order to end the gathering. If the number of nodes occupied in T_1 is equal to that in T_2, it is always possible to determine which subtree is less than the other with respect to a natural ordering on labeled trees (see [1, 7]). To define the smaller tree as the one with the robots closer to the root, we associate label 1 to empty nodes, and label 0 to nodes occupied by robots. Then the algorithm would exploit this ordering in order to detect the robots to move from one subtree towards the root of the other one. If a robot moves over a node already occupied, the number of occupied nodes in the original subtree decreases. As soon as one robot moves towards the other subtree, the number of robots in the two subtrees is no longer equal and the previous strategy can be applied. Similarly to what happened in the case of even × odd grids of Sect. 13.3.2, the occurrence of multiplicities does not affect the proposed algorithm.

13.5 Conclusion

In this chapter, we surveyed recent results about the gathering problem under the Look-Compute-Move model in various graph topologies.

For most of the cases under investigation, it turned out that the problem has been fully characterized. For trees and grids, the multiplicity detection capability does not strengthen the model, that is, all the cases which admit gathering are still solvable without such a capability. The only exception is provided by the very specific case of three robots on a 2 × 2 grid. However, the multiplicity detection capability can strengthen the model in the case of ring topologies. In the literature, the case that assumes global-weak multiplicity detection has been fully characterized while that assuming local-weak multiplicity detection still lacks of a unified algorithm, and there are some open cases that deserve further investigations.

The study of different topologies has required very different and sometimes opposite approaches that stimulate main advances in robot-based computing systems. On the other hand, some of the techniques described for different topologies share common ideas that can be, therefore, used in other topologies. Infinite grids, tori, and hypercubes might represent just a sampling set.

Another challenging direction would be that of investigating the minimum number of steps required by the robots to accomplish the gathering task. So far, the research has mainly focused on the feasibility of the gathering, while few results concern the minimization of the robots' movements. Similarly, low effort has been spent in order to increase the opportunity to parallelize movements. As we have seen for ring networks, at most two robots are allowed to move concurrently unless robots composing multiplicities must move. Whereas, on grids and trees, less restrictions are imposed on the robots' movements.

It would be interesting to investigate how the proposed techniques may affect the resolution of different tasks, as well as how different assumptions on the capability of the robots may change the required strategies.

References

1. Aho, A., Hopcroft, J., Ullman, J.: Data Structures and Algorithms. Addison Wesley (1983)
2. Alpern, S.: Rendezvous search: A personal perspective. Operations Research **50**(5), 772–795 (2002)
3. Alpern, S., Gal, S.: The Theory of Search Games and Rendezvous, *International Series in Operations Research & Management*, vol. 55. Springer (2003)
4. Baldoni, R., Bonnet, F., Milani, A., Raynal, M.: On the solvability of anonymous partial grids exploration by mobile robots. In: Proceedings of the 12th International Conference On Principles Of Distributed Systems (OPODIS), *Lecture Notes in Computer Science*, vol. 5401, pp. 428–445. Springer-Verlag (2008)
5. Blin, L., Milani, A., Potop-Butucaru, M., Tixeuil, S.: Exclusive perpetual ring exploration without chirality. In: Proceedings of the 24th International Symposium on Distributed Computing (DISC), *Lecture Notes in Computer Science*, vol. 6343, pp. 312–327. Springer (2010)
6. Bonnet, F., Milani, A., Potop-Butucaru, M., Tixeuil, S.: Asynchronous exclusive perpetual grid exploration without sense of direction. In: Proceedings of the 15th International Conference On Principles Of Distributed Systems (OPODIS), *Lecture Notes in Computer Science*, vol. 7109, pp. 251–265. Springer (2011)
7. Buss, S.: Alogtime algorithms for tree isomorphism, comparison, and canonization. In: Kurt Gödel Colloquium, *Lecture Notes in Computer Science*, vol. 1289, pp. 18–33. Springer (1997)
8. Czyzowicz, J., Gasieniec, L., Pelc, A.: Gathering few fat mobile robots in the plane. Theoretical Computer Science **410**(6–7), 481–499 (2009)
9. D'Angelo, G., Di Stefano, G., Navarra, A.: Gathering of six robots on anonymous symmetric rings. In: Proceedings of the 18th International Colloquium on Structural Information and Communication Complexity (SIROCCO), *Lecture Notes in Computer Science*, vol. 6796, pp. 174–185 (2011)
10. D'Angelo, G., Di Stefano, G., Klasing R., Navarra A.: Gathering of robots on anonymous grids without multiplicity detection. In: Proceedings of the 19th International Colloquium on Structural Information and Communication Complexity (SIROCCO), *Lecture Notes in Computer Science*, vol. 7355, pp. 327–338 (2012)
11. D'Angelo, G., Di Stefano, G., Navarra, A.: How to gather asynchronous oblivious robots on anonymous rings. In: Proceedings of the 26th International Symposium on Distributed Computing (DISC), *Lecture Notes in Computer Science*, vol. 7611, pp. 330–344 (2012)
12. Devismes, S., Petit, F., Tixeuil, S.: Optimal probabilistic ring exploration by semi-synchronous oblivious robots. In: Proceedings of the 16th International Colloquium on Structural Information and Communication Complexity (SIROCCO), *Lecture Notes in Computer Science*, vol. 5869, pp. 195–208 (2009)
13. Flocchini, P., Ilcinkas, D., Pelc, A., Santoro, N.: Computing without communicating: Ring exploration by asynchronous oblivious robots. Algorithmica. To appear.
14. Flocchini, P., Ilcinkas, D., Pelc, A., Santoro, N.: Remembering without memory: Tree exploration by asynchronous oblivious robots. Theoretical Computer Science **411**(14–15), 1583–1598 (2010)
15. Flocchini, P., Prencipe, G., Santoro, N., Widmayer, P.: Gathering of asynchronous robots with limited visibility. Theoretical Computer Science **337**, 147–168 (2005)
16. Haba, K., Izumi, T., Katayama, Y., Inuzuka, N., Wada, K.: On gathering problem in a ring for 2n autonomous mobile robots. In: Proceedings of the 10th International Symposium on Stabilization, Safety, and Security of Distributed Systems (SSS), poster (2008)
17. Izumi, T., Izumi, T., Kamei, S., Ooshita, F.: Mobile robots gathering algorithm with local weak multiplicity in rings. In: Proceedings of the 17th International Colloquium on Structural Information and Communication Complexity (SIROCCO), *Lecture Notes in Computer Science*, vol. 6058, pp. 101–113 (2010)
18. Kamei, S., Lamani, A., Ooshita, F., Tixeuil, S.: Asynchronous mobile robot gathering from symmetric configurations. In: Proceedings of the 18th International Colloquium on Structural Information and Communication Complexity (SIROCCO), *Lecture Notes in Computer Science*, vol. 6796, pp. 150–161 (2011)

19. Kamei, S., Lamani, A., Ooshita, F., Tixeuil, S.: Asynchronous mobile robot gathering from symmetric configurations without global multiplicity detection. In: Proceedings of the 37th International Symposium on Mathematical Foundations of Computer Science (MFCS), vol. 7464, pp. 542–553. Springer-Verlag (2012)

20. Kempkes, B., Meyer auf der Heide, F.: Continuous local strategies for robotic formation problems. In: Proceedings of the 11th International Symposium on Experimental Algorithms (SEA), *Lecture Notes in Computer Science*, vol. 7276, pp. 9–17. Springer-Verlag (2012)

21. Klasing, R., Kosowski, A., Navarra, A.: Taking advantage of symmetries: Gathering of many asynchronous oblivious robots on a ring. Theoretical Computer Science **411**, 3235–3246 (2010)

22. Klasing, R., Markou, E., Pelc, A.: Gathering asynchronous oblivious mobile robots in a ring. Theoretical Computer Science **390**, 27–39 (2008)

23. Koren, M.: Gathering small number of mobile asynchronous robots on ring. Zeszyty Naukowe Wydzialu ETI Politechniki Gdanskiej. Technologie Informacyjne **18**, 325–331 (2010)

24. Kranakis, E., Krizanc, D., Markou, E.: The Mobile Agent Rendezvous Problem in the Ring. Morgan & Claypool (2010)

25. Kranakis, E., Krizanc, D., Rajsbaum, S.: Mobile agent rendezvous: a survey. In: Proceedings of the 13th International Colloquium on Structural Information and Communication Complexity (SIROCCO), *Lecture Notes in Computer Science*, vol. 4056, pp. 1–9. Springer-Verlag (2006)

26. Pelc, A.: Deterministic rendezvous in networks: A comprehensive survey. Networks **59**(3), 331–347 (2012)

27. Prencipe, G.: Impossibility of gathering by a set of autonomous mobile robots. Theoretical Computer Science **384**, 222–231 (2007)

28. Santoro, N.: Design and Analysis of Distributed Algorithms. John Wiley & Sons (2007)

29. Suzuki, I., Yamashita, M.: Distributed anonymous mobile robots: Formation of geometric patterns. SIAM J. Comput. **28**(4), 1347–1363 (1999)

Chapter 14
Ten Open Problems in Rendezvous Search

Steve Alpern

Abstract The rendezvous search problem asks how two (or more) agents who are lost in a common region can optimize the process by which they meet. Usually they have restricted speed (unit speed in the continuous time context; moves allowed to adjacent nodes in discrete time). In all cases the agents are not aware of each other's location. This chapter is concerned with the 'operations research' version of the problem – where optimization of the search process is interpreted as minimizing the expected time to meet, or possibly maximizing the probability of meeting within a given time. The deterministic approaches taken by the theoretical computer science community will not be considered here.

14.1 Introduction

A precursor of the rendezvous problem is the work of Schelling [32] on the coordination of meeting places (focal points). However these one-shot games lack the main ingredient of search, namely that the process continues after each failure to meet until (hopefully) the rendezvous is achieved. The rendezvous search problem was first proposed by the author at the end of a talk on search games given in Vienna in 1976 [1].

We shall be mainly concerned with the so called *symmetric*, or *player–symmetric* form of the problem, where both players (agents) must adopt the same rendezvous strategy, though when using mixed strategies they must randomize independently. This version is said to have *indistinguishable agents*. For example if two players, Tom and Mary, were trying to rendezvous on a circle drawn on the plane, we would not allow the solution where Tom proceeds clockwise and Mary counterclockwise. The symmetric solution could be written in a book on rendezvous so that

S. Alpern (✉)
ORMS Group, Warwick Business School, University of Warwick, Coventry CV4 7AL, UK
e-mail: steve.alpern@wbs.ac.uk

S. Alpern et al. (eds.), *Search Theory: A Game Theoretic Perspective*,
DOI 10.1007/978-1-4614-6825-7_14, © Springer Science+Business Media New York 2013

when two people find themselves faced with the rendezvous problem they could read what to do in this book, without having decided beforehand on which roles (e.g. clockwise or counter-clockwise) to play. The asymmetric version allows Tom and Mary to agree beforehand on which roles to take: for example one could wait while the other searches (the *Wait for Mommy* strategy).

14.2 The Problems

We now list some rendezvous problems which are unsolved. When partial solutions, or solutions to special cases have been found, we will mention this afterwards.

1. **The astronaut problem.** Two unit speed astronauts land simultaneously on a small smooth sphere. They walk at the same speed and can detect each other when they are a given distance apart. What is the least expected meeting time they can guarantee, and what strategies should they use to achieve this time?

As far as we are aware, no significant progress has been made on this problem. An easier version would spin the sphere, so that the two poles would be focal points. Even the latter problem seems to be open, though some sort of randomized oscillation between the poles would seem to be useful in the latter problem. But how long should one wait before heading for the other pole? A lower dimension analog of this problem is given next.

2. **Rendezvous on a circle.** Two players are uniformly and independently placed on a circle, without any common sense of direction (or of *up*, if the circle is drawn on a plane). They move at unit speed and must use the same mixed strategy. Determine the least expected meeting time.

It has been conjectured by the author that the solution of this problem is for both players to use the so called cohato (coin half tour) strategy: oscillate between your starting point and its antipode, each time choosing equiprobably and independently of prior choices between your clockwise and your counter-clockwise (or simply choose two directions and do one if Heads and the other if Tails on the coin). Simpler versions of this problem, where the players have some additional information or common notions, have been solved by Howard [22] and Alpern [4, 6].

3. **Rendezvous on the infinite line (agent-symmetric form).** In this version
   of the rendezvous problem, the two players are randomly placed on a common line and try to find each other. Even for the apparently simpler version
   where the initial distance (taken for simplicity as 2) is known, the problem
   is open.

It has been shown by Lim, Alpern and Beck [27] and Qiaoming, Du, Vera and
Zuluaga [31] that the players can optimally restrict their mixing to strategies
which move at unit times along the integer lattice determined by their starting
point. So anyone trying to solve this problem can restrict their search to these
simple strategies.

The author has recently established by a compactness argument that the least
expected time and optimal strategies exist (not just ε−optimal ones). The author initially [2, 3] suggested the strategy where in each time period of length
3 the players independently chose a forward direction and then move forward,
back, back. This strategy is easily seen to have expected meeting time 5/2. It
has been bettered by strategies using longer sequences of moves, by Anderson
and Essegaier [15], Baston [17], Uthaisombut [34], Alpern [7] and Han et al.
[31]. Currently the best estimates combine to give $2.091 \leq R^s \leq 2.1287$.

4. **Rendezvous on the infinite line (agent-asymmetric form).** In this version
   of the problem, the cumulative distribution function $F(d)$ of the initial distance d between the players is given. In general, this is an easier problem,
   though a solution for general F is not known.

In the case where the initial distance d is given, the solution was found by
Alpern and Gal [11], and the least expected meeting time is $13d/8$. Further
progress was made when Alpern and Howard [13] showed that the problem
was equivalent to a single agent problem where a single searcher seeks to find
an object hidden at one of two possible locations, where each location must
be searched in a given order and alternation between locations is costless (the
alternating search problem). Using this approach Alpern and Beck [10] were
able to solve the problem for the case where d has a convex distribution on an
interval $[0, D]$. In these cases the solution is a variation on the so called *Wait For
Mommy* strategy, where one player (Baby) waits while the other carries out an
optimal search for an immobile hider. In the variation, Mommy doesn't change
her strategy, but Baby moves to meet Mommy. These solutions don't apply to
the symmetric problem because they require coordination in the assignment of
roles (Baby, Mommy) to the players.

NOT BEING SPECIFIC ENOUGH IN
SALZBURG ... OR VIENNA ?

5. **Mozart Café problem.** Two friends agree to meet for lunch on the first day of the millennium year 2,000 at the Mozart Café in Vienna. When the time comes, each arrives at Vienna Airport and asks a cab to take them to the Mozart Café. Each is troubled to hear the answer 'There are four of them; Which one do you want?' On the first day the four cafes are indistinguishable so each can do no better than picking one at random. If they don't meet on the first day (1 January) then they can choose to return on day 2 to the same cafe, or to go to a random new one. And so on. What is the best strategy, assuming both players use the same one (with independent randomization)?

This is the first discrete time rendezvous problem. It remains open if there are at least $n = 4$ cafes. If there are only $n = 2$ cafes, Anderson and Weber [16] showed that the random strategy (pick randomly every day) is optimal. They proposed a general strategy $AW(n)$ for the case of n cafes: If you haven't met on day 1, do the following in every successive interval of $n - 1$ days; with probability p_n, search the other $n - 1$ cafes in random order; with probability $1 - p_{n-1}$, stay where you are for another $n - 1$ days. Weber [35] has recently used an elegant but elaborate argument to show that $AW(3)$ is indeed optimal for the 3 cafe problem, but that $AW(4)$ is not optimal for $n = 4$.

6. **Mozart Cafe problem with river.** Suppose there are $2n$ cafes, with n of them on each side of the Danube. The problem has changed, because after not meeting on day 1 for example, a player has three choices for the next day: the same cafe, a different one on the same side of the river, a random cafe on the other side of the river.

Nothing is known about this version of the problem. The players can ignore the additional information, so they should be able to meet in the same expected time as the regular $2n$ problem. But can they do better? More generally, we could have $n = mr$ cafes partitioned into m sets of r. (Unequal partitions could allow coordination on certain sets.)

7. **Multiple agent rendezvous.** The rendezvous problems can be modified so that the goal is for n players to all meet at the same location. For example, n astronauts could land independently with the same distribution on the sphere.

For these problems it could be assumed that players who meet must stick together (sticky) or not. One could also look at problems where there are n players but only k of them have to meet (say, to carry out some task). Some references to various versions of this problem (including some in the computer science literature which we have been otherwise excluding) are Alpern [6], Lin, Morse, and Anderson [28, 29], Baston [17], Lim, Alpern and Beck [27], Alpern and Lim [14, 26], Dessmark, Fraigniaud, Kowalski and Pelc [20], Marco, Gargano, Kranakis, Krizanc and Pelc [30], Kranakis, Krizanc and Rajsbaum [25],Kowalski and Malinoski [24], Dobrev, Flocchini, Prencipe and Santoro [21].

8. **Asynchronous rendezvous.** All of the above problems assume that the two (or more) players enter the search region, and begin their search, at the same time. This allows some coordination.

This problem has received some attention in unpublished work of V. Baston and A. Beck, and in the alternative optimization criteria of the theoretical computer science approach of Marco, Gargano, Kranakis, Krizanc and Pelc [30]. See also Lin, Morse and Anderson [29].

9. **Rendezvous without proximity.** The classical form of the rendezvous search problem assumes that the problem is solved (rendezvous achieved) when the two players meet spatially. That is, when they come within a specified distance or, in one dimensional scenarios, when they have the same location. However, other end conditions are possible. More generally we could posit a subset R of $Q \times Q$, where Q is the search region, where the game ends when the locations of the two players form a pair in R. The classical version constitutes the diagonal.

The only problem of this type that has been explored in the literature is by the author [8], who considered a rendezvous problem in Manhattan, where the two players rendezvous when they arrive at a common street or avenue (and thus can see each other without buildings coming between them). It would be useful to develop a general theory, or perhaps simply explore another example.

10. **When to give up: rendezvous with failure.** As stated in the Introduction,
what distinguishes rendezvous search from Schelling's coordination problems is that the search continues until coordination (meeting) is achieved.
However it is common that when two people agree to meet in a large area,
one or more may eventually give up, assuming the other didn't come to the
area. Similarly, searches for missing people are eventually terminated.

There are two ways we could incorporate giving up the search: we could keep
the classical formulations but change the cost function; or we could put in new
ingredients to the problem. For the former, we could postulate a cost function
$c(T) + f \cdot C$, where c is an increasing convex function of search time T and C
is a cost of failure that is incurred if the search is terminated at a time T without meeting (in this case $f = 1$; otherwise $f = 0$). Or we could have a value of
meeting which decays exponentially, while the cost is still the search time T.
The later version involves changing the ground rules of the rendezvous problem
itself. We could have a given probability that each player simply does not turn
up to the game. Or this could be implicit: for example the initial distribution of
the players could be uniform on the union of disjoint complete graphs of size 8
and 2. Presumably one would start off in the hope that the other is in the same
component and play optimally within your component. At some point (earlier
if you are in the small component), you have a sufficiently low updated probability that rendezvous is possible, and you would stop. If the players stop at
different times it is not clear how to allocate costs to search times.

14.3 Further Comments

For more results in the field of the operations research aspects of rendezvous, see
the monograph of the author and S. Gal [12] and the survey of the author [5]. For a
broader approach to search games and search problems, see the monograph of Stone
[33] and the survey of Benkoski, Monticino and Weisinger [18].

References

1. Alpern, S. Hide an seek games. Seminar, Institut fur Hohere Studien, Vienna, 26 July 1976.
2. Alpern, S. The rendezvous search problem. *LSE CDAM Research Report* 1993 **53,** London
   School of Economics.
3. Alpern, S. The rendezvous search problem. *SIAM J. Control & Optimization* 1995;
   33:673–683.
4. Alpern, S. Asymmetric rendezvous search on the circle. *Dynamics and Control* 2000;
   10:33–45

5. Alpern, S. Rendezvous search: a personal perspective. *Operations Research* 2002; **50** (5):772–795.
6. Alpern, S. Rendezvous search on labelled networks. *Naval Research Logistics* 2002; **49**:256–274.
7. Alpern, S. Rendezvous search with revealed information: applications to the line. *Journal of Applied Probability* 2007; 44 (1):1–15.
8. Alpern, S. Line-of-sight rendezvous. *European Journal of Operational Research* 2008; 188 (3): 865–883.
9. Alpern, S. Rendezvous Search Games, Wiley Encyclopedia of Operations Research and Management Science 2012, John Wiley & Sons, Inc.
10. Alpern, S, and Beck, A. Asymmetric rendezvous on the line is a double linear search problem. *Mathematics of Operations Research* 1999; **24** (3): 604–618.
11. Alpern, S and Gal, S. Rendezvous search on the line with distinguishable players. *SIAM J. Control & Optimization* 1995; **33**:1270–1276.
12. Alpern, S and Gal, S. Search Games and Rendezvous Theory, Springer, 2003.
13. Alpern, S and Howard, JV. Alternating search at two locations. *Dynamics and Control* 2000; **10**:319–339.
14. Alpern, S and Lim, WS. Rendezvous of three agents on the line. *Naval Research Logistics* 2002; **49**:244–255. (6):2233–2252.
15. Anderson, EJ and Essegaier, S. Rendezvous search on the line with indistinguishable players, *SIAM Journal of Control and Optimization* 1995; **33**:1637–1642.
16. Anderson, EJ. and Weber, R. The rendezvous problem on discrete locations. *Journal of Applied Probability* 1990; **27**:839–851.
17. Baston, VJ. Two rendezvous search problems on the line, *Naval Research Logistics* 1999; **46:335–340.**
18. Benkoski, S, Monticino, M and Weisinger, J. . A survey of the search theory literature. *Naval Research Logistics* 1991; **38**:469–494.
19. Chester, EJ and Tutuncu, RH. Rendezvous search on the labelled line *Operations Research* 2004; **52**(2):330–334. *Operations Research* 1999; **47** (6):849–861.
20. Dessmark, A, Fraigniaud, P, Kowalski, D and Pelc, A. Deterministic rendezvous in graphs. *Algorithmica* 2006; **46**:69–96.
21. Dobrev,S, Flocchini,P, Prencipe,G and Santoro, N. Multiple agents rendezvous in a ring in spite of a black hole. *Lecture Notes in Computer Science* 2004: Volume 3144/2004, 34–46
22. Howard, JV. Rendezvous search on the interval and circle. *Operations Research* 1999; **47** (4):550–558.
23. Kikuta, K and Ruckle, W. Rendezvous search on a star graph with examination costs. *European Journal of Operational Research* 2007; **181** (1):298–304.
24. Kowalski, D. and Malinoski, A. How to meet in anonymous network. *Theoretical Computer Science* 2008; **399**:141–156.
25. Kranakis, E, Krizanc, D and Rajsbaum, S. Mobile agent rendezvous: a survey. Lecture Notes in Computer Science 2006, Volume 4056/2006, Springer Berlin / Heidelberg.
26. Lim, WS and Alpern, S. Minimax rendezvous search on the line. *SIAM J. Control Optim.* 1996; **34**:1650–1665.
27. Lim, WS, Alpern, S and Beck, A. Rendezvous search on the line with more than two players. *Operations Research* 1997; 45 (3):357–364.
28. Lin, J Morse, AS and Anderson, BDO. The Multi-Agent Rendezvous Problem. Part 1: The synchronous Case. *SIAM J. Control Optim* 2007; **46**(6):2096–2119.
29. Lin, J, Morse, AS and Anderson, BDO. The Multi-Agent Rendezvous Problem. Part 2: The asynchronous Case. *SIAM J. Control Optim* 2007; **46**(6):2120–2147.
30. Marco,G, Gargano, Kranakis,E, Krizanc,D and Pelc, A. Asynchronous deterministic rendezvous in graphs. *Theoretical Computer Science* 2006; **355**:315–326.
31. Qiaoming H, Du, D, Vera,J and Zuluaga, LF. Improved Bounds for the Symmetric Rendezvous Value on the Line *Operations Research* 2008; 56 (3):772–782.
32. Schelling, T. *The Strategy of Conflict.* Harvard University Press, Cambridge, 1960.

33. Stone, LD. *Theory of Optimal Search*, 2nd edition. Operations Research Society of America, Arlington, VA, 1989.
34. Uthaisombut, P. Symmetric rendezvous search on the line using moving patterns with different lengths. Department of Computer Science, University of Pittsburgh, 2006.
35. Weber, R. Optimal Symmetric Rendezvous Search on Three Locations. *Mathematics of Operations Research* 2012; **37** (1): 111–122.

Part IV
Search in Biology

Chapter 15
Interactions Between Searching Predators and Hidden Prey

Mark Broom

Abstract Predator-prey interactions are among the most fundamental in nature. In this chapter we look at some existing models of the interaction between predators and prey, where prey and/or predators have important strategic choices to make. Firstly we consider situations where both predator and prey are aware of each other, and the predator is approaching the prey. When should the prey flee? Then we consider a moving predator which is unaware of a hidden prey individual. Should the prey stay where it is, or should it flee, and if so when? Finally we consider a new model of a stationary predator searching for a hidden prey (which may not be present), where both prey and predator have important decisions to make. Should the prey flee, and if so when? How long should the predator search before giving up, and moving elsewhere?

15.1 Introduction

Avoiding predation is a central component of the lives of the majority of animals. Depending upon the type of prey and predator, there are a number of potential methods that can be used by the prey to avoid predators. These include physical defences such as spines or toxins, and such individuals may be able to fight off or deter a predator attack. Often, as in the cases we shall consider, predators are overwhelmingly stronger than their prey, and there are two main defences, to run or to hide.

Consider a prey that has just discovered a predator. Sometimes the optimal strategy can be to flee immediately, for instance if a ground feeding bird is being

M. Broom (✉)
Department of Mathematical Science, City University London, Northampton Square, London, EC1V 0HB, UK
e-mail: mark.broom.1@city.ac.uk

S. Alpern et al. (eds.), *Search Theory: A Game Theoretic Perspective*,
DOI 10.1007/978-1-4614-6825-7_15, © Springer Science+Business Media New York 2013

attacked by a hawk [7]. However, there are also important situations where this is not the case. The classic case is on open plains where predators are very often present and where both prey and predators can be clearly visible to each other. To flee on all

PREY AVOIDING A PREDATOR

occasions would disrupt foraging to an unsustainable extent. Rather the prey must decide when the level of risk is sufficiently high to make fleeing the best strategy, and when to continue foraging [12]. A second scenario is when a prey individual has not yet been observed by the predator, for example an ungulate calf hiding in long grass. To run will alert the predator to its presence, so it may be best to remain still if there is sufficient chance that the predator will not discover the individual. Thus the key decision for the prey here again, is when to run and when not to [2]. These two important cases are related, and we shall look at existing models of both, although the central focus of this chapter is on the second case. In particular we shall then introduce a new model of an interesting variant of the second case where the predator also has key strategic decisions to make.

15.2 Interactions Between Visible Prey and Predators

Avoiding and fleeing from predators is one of the central activities of many animals. It has been postulated that this task is so important in comparison to foraging (the threat of death versus an incremental gain in resources) that when there is a predator threat this always takes precedence over foraging, see for example [9]. Thus when an approaching predator is detected by a prey individual this should lead to some immediate anti-predator response, such as flight. Ydenberg and Dill [14] suggested that this view was incorrect and that detection and response were often not so directly associated. They introduced an "economic" model, with benefits and costs associated with fleeing from predators or continuing to forage (with the potential for later flight).

15.2.1 The Model of Ydenberg and Dill

The argument of [14] is as follows. Suppose that a prey individual has seen a predator approaching it. Thus both are aware of each other's location. If the prey continues foraging the predator will continue its approach and an attack will occur. If the prey flees it will lose some foraging time. The earlier it flees, the more foraging time will be lost. The earlier the prey flees, the more chance it will have of escaping the predator. There is thus a benefit from fleeing B which decreases with predator distance, and a cost C which increases with predator distance. They contend that at every point the prey chooses to flee if the benefit of doing so outweights the cost, i.e. $B > C$. $B - C$ decreases with distance, and so increases with time (the distance decreases with time as the predator approaches), and so the prey should flee at the critical point when $B = C$.

It should also be noted that instead of just not fleeing, prey often give responses short of flight when detecting a predator. This includes lifted tails in squirrels, standing posture in hares, primate alarm calls and stotting by gazelles. Stotting is where the gazelle jumps high into the air when faced by a potential predator, and is believed to be a signal to the predator that the prey is healthy and so should not be chased, see [4]. These can be signals to the predator that it has been detected, and that it consequently has low success probability, and so should not waste its time and energy in an attack.

The model of [14] makes clear predictions which match real observations well, and consequently has been very influential. The key predictions of the model are as follows:

(i) Flight distance should increase with the risk of capture.

(ii) Flight distance should decrease with increased cost of fleeing.

(iii) Flight distance changes with the effectiveness of alternative defence tactics such as crypsis. If a prey is cryptic then perhaps an approaching predator has not seen it and the level of crypsis may affect flight distance (see the model of [2] described in Sect. 15.3).

(iv) Flight distance varies with the fitness benefit associated with group membership. Larger group size may reduce predation risk (but also possibly foraging efficiency) and so may reduce flight distance.

15.2.2 The Model of Cooper and Frederick

In the [14] model prey flee when the cost of fleeing precisely matches the benefit. The concept of benefits and costs here are rather abstract. The cost and benefit of fleeing immediately can only be evaluated if we know what happens if the prey does not flee immediately, which in turn depends upon when it does eventually flee. Similarly between two potential fleeing distances, the actions of the prey may affect subsequent costs and benefits. A more pertinent consideration is, what is the optimal time to flee for a prey individual to maximise its fitness?

Cooper and Frederick [3] modelled this by developing a model of the foraging scenario in [14] using an explicit fitness function. In their model again both prey and predator can see the other, and the fitness of an individual, if it survives, depends upon its resource level. It is assumed that the predator approaches the prey at a constant speed, so that there is a simple relationship between the time since the start of the encounter and the distance between prey and predator, and the resource level is given by

$$F_0 + B(d) - E(d), \tag{15.1}$$

where d is the distance of the predator from the prey, F_0 is the fitness at the start of the encounter (when $d = d_d$), $E(d)$ is the energetic cost of escaping at distance d, and $B(d)$ is the benefit of waiting from the start of the encounter until the prey has reached distance d, which is increasing with d (so $B(d_d) = 0$). We note that this benefit is achieved through extra foraging opportunity, and so more properly depends upon the time since the start of the encounter rather than the distance. In this instance, since there is a deterministic relationship between time and distance this is not problematic, but a more realistic model (e.g. with variable predator speed) would need to contain time as a separate factor.

When the prey flees is has probability of survival $P_s(d)$ which increases with d. The total fitness at distance d is thus

$$F(d) = \{F_0 + B(d) - E(d)\}P_s(d). \tag{15.2}$$

We note that in reality P_s may also increase with the level of resources, which increases with time (and so decreases with d), and so it is possible that in some situations P_s might not increase monotonically with d (although this would likely have to be associated with a very slow predator approach).

Cooper and Frederick [3] used the following example functions

$$B(d) = B^* \left(1 - \frac{d}{d_d}\right)^n, E(d) = \frac{f}{d^m}, P_s(d) = 1 - e^{-cd}, \tag{15.3}$$

giving the fitness as

$$F(d) = \left(F_0 + B^* \left(1 - \frac{d}{d_d}\right)^n - \frac{f}{d^m}\right)(1 - e^{-cd}). \tag{15.4}$$

Cooper and Frederick [3] developed a more realistic model, which nevertheless agreed with the main predictions of the model of [14]. The extra flexibility of their model meant that they were also able to make new predictions based upon the different parameters that they introduced, that we see in (15.4). For instance flight distance should increase with initial fitness F_0, as the fitter the individual initially, the more it has to lose and the costs associated with predator attack increase, whereas the gain associated with extra forgaging is unchanged. Conversely if the benefit of foraging B^* increases, then the gain associated with remaining increases, but the cost is unchanged, and so the optimal flight distance is reduced. For each parameter there was a clear prediction of the effect that altering its size would have on the flight distance. We should note also that it is of course easy to replace the functions used with alternative ones if they represent a particular scenario better, but that effectively the same qualitative behaviour is likely to result from most plausible functions.

15.3 Interactions Between Cryptic Prey and a Mobile Visible Predator

Broom and Ruxton [2] consider prey that are initially stationary and to some extent cryptic in their environment, such that predators cannot detect them at a distance. Examples include ungulate calves hiding in long grass, many cryptically coloured ground-nesting birds and flat fish lying on the sea bottom.

15.3.1 The Predator-Prey Interaction

In their model, a predator-prey encounter begins when the prey detects an approaching predator. Since the predator has yet to detect the prey when the interaction begins, there is no reason to expect that its trajectory will be taking it directly towards the prey. The closer the predator gets to the prey, the more likely it is to detect it. The closer the predator is when the prey is detected, the more likely it is that the ensuing attack by the predator will be successful. Fleeing from the predator will in most cases alert the predator to the presence of the prey individual. Hence, there may be a countervailing pressure for the prey to sit tight, rely on its crypsis, and only flee if it perceives that the predator has detected it, and is attacking. A key difference between this model and that of [14] is in the cost to the prey associated with fleeing early; in the model of [14] this cost is the opportunity cost associated with reduced time spent feeding.

The predator starts at a distance r from the prey, initially at an angle θ to the prey (so $\theta = 0$ means it is heading straight for the prey). The predator is assumed to move in an undeviating straight line at a constant speed, unless it detects the prey individual. Thus, as in Sect. 15.2.2, there is a deterministic relationship between time and the position of the predator, and we again focus on the predator position.

This initial position on the trajectory is denoted $v = -1$, the position when predator and prey are closest is position $v = 0$, and the position when the predator is again at distance r from the prey, and it is assumed that the potential encounter finishes, is position $v = 1$. The length of the trajectory is $2r\cos\theta$ and the minimum distance between predator and prey (occurring when $v = 0$) is $r\sin\theta$. At any position v on the trajectory, the distance between predator and prey can be found by simple triangular geometry, see Fig. 15.1.

Parameter	Meaning
r	The maximum distance at which prey can detect predators
θ	The angle between the predator's trajectory and the prey direction
s	The speed of movement of the predator prior to an attack
$d(v)$	The distance from predator to prey when the predator is at position v
Δ	The distance advantage the prey gets from initiating a chase itself
t	The time taken by the predator to reach position v
$f(d)$	The probability of the predator catching the prey from distance d
$g(d)$	The rate that the predator detects the prey at distance d
$A(v)$	The probability that the prey has been detected by position v
c	The multiplicative cost of surviving through outrunning a predator

Table 15.1 The parameters for the model of Broom and Ruxton [2]

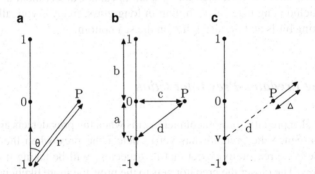

Fig. 15.1 The prey hides at point P. (**a**) The interaction begins when the predator enters the area at point $v = -1$, at distance r from the prey, moving in a direction at an angle θ to the most direct route to the prey. (**b**) The distance d of the predator from the prey at point v follows from Pythagoras' theorem using the distances a and c, with distances $a = vr\cos\theta$, $b = r\cos\theta$, $c = r\sin\theta$. (**c**) If the prey flees, it gains an extra distance Δ ahead of the predator in the pursuit

Detection of the prey occurs at rate $g(d(v))$, and the probability that the prey has been detected by position v is denoted by $A(v)$. Note that the rate of detection is independent of the speed of the predator, but this speed still affects the probability of the prey being spotting because it affects how long it will be in range of the prey. Hence

$$1 - A(v) = exp\left(-\int_0^t g(d(v))dt\right) \Rightarrow \tag{15.5}$$

$$-\ln(1-A(v)) = \frac{r\cos\theta}{s}\int_{-1}^{v}g(d(v))dv. \qquad (15.6)$$

Broom and Ruxton [2] considered three cases, and we shall briefly look at two of them. A summary of the important model parameters is given in Table 15.1. Rewards to both predator and prey from the three different possible types of interaction are shown in Table 15.2.

15.3.2 The Predator Attacks As Soon As the Prey Is Observed

In this version of the model, the predator must attack as soon as it detects the prey, and can see behind itself (and so may still attack after the closest position).

The payoff for never fleeing unless attacked is given by

$$R(N) = \int_{-1}^{1}\{1-f(d(v))\}(1-c)\frac{dA(v)}{dv}dv + 1 - A(1). \qquad (15.7)$$

Strategy V is to flee if the predator attacks or if the predator reaches position $v = V$ without attacking, for some position $V \in [-1, 1]$. The payoff from this strategy is

$$R(V) = \int_{-1}^{V}\{1-f(d(v))\}(1-c)\frac{dA(v)}{dv}dv + (1-A(V))(1-c)\{1-f(d(V)+\Delta)\},$$
$$(15.8)$$

where Δ is the extra distance that the prey can move when it initiates a chase before the predator is aware of the fleeing prey (so that effectively the chase starts with a distance $v + \Delta$ between the two individuals, see Fig. 15.1c).

The case of fleeing immediately is simply $V = -1$ in the above i.e.

$$R(-1) = (1 - f(r+\Delta))(1-c). \qquad (15.9)$$

For any $V > -1$, $R(V)$ is a weighted average of terms of the form $(1-c)(1-f(d))$ that are never bigger than $R(-1)$, and so $R(-1) > R(V)$.

Situation	Prey's payoff	Predator's payoff
No chase	1	0
Attack-initiated chase	$(1-c)(1-f(d(v)))$	$f(d(v))$
Flight-initiated chase	$(1-c)(1-f(d(v)+\Delta))$	$f(d(v)+\Delta)$

Table 15.2 The possible interactions and associated payoffs for the model of Broom and Ruxton [2]

Thus the optimal strategy is either to run immediately (with payoff $R(-1)$) or only to run when the predator initiates an attack (with payoff $R(N)$).

15.3.3 The Predator Can Delay Its Attack

In this case, the predator can still see behind it, but need not attack as soon as it has seen the prey. Now the prey can play the same strategies as considered above, and the predator plays a strategy U, which involves delaying an attack until position $U \in [-1,1]$ if the prey is detected before this position, or attacking immediately on detecting the prey, if detection occurs after it reaches this position.

If $V \leq U$, then the prey will always flee before a predator attacks, and will receive payoff

$$R(V,U) = (1-c)\{1 - f(d(V) + \Delta)\}. \tag{15.10}$$

The predator receives payoff

$$P(V,U) = f(d(V) + \Delta). \tag{15.11}$$

If $V > U$ then either the predator will detect the prey before position U and will initiate an attack at position U, or the predator will detect the prey and thus initiate an attack at some position between U and V, or the predator will not detect the prey before position V and the prey will flee at position V. These three possibilities (respectively) lead to three terms in the payoffs to prey and predator:

$$R(V,U) = A(U)(1-c)\{1 - f(d(U))\} + \int_U^V (1-c)\{1 - f(d(v))\}\frac{dA(v)}{dv}dv$$

$$+ (1 - A(V))(1-c)\{1 - f(d(V) + \Delta)\}, \tag{15.12}$$

$$P(V,U) = A(U)f(d(U)) + \int_U^V f(d(V))\frac{d(A(v))}{dv}dv + (1 - A(V))f(d(V) + \Delta). \tag{15.13}$$

Similarly if the prey plays strategy N, the rewards are

$$R(N,U) = A(U)(1-c)\{1 - f(d(U))\} + \int_U^1 (1-c)\{1 - f(d(v))\}\frac{dA(v)}{dv}dv$$

$$+ \{1 - A(1)\}, \tag{15.14}$$

$$P(N,U) = A(U)f(d(U)) + \int_U^1 f(d(V))\frac{d(A(v))}{dv}dv. \tag{15.15}$$

By a similar argument to that used before, $R(-1,U)$ is greater than the weighted average of similar terms that constitute $R(V,U)$ for all $V > -1$. This is true for any U, so whatever the predator's strategy, the prey's best strategy will be to either never flee unless attacked (Strategy N) or flee immediately on detecting the predator (strategy -1).

If strategy -1 is adopted, then the payoffs to both parties are independent of the predator's strategy. For the case where the prey picks strategy N, It was shown in [2] that the value of U which maximises the predator's payoff $P(N,U)$ is $U = 0$. Thus the predator should always wait until the closest position, and the prey should either flee immediately if $R(-1,0) > R(N,0)$, or never flee otherwise.

15.3.4 Summary

The optimal strategy for the prey is thus either to run immediately or never to initiate a chase. This is true whether the predator attacks immediately on discovering the prey or whether the predator delays its attack until it has closed the distance to the prey. We note that in our different models either the predator had no choice, or there was a clear predator optimal strategy, regardless of whether the prey runs immediately or never. This reduces the prey's strategy to a simple optimisation problem, and thus effectively rules out a mixed strategy. The strategy of immediate flight is associated with slow predator search speed, a low non-predation cost to running, a large advantage to the prey in initiating chases, limited ability to spot the predator at distance, a high ability to spot prey by the predator, and a high probability that chases will be successful. Note that it was assumed that searching predators never spontaneously change direction, or do so sufficiently rarely that this situation can be ignored. If many such directional changes are made, this will affect optimal prey and predator strategy.

15.4 Interactions Between Cryptic Prey and a Stationary Visible Predator

In this section we consider a model of a related scenario to the above, where an ambush predator appears close to a concealed prey individual. The predator does not move, but begins to scan the environment for hidden prey. An example is a bird of prey such as an eagle settled at a high vantage point. The prey must choose when (if at all) to run. This model was developed following experiments by Martin et al. [8] which we describe below.

15.4.1 Experiments with Hidden Rock Lizards

Martin et al. [8] conducted experiments in central Spain with hidden rock lizards in their natural habitat, and humans as simulated predators. The experimenter walked around the study area until a lizard was spotted. Then the experimenter carried out one of two procedures. In the first case they approached the lizard slowly without obviously looking at it, and stopped at a short distance from the lizard. The experimenter then timed how long it took for the lizard to flee to a refuge (usually a crevice in a rock). In the second procedure, the experimenter proceeded as above, but after stopping ran towards the lizard, simulating an attack. The lizard then fled to a refuge, after which the experimenter stayed in the local area until the lizard next left its refuge. When the lizard left, the first procedure was again followed. In each case the human predators did not discontinue their search until the prey had fled.

Thus in the experiments, two levels of predator threat were displayed. A lower level threat where the predator had just arrived in the area, was clearly searching but had not discovered the lizard, and a higher level one where the prey was hiding as a direct result of a chase, so it was clear that the predator knew it was in the vicinity. Martin et al. [8] found that the lizards fled after some time, typically of the order of 3 min or less. This time depended upon the level of danger that they were in. In the high risk case the time was significantly shorter than in the low risk case. There were differences between males and females, with males generally fleeing earlier (males are more brightly coloured so that crypsis is a less reliable defence).

15.4.2 A Model of the Interaction Between a Stationary Searching Predator and a Hidden Prey

Here we describe a model of this predator prey interaction. A predator arrives at a location with cover that may conceal hidden prey, and a good vantage point for searching. The predator begins to search the cover area for prey. It may be that at some point during the search a hidden prey individual flees. We shall assume that the cover area does not contain more than one prey individual. It is possible that there are no prey present. Eventually, if it does not find prey, the predator will give up and try a new location. The predator must decide when to give up searching if no prey is found.

A prey individual is in cover when it observes a predator arrive. The prey knows that there are no other prey nearby, and that the predator is not aware of its presence. It also knows that the predator is there to hunt and is searching to find prey, and that eventually if the predator finds nothing it will give up and move elsewhere. The prey must decide how long to stay in hiding and when to run, if ever.

When the predator arrives, the probability that there is a prey individual present is $0 < \alpha < 1$. If the predator finds the prey it will attack it, the prey surviving with probability γ. If the prey runs before the predator sees it, it will alert the predator to its presence, but have an increased chance of survival β, where $0 \leq \gamma < \beta < 1$. The rate that the predator spots a prey individual (conditional on it being there) is given by $g(t)$ which depends upon how long the predator has been searching, and so the probability that the prey has been seen by time t is

$$G(t) = \int_0^t g(s)ds. \tag{15.16}$$

The predator pays an opportunity cost λ per unit time it spends looking for the prey i.e. if it leaves, the expected reward it would get elsewhere per unit time is λ. The model parameters are summarised in Table 15.3.

Parameter	Meaning
α	The probability that a prey individual is present
β	The probability of survival if the prey breaks cover and runs
$\gamma (< \beta)$	The probability of prey survival if the predator spots the prey
$G(t)$	The probability that the predator finds the prey (if present) by time t
$g(t)$	The rate of search success of the predator, the derivative of $G(t)$
λ	The cost paid by the predator per unit time during the search
v	The predator's discovery rate at $t = 0$

Table 15.3 The parameters for the stationary predator and hidden prey model

The prey's strategy is the time s after the predator's arrival when it will run. The predator's strategy is the time t after its arrival to give up if no prey has been found.

For given values of s and t, the prey reward (simply its survival probability) is $R(s,t)$ where

$$R(s,t) = \begin{cases} G(t)\gamma + (1 - G(t)) & s > t \\ G(s)\gamma + (1 - G(s))\beta & s \leq t \end{cases} \tag{15.17}$$

The predator reward $P(s,t)$ is its probability of catching the prey minus λ times the expected time spent in the search. This is

$$P(s,t) = \begin{cases} \alpha\{G(t)(1-\gamma)\} - \lambda\{(1 - \alpha G(t))t + \alpha \int_0^t g(x)x\,dx\} & s > t \\ \alpha\{G(s)(1-\gamma) + (1 - G(s))(1 - \beta)\} - \lambda\{(1 - \alpha)t \\ \quad + \alpha(\int_0^s g(x)x\,dx + (1 - G(s))s)\} & s \leq t \end{cases} \tag{15.18}$$

Our game is a type of asymmetric generalized war of attrition. The longer the predator waits, the greater the conditional probability that in fact there is no prey in cover, so the predator must eventually stop searching. Thus under most circumstances the expected incremental payoff to the predator of waiting extra time will decrease and there is likely to be a unique point at which it is best for the predator to leave. We suppose that the predator chooses pure strategy t.

For the prey the risk is high early on, but it knows that it can outwait the predator. A natural strategy for the prey is to either run immediately, or not at all. In fact, similarly to the game of [2], there is no point it waiting a short time, and exposing itself to the risk of being attacked, only then to flee before the predator gives up. The only possible choices are thus to flee immediately or to choose a leaving time which is longer than the predator is prepared to wait (and any such time is equivalent to never to flee). We suppose that the prey chooses $s = 0$ with probability p, and $s = \infty$ with probability $1 - p$. We denote this strategy by p.

We thus have a game where the predator chooses a time t, and the prey chooses a probability p. We note that we have not ruled out the possibility of other strategy combinations associated with a mixed predator strategy. If the predator used a mixed strategy involving at least two different strategies t_1 and $t_2 > t_1$, then the payoffs to playing either of these as a pure strategy would have to be equal in any population in equilibrium. This in turn means that there must be some probability of the prey fleeing between times t_1 and t_2. Why would a prey individual flee at such intermediate times, having exposed itself to initial risk of being attacked (similarly to the argument above)? One reason could be that the predator gets better at searching as it aclimatises to the location. This cannot be ruled out, so it may be that such mixed strategies occur (although it is not obvious that they can). As we shall see, under reasonable assumptions (including that the predator does not get better at searching the longer it waits) we find a unique strategy of this type for all parameters. We also assume that if the predator and prey choose to leave at the same time, then an attack initiated by the prey is the result (i.e. as it prepares to go the predator sees the prey flee and launches an attack).

The payoffs for prey and predator become

$$R(p,t) = p\beta + (1-p)G(t)\gamma + (1-p)(1-G(t)), \qquad (15.19)$$

$$P(p,t) = \alpha \{p(1-\beta) + (1-p)G(t)(1-\gamma)\} - \lambda\{(1-\alpha)t + (1-p)\alpha(\int_0^t g(x)x\,dx + (1-G(t))t)\}. \qquad (15.20)$$

Against a given t the optimal prey strategy is the value of p which maximises the expression for $R(p,t)$ in (15.19), which is $p = 0$ ($p = 1$) when

$$G(t) < (>) \frac{1-\beta}{1-\gamma}, \qquad (15.21)$$

and all values are equivalent if (15.21) becomes an equality.

The optimal predator strategy occurs where $P(p,t)$ achieves its maximum value in (15.20). A local maximum of this expression occurs when

$$\frac{dP(p,t)}{dt} = \alpha(1-p)(1-\gamma)g(t) - \lambda(1-\alpha) - \lambda(1-p)\alpha(1-G(t)) = 0. \quad (15.22)$$

For most reasonable functions $G(t)$ this decreases with t, so there will be at most one such value. If there is such a value this is the optimal predator choice of t, otherwise (if the right hand side of (15.22) is negative at $t = 0$) then $t = 0$ is optimal.

We thus look for combinations of t and p which satisfy (15.21) and (15.22). It is easy to see that there is no stable solution with $p = 1$, since this means that it is impossible for prey to be in hiding after time 0, so that $t = 0$ maximises (15.20), but for $t = 0$ it is best for the prey to wait (i.e. $p = 0$ maximises (15.19)). Similarly no stable solution can exist involving a mixed strategy $0 < p < 1$ and $t = 0$, again since $p = 0$ maximises (15.19) when $t = 0$. This leaves three possibilities, $p = 0, t = 0$; $p = 0, t > 0$ and $0 < p < 1, t > 0$.

15.4.3 An Example Payoff Function

We shall consider the function $g(t) = v\exp(-vt)$. This represents a predator who spots a prey individual at constant rate v, conditional on the fact that there was initially a prey individual to be spotted, and that it has not already fled or been spotted by the predator. We obtain the following results, which are illustrated for a particular choice of parameters in Fig. 15.2.

1. $p = 0, t = 0$ defines an equilibrium if $t = 0$ maximises (15.20) when $p = 0$, (note that $p = 0$ automatically maximises (15.19) for $t = 0$). Using (15.22), this gives

$$\alpha(1 - \gamma)v - \lambda(1 - \alpha) - \lambda\alpha < 0 \Rightarrow \tag{15.23}$$

$$v < \frac{\lambda}{\alpha(1 - \gamma)}. \tag{15.24}$$

2. $p = 0, t > 0$ defines an equilibrium if (15.22) is satisfied for a value of $t > 0$ when $p = 0$, and for this t, $p = 0$ is optimal. For $p = 0$ (15.22) becomes

$$\alpha(1 - \gamma)ve^{-vt} - \lambda(1 - \alpha) - \lambda\alpha e^{-vt} = 0 \Rightarrow \tag{15.25}$$

$$t = \frac{1}{v}\ln\frac{\alpha(1 - \gamma)v - \lambda\alpha}{\lambda(1 - \alpha)}. \tag{15.26}$$

The value of t in (15.26) is only positive if (15.24) does not hold. $p = 0$ is optimal if (15.21) holds i.e.

$$1 - e^{-vt} < \frac{1 - \beta}{1 - \gamma}. \tag{15.27}$$

These two conditions thus yield

$$\frac{\lambda}{\alpha(1 - \gamma)} < v < \lambda\left(\frac{1}{1 - \gamma} + \frac{1 - \alpha}{\alpha}\frac{1}{\beta - \gamma}\right). \tag{15.28}$$

3. $p > 0, t > 0$ defines an equilibrium if (15.21) and (15.22) are satisfied with equality. These yield

$$t = \frac{1}{v}\ln\frac{1 - \gamma}{\beta - \gamma} \tag{15.29}$$

and

$$p = 1 - \frac{\lambda(1 - \alpha)(1 - \gamma)}{\alpha(\beta - \gamma)((1 - \gamma)v - \lambda)}, $$

the second of which requires

$$\lambda\left(\frac{1}{1 - \gamma} + \frac{1 - \alpha}{\alpha}\frac{1}{\beta - \gamma}\right) < v. \tag{15.30}$$

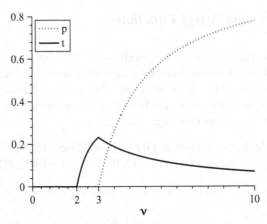

Fig. 15.2 Optimal values of p and t for various values of v in the stationary predator, hidden prey game. Other parameters are $\lambda = 1$, $\alpha = 0.5$, $\gamma = 0$, $\beta = 0.5$

Thus for different values of search efficiency v we have:

v small: the prey does not move and the predator gives up immediately (e.g. the cover is too dense, opportunities are better elsewhere than here).

v medium: the predator will search for some time before leaving, but the prey will never run if not found.

v large: the predator will search for some time before leaving, and the prey will play a mixed strategy, fleeing immediately with some probability, and otherwise never running unless attacked.

15.5 Discussion

Initially we looked at existing models of two situations where prey must decide whether to flee. Sometimes elements of both models will be present. Although the cost of fleeing considered by Broom and Ruxton [2], energetic costs and risk of capture, and that considered by Ydenberg and Dill [14] and Cooper and Frederick [3], loss of foraging time, are not mutually exclusive, we can make predictions about the relative importance of the two costs.

In cases where predators can see prey from a distance then the Cooper and Frederick [3] model is better. An example is an adult zebra or gazelle grazing on the savannah during the day, see [13]. This is particularly true if attempted predation is common, so that costs of flight are significant to the daily energy or time budget of the prey, or if fleeing causes the prey to lose a valuable food item (e.g. a cheetah being driven from a kill by approaching lions). There are many instances of prey not fleeing as soon as a predator is seen; for example [14] gives a number of situations where flight distance is affected by the speed of a predator's approach, and many prey become alert when a cheetah is spotted, but do not take flight until it is sufficiently close [6]. However, if the prey is cryptic such that predators can pass reasonably close to it without detecting it (e.g. a juvenile gazelle lying motionless in long grass) then the Broom and Ruxton [2] model is better. This will especially apply to cases where predation attempts (or at least proximate predators) are less frequent, so that avoiding predation makes up a small part of an animal's time or energy budget or if the prey animal can quickly return to its previous feeding behaviour as soon as the predator has passed (e.g. many grazers). It may be that more sophisticated models that take into account different costs to the prey are needed.

We then looked at a new model similar to that developed in [2], inspired by experiments carried out in [8]. In the new model, the predator remains still and has to decide how long to search for prey before giving up; the prey has to decide whether to run, and if so when. We found some circumstances, where a particular area was just too difficult to search, where the predator should give up immediately (and of course the prey does not run). As the ease of search in a particular area improves, there comes a critical point when it is worthwhile for the predator to search, but where the prey should still always stay in cover. As searching becomes easier the length of time the predator should stay increases. This continues until a point where it becomes optimal for the prey to play a mixed strategy, sometimes staying in cover, but sometimes running immediately. As searching becomes easier still, the probability that the prey runs immediately should increase, and the search time of the predator starts to decrease again. This is because, given that prey often run immediately, conditional on not having found prey or seen them run, the probability of there actually being no prey to find increases with ease of search through the increased chance of early running.

Why did we not get the results of [8], that the prey sometimes wait before running? The most obvious reason is that we focused on finding a different kind of solution, which we considered to be the natural solution to the problem that we considered. But why did they not find such solutions, but instead have prey that ran after some time, rather than immediately or not at all? One possible explanation is that the lost foraging cost of [14] or [3] is sufficiently important, and that the lizards cannot afford to wait it out.

Interestingly, the "predators" of [8] effectively carried on searching forever, so did not behave like strategic predators. This provides another possible explanation. In our model, if prey knew there were a small fraction ε of predators that waited forever, then if a predator stays beyond time t, they would know that one of the new predators was present, thus fleeing is then optimal.

Alternatively, what if prey can observe which area is being searched? Thus for the prey the search rate may be v_H or $v_L < v_H$, depending on whether the predator is searching in its vicinity. There would be transitions between these two rates, as the predator's search continued. If these transition rates were r_{HL} and r_{LH} from high risk to low risk and low risk to high risk respectively, then we can show that the mean search rate v is found from

$$v = \frac{r_{HL}}{r_{HL} + r_{LH}} v_L + \frac{r_{LH}}{r_{HL} + r_{LH}} v_H. \qquad (15.31)$$

Thus it may be that the optimal strategy for a prey individual may be to run if it is in the more dangerous search phase, and so we simply observe a prey individual run the first time that this phase is entered.

We note that in the above work the strategic choices of the predators have been limited i.e. to attack immediately or wait, or when to cease searching. However, we have not considered the best active searching strategy of the predator. Some important work on predator searching strategies has been carried out, with examples including [1, 10, 11] and [5]. The choices of predators in general can of course be important, and in turn influence the choices of the prey, as we have seen in our model of the stationary predator and hidden prey. This is likely to be true for a variety of scenarios, and so the choices of both prey and predators should be considered in our models.

References

1. Alpern,S. Fokkink,R. Timmer,M. and Casas,J. (2011) Ambush frequency should increase over time during optimal predator search for prey. Journal of the Royal Society Interface 8 1665–1672.
2. Broom,M. & Ruxton,G.D. (2005) You can run - or you can hide: optimal strategies for cryptic prey against pursuit predators. Behavioral Ecology 16 534–540.
3. Cooper Jr, W.E., Frederick, W.G. (2007) Optimal flight initiation distance. Journal of Theoretical Biology 244: 59–67.
4. Caro,T.M. (1986) The functions of stotting in Thomson's gazelles: Some tests of the predictions. Animal Behaviour 34 663–684.
5. Desouhant,E., Lucchetta,P. Giron,D. and Bernstein,C. (2010) Feeding activity pattern in a parasitic wasp when foraging in the field. Ecological Research 25 419–428.
6. Ewer,R.F. (1968) Ethology of Mammals. Elek Science, London.
7. Kenward,R.E. (1978) Hawks and doves: factors affecting success and selection in goshawk attacks on woodpigeons. Journal of Animal Ecology 47 449–460.
8. Martin, J., Luqu-Larena, J.J., Lopez, P. (2009) When to run from an ambush predator: balancing crpysis benefits with costs of fleeing in lizards. Animal Behaviour 78 1011–1018.
9. Myers,J.P. (1983) Commentary. In "Perspectives in Ornithology" (A.H.Brush and G.A.Clark Jr, eds) pp 216–221. Cambridge University Press.
10. Pitchford,J.W., James,A. and Brindley,J. (2005) Quantifying the effects of individual and environmental variability in fish recruitment. Fisheries Oceanography 14 156–160.
11. Preston,M.D., Pitchford,J.W. and Wood,A.J. (2010) Evolutionary optimality in stochastic search problems. Journal of the Royal Society Interface 7 1301–1310.
12. Rowe-Rowe,D.T. (1974) Flight behaviour and flight distance of blesbok. Zeitschrift fur Tierpsychologie 34 208–211.
13. Walther,R.F. (1969) Flight behaviour and avoidance of predators in Thomson's gazelle (Gazella thomsoni Guenther 1884). Behaviour 34 184–221.
14. Ydenberg, R.C., Dill, L.M. (1986) The Economics of Fleeing from Predators. Advances in the Study of Behaviour 16 229–249.

Chapter 16
A Discrete Search-Ambush Game with a Silent Predator

Robert Arculus

Abstract We investigate the problem faced by a searcher attempting to capture a mobile hider hidden in one of K discrete locations. In each time period the searcher may either inspect a location or ambush. The hider can attempt to move to a random location at any time, but will be captured during the move if the searcher is ambushing. This game was invented by Steve Alpern, and is an outgrowth of the recent studies of Alpern et al. (J R Soc Interface 8, 2011), which assume a 'noisy predator', where the prey always knows the fraction of the search area that has been inspected. The distinguishing feature of the problem addressed here is that the predator is 'silent', so that the prey can only surmise the extent of the current exploration. The prey, however, remains 'noisy', in that the predator is informed if a successful move occurs. To make the game tractable via computation, the searcher is constrained to inspect the entire space before some chosen number of periods elapse. Here, we present general upper and lower bounds on the expected time until capture, and state solutions to the game for some simple cases; among other results, we observe that when the search space consists of two locations and the searcher is unconstrained, the expected capture time is equal to the square of the golden ratio. We also provide some conjectures concerning optimal behaviour and bounds on the value of the game for general K with an unconstrained searcher. Our numerical results are mentioned but not discussed in detail, though they have potential to provide further insight into the structure of the game.

R. Arculus (✉)
Department of Mathematics, Columbia House, London School of Economics,
Houghton Street, London WC2A 2AE, UK
e-mail: rarculus@gmail.com

S. Alpern et al. (eds.), *Search Theory: A Game Theoretic Perspective*,
DOI 10.1007/978-1-4614-6825-7_16, © Springer Science+Business Media New York 2013

16.1 Introduction and Terminology

Consider a zero-sum game played between a hider and a searcher on a search space consisting of K discrete cells, with the searcher and hider minimising and maximising, respectively, the time elapsed until capture. In every unit of time the searcher chooses either to inspect one of the cells, or to *ambush* in some central location. Similarly, in each time unit the hider chooses either to remain motionless, or to attempt to move from their current cell to a randomly selected cell (note this includes the possibility that they may return to the cell they started the period in). If such a move is attempted while the searcher is ambushing, then the hider is captured; otherwise, the searcher loses all information gained so far regarding which cells do not contain the hider.

We adopt the following terminology: a *period* is one discrete unit of time; a *round* is a run of periods which ends when the hider is either caught or successfully randomises their position; a *game* consists of a sequence of rounds which ends whenever there is a round in which the hider is caught. We include a second parameter, L, which is a limitation on the behaviour of the searcher, such that in any round the searcher must commit to having searched all K cells by the time L periods have elapsed. Thus while L limits the length of any round, any particular game can potentially continue indefinitely. This parameter serves to make the game tractable for computation purposes.

We assume that the searcher is *silent*, meaning the hider does not receive additional information after the start of a round regarding the proportion of cells that have been inspected. We further assume that the hider is *noisy*, in that, if capture does not occur in a period, the searcher is informed whether the hider moved or remained motionless. The game can thus be modelled using a matrix representing a simultaneous strategy commitment by both hider and searcher, where a hider strategy involves picking a period in which to move, and a searcher strategy involves picking K out of the first L periods in which to search. For a game to be non-trivial and well-defined, we require $L > K$; if $L = K$ the hider guarantees an infinite payoff by moving in the first period of every round, and if $L < K$ the searcher's constraint cannot be satisfied. We refer to the game in general as the *silent search-ambush game*.

We suppose that moving while search is taking place always allows the hider to escape, even if the searcher is inspecting the cell that the hider was occupying at the end of the previous period. Note also that when capture occurs, for the purposes of measuring payoffs it is assumed to take place at the end of the relevant period.

As space available is abbreviated, some proofs will be omitted. Note also that many of the results here were inspired by numerical solutions obtained computationally; we make occasional reference to these results, particularly when discussing our conjecture in Sect. 16.7, but they are not addressed in as much detail as they might be.

16.2 Literature Review

The silent search-ambush game was invented by Steve Alpern, and is an outgrowth of recent work by Alpern et al. [3] on the solution to a very similar game, with two distinguishing characteristics: first, their search space is continuous, with total area equal to unity; and second, more crucially, their hider is always aware of what proportion of the search space has been inspected (i.e., the searcher is "noisy"). A key result in their game is the "square root law of predation search": that the searcher's optimal probability of searching is equal to the square root of the fraction of the search space that remains uninspected. Qualitatively, this implies the pace of search should be rapid at the start of a game, slowing in favour of ambush as the game progresses.

The issue of the potential usefulness of ambush (in the sense of the searcher remaining motionless in the hope that the hider will move into them) extends further back into the search game literature, arising naturally in questions of optimal search for a mobile hider on a network, especially where the network in question has the form of a figure-eight or n-leaf clover, or star network: here, it is reasonable to suspect that the searcher may wish to occasionally remain still at the central node. This issue was considered in detail by Alpern and Asic [1, 2], finding that such "loitering" strategies are indeed sometimes advantageous (a useful overview of this and related issues is provided in Alpern and Gal [4]).

Our game is similar but distinct to games that model the problem of pursuit-evasion in a continuous space with differential equations or games that consider the geometric problem of optimal ambush locations. Example of this latter type include a variety of games based on an agent attempting to defend some discrete or continuous search space (see Ruckle et al. [15]; Joseph [12]; Washburn [17] or Baston and Kikuta [5]). A case which has some interesting parallels to ours is the one-dimensional evasion game first described by Gal [8] and solved by Lalley [13] (with several generalisations of the game investigated by other authors); here, an infiltrator starts in a safe spot but then must progress incrementally through M discrete sites within N periods of time without being detected by a defender. One result is that a min-max strategy for the infiltrator is to play what Lalley terms "Admiral Farragut"[1] strategies: waiting for some period of time before progressing through all sites as rapidly as possible (in Sect. 16.6, we will, less evocatively, term our equivalent of these *wait-then-exhaustive* strategies).

Trade-offs between searching and ambushing behaviour can be seen in a variety of real-world contexts. The issue is particularly relevant in biology, with much modern research following from the observations of different foraging modes made by Schoener [16]. Some of many possible examples of such behaviour include brook trout (Grant and Noakes [9]), mantids (Inoue and Marsura [11]), and between different species of desert-dwelling lizards (Huey and Pianka [10]). Possible explanations offered for such behavioural variations include environmental conditions, the threat

---

[1] The reference is to the US Navy flag officer who, among other things, is remembered for the paraphrased exclamation "Damn the torpedoes, full steam ahead!".

of predation on the predator, available food (Olive [14]), and the necessity of either overwhelming or not alarming the prey (Casas et al. [7]). The suggestion from game theory is that such variation could represent optimal play if the interaction is modelled as a zero-sum game; this follows from [3], and is also considered elsewhere, such as in Zoroa et al. [18].

Our game is also somewhat connected with the question of timing when to flee from an approaching predator, where the probability of successful escape is increasing the further away the predator is and the quicker they are moving, with an assumption that the predator's speed comes at the cost of accuracy; see Broom and Ruxton [6] for more discussion of this point. This can be linked to our condition that the hider always escapes against a searching predator, even if the predator inspects the cell the hider started the round in: this is reasonable if we assume that a moving hider is "too difficult a target" for a moving searcher to effectively intercept.

16.3 Constructing Payoff Matrices

We start with some additional terminology. We denote the searcher strategy set S, and the hider strategy set H, with elements of these sets denoted by lower-case s and h. Specific hider strategies are denoted h^i, where $i \in \{1 \ldots L, L+\}$; thus $|H| = L+1$. Given the searcher's constraint, the strategy h^{L+} (which is intended to refer to moving in any period greater than L) is equivalent to not moving at all. It is straightforward to show that this is never optimal for the hider – at a minimum, if period L is reached, it must be sensible to move, since in that period the searcher will be obligated to search – and thus after this section such strategies will be excluded from our payoff matrices.

Since every searcher strategy is constructed by choosing K periods out of a total of L periods, we have $|S| = \binom{L}{K}$. When $K = 2$, we have $|S| = \frac{L(L-1)}{2}$, and thus $|S| = O(L^2)$. In general, for any fixed K, we have $|S| = O(L^K)$.

The expected time until capture based on a pair of strategies s and h for a game defined by K and L is denoted $T_{L,K}(s,h)$.

The *value* of a game, i.e. the expected time until capture under optimal play, is denoted $\hat{T}_{L,K}$ (see [4] for more discussion of this concept of value in search games).

The optimal (max-min and min-max) hider and searcher strategies are denoted $\hat{s}$ and $\hat{h}$; we then have $\hat{T}_{L,K} = T(\hat{s}, \hat{h})$, which again may be abbreviated to $\hat{T}$. The subscripts $_L$ and $_K$ will be dropped when they are clear from the context.

We now turn to constructing game matrices, starting by describing how the expected time until capture (less precisely referred to as the payoff) is calculated from any pair $s \in S$ and $h \in H$. Searcher strategies in general can be difficult to denote neatly; for this section, we will use the game where $K = 2$, $L = 4$ as an illustrative example, and a search strategy that involves searching in, say, the second and fourth periods will be denoted $s^{2,4}$. In general, a searcher strategy s can be interpreted as a binary vector of length L, where a 1 in the n-th position indicates searching in the n-th period, while a 0 indicates ambushing. Thus $s^{2,4} = \{0,1,0,1\}$. A subscript on the s represents the relevant vector entry; hence $s_2^{2,4} = 1$, $s_3^{2,4} = 0$. Our constraint on the searcher requires $\sum_{l=1}^{L} s_l = K$. Any pure hider strategy, h^i, could also be represented by a binary vector, but this rarely proves necessary.

Our notation provides us with a natural ordering for the searcher strategies: concatenate the relevant binary vectors to create binary numbers, then sort in descending numerical order. Under this approach, strategies that emphasise relatively "early" search will be listed ahead of those that emphasise "late" search.

To provide a compact formula for the expected payoffs, we define the vector W^m of length L, where $W_j^m = j$ if $j < m$ and 0 otherwise; as an example, $W^3 = \{1,2,0,0\}$. Define $W^{L+} = W^{L+1}$. The expected time until capture from any combination of pure hider and searcher strategies is:

$$T(s,h^i) = \frac{1}{K} sW^i + \frac{K - \sum_{l=1}^{i-1} s_l}{K} (Ts_i + i), \qquad (16.1)$$

Vector multiplication is assumed to proceed appropriately, producing a scalar in every case, without introducing notation to distinguish row and column vectors.

The interpretation of this formula is not obvious, so some explication follows. We will first consider the case where the hider plays the suboptimal stratyegy of not moving before time L, that is, $i = L+$. In that case the expected time until capture is $1/K$ multiplied by the search strategy, s, multiplied by W, which will just be a vector of integers counting from 1 to L (we will have that $\sum_{l=1}^{i-1} s_l = K$, so the second term in the formula is zero). Thus, against the strategy $\{0,1,0,1\}$, the expected capture time in our example game would be $2(\frac{1}{2}) + 4(\frac{1}{2}) = 3$.

Consider another case, where the hider moves before the first period in which the searcher searches. In that case the first term is zero (as $sW^i = 0$), while $\sum_{l=1}^{i-1} s_l = 0$. Since in this case $s_i = 0$, the entire equation is simply equal to i. If, however, the hider moves in the first period in which the searcher searches (so that $s_i = 1$), the payoff is $T + i$, where T is the payoff obtained by playing the game again assuming the hider and searcher hold their probability distributions over their strategies constant.

The remaining and most complex case is thus when the hider moves after some cells have been searched, but not all of them. In this case the term $\frac{1}{K} sW^i$ represents the proportion of the expected payoff due to the possibility of capture *before* the period in which the hider moves. $\frac{K - \sum_{l=1}^{i-1} s_l}{K}$ represents the probability that the hider will not have been captured by time i, and thus the probability of obtaining the payoff $T + i$ (if movement occurs during search) or i (if it does not).

Thus: against $\{0,1,0,1\}$, h^1 provides a payoff of 1; h^2 of $2 + T$; h^3 of $2(\frac{1}{2}) + 3(\frac{1}{2}) = \frac{5}{2}$, and h^4 of $2(\frac{1}{2}) + (4+T)(\frac{1}{2}) = 3 + \frac{T}{2}$.

For a particular game, the payoff matrix is denoted $A^{L,K}$. The searcher is taken as the column player (with strategies ordered as described previously) while the hider is the row player. For our example, we have:

$$A^{4,2} = \begin{array}{c} h^1 \\ h^2 \\ h^3 \\ h^4 \\ h^{4+} \end{array} \begin{matrix} s^{1,2} & s^{1,3} & s^{1,4} & s^{2,3} & s^{2,4} & s^{3,4} \end{matrix} \left(\begin{matrix} 1+T & 1+T & 1+T & 1 & 1 & 1 \\ {(3+T)}/{2} & {3}/{2} & {3}/{2} & 2+T & 2+T & 2 \\ {3}/{2} & {(4+T)}/{2} & 2 & {(5+T)}/{2} & {5}/{2} & 3+T \\ {3}/{2} & 2 & {(5+T)}/{2} & {5}/{2} & {(6+T)}/{2} & {(7+T)}/{2} \\ {3}/{2} & 2 & {5}/{2} & {5}/{2} & 3 & {7}/{2} \end{matrix} \right)$$

By definition, where $\mathbf{q}$ represents the vector of searcher probabilities and $\mathbf{p}$ the vector of hider probabilities, both in row form, we have:

$$T_{L,K}(s,h) = \mathbf{p}A^{L,K}\mathbf{q}^{\mathbf{T}}, \tag{16.2}$$

16.4 Some Rudimentary Bounds

We begin by presenting some upper and lower bounds. These bounds are of interest in their own right, but also serve as a check on our numerical results for large L and K, especially given that they are substantially easier to compute. First, by considering the payoff that would result if the hider moves with equal probability in each of the periods from 1 to L, the folowing can be shown:

Theorem 1. *There exists a hider strategy which guarantees the lower bound:*

$$\hat{T}_{L,K} \geq \frac{2}{K(2L-K-1)}\left(\frac{K^3}{3} + \frac{K^2}{2} + \frac{K}{6}\right) + \frac{(L-K)(K+1)}{(2L-K-1)}, \tag{16.3}$$

Second, consideration of the analogous seacher strategy, where the searcher gives equal probability to each of their possible strategies, leads to:

Theorem 2. *There exists a searcher strategy which guarantees the upper bound:*

$$\hat{T}_{L,K} \leq \max_{1\leq n\leq K+1, n\in\mathbb{N}} \left\{ \frac{nK(2K+3-n)}{2+2n(K-1)} \right\}, \tag{16.4}$$

Note that the parameter L does not influence the value of this bound.

A final bound can be obtained by considering the game as a restricted version of search on a network structured so that there are K directed ("one-way") loops from a central node, i.e. a K-leaf clover. The central node is then taken as the searcher's ambush point. We will not go into detail, but using existing techniques from the search game literature it is possible to conclude the following:

Theorem 3. *When $L = \infty$, there exists a searcher strategy which guarantees the upper bound:*

$$\hat{T}_{\infty,K} \leq K+1, \tag{16.5}$$

16.5 Ambush in a Single Cell

We will now present some analytical solutions to versions of the silent search-ambush game, starting with the case where $K = 1$ (that is, there is only a single cell). In this case, movement by the hider does not involve a meaningful randomisation of the hider's position from the searcher's point of view, since the searcher effectively "knows" which cell the hider is occupying at all times. Rather, a successful move simply allows the hider to evade capture for one additional period. The "silent searcher" condition retains its meaning in the sense that strategies are not changed while a round is in progress, though it is no longer even possible for the hider to be aware of some fraction of the space search having been searched (either none of the search space has been searched, or the entirety has. In the latter case the hider has been caught, and, under the predator-prey interpretation, consumed, rendering the content of the hider's information set a moot issue).

Geometrically, some intuition can be gained by imagining the $K = 1$ case as a search game on a one-way circular graph that takes unit time to traverse, with hiding and ambush points placed at antipodean points. If the hider moves while the searcher inspects, they "chase each other's tails" around the circle, coming back to their starting points (this is the simplest instance of the K-leaf clover network referenced in Sect. 16.4).

16.5.1 The Game with Two Periods

As before, a strategy for the hider is denoted $h^i \in H$. Ignoring h^{L+}. We have $i \in \{1 \ldots L\}$. Similarly, a strategy for the searcher will be denoted $s^j \in S$ where $j \in \{1 \ldots L\}$ represents the period of search. The payoffs are defined as follows:

$$T_{L,1}(h^i, s^j) = \begin{cases} \frac{i+j}{2} + \hat{T}_{L,1} & \text{if } i = j \\ \min\{i, j\} & \text{if } i \neq j \end{cases}, \tag{16.6}$$

That is, if the hider successfully restarts the game, the payoff is equal to the period in which this occurred plus the value of the restarted game; otherwise, the payoff is simply in the period in which either the searcher searches before the hider moves, or the hider moves while the searcher is ambushing. The simplest possible interesting form of the general game with finite L is where $K = 1, L = 2$, which we will consider now.

The relevant payoff matrix is symmetric with the form:

$$A^{2,1} = \begin{array}{c} \\ h^1 \\ h^2 \end{array} \begin{array}{c} s^1 \quad\quad s^2 \\ \begin{pmatrix} 1+T & 1 \\ 1 & 2+T \end{pmatrix} \end{array}$$

For instance, suppose the searcher inspects in the first period (that is, plays s^1), while the hider moves (plays h^1). No capture occurs in that period; instead the hider escapes but, at the end of the period, returns to the only hiding place available, while the searcher returns to their ambush position. The searcher is informed of the hider's escape and a new round starts, so the payoff is $1 + T$.

This game can be solved using basic game theory techniques, with the only complicating factor being that the value of the game is unknown and is included in the payoffs. Given the interpretation of the game, any payoff must be strictly positive, and we therefore have $T > 0$. As would be expected, there is no equilibrium in pure strategies. To solve in mixed strategies: let $q(s^i)$ represent the probability attached by the searcher to searching in the i-th period; for simplicity, in this section we will shorten this to simply q_i. For the searcher to make the hider indifferent between their two remaining strategies, we require:

$$q_1(1+T) + (1-q_1) = q_1 + (2+T)(1-q_1), \tag{16.7}$$

$$\Rightarrow q_1 = \frac{1+T}{2T+1}, \tag{16.8}$$

By symmetry, an equivalent equation will hold for the probability of the hider moving in the first period. To solve for T, we can substitute (16.8) into:

$$q_1(1+T) + (1-q_1) = T, \tag{16.9}$$

which represents the payoff the hider will receive in equilibrium. This provides us with an expression solely in terms of T:

$$\frac{(1+T)^2}{2T+1} + (1 - \frac{1+T}{2T+1}) = T, \tag{16.10}$$

$$\Rightarrow \hat{T}_{2,1}(\hat{h}, \hat{s}) = 1 + \sqrt{2}, \tag{16.11}$$

where we have eliminated the negative root of the quadratic due to T being positive. Substituting into (16.8), we conclude that the searcher inspects in the first period with probability:

$$q_1 = \frac{2+\sqrt{2}}{3+2\sqrt{2}} = \frac{\sqrt{2}}{1+\sqrt{2}} = 2 - \sqrt{2}, \tag{16.12}$$

which by the symmetry of the game is likewise the equilibrium probability that the hider moves in the first period; and thus the probability of the searcher searching or the hider moving in the second period is[2]:

$$(1-q_1) = \frac{1}{1+\sqrt{2}} = \sqrt{2} - 1, \tag{16.13}$$

---

[2] For incidental interest, $1/(1+\sqrt{2})$ has been known as the silver ratio, which, in conjunction with the results of Sect. 16.7, seems worth mentioning if only to identify a theme of metallurgical nomenclature through this chapter.

16.5.2 The Game with an Arbitrary Number of Periods

For an undefined but finite L, the payoff matrix has the following appearance:

$$A^{L,1} = \begin{array}{c} \\ h^1 \\ h^2 \\ h^3 \\ \vdots \\ h^L \end{array} \begin{array}{ccccc} s^1 & s^2 & s^3 & \ldots & s^L \\ \left(\begin{array}{ccccc} 1+T & 1 & 1 & \ldots & 1 \\ 1 & 2+T & 2 & \ldots & 2 \\ 1 & 2 & 3+T & \ldots & 3 \\ \vdots & \vdots & \vdots & \ddots & \vdots \\ 1 & 2 & 3 & \ldots & L+T \end{array}\right) \end{array}$$

We claim that in equilibrium every hider and searcher strategy must have some positive weight. If this is the case then every hider strategy must provide the same expected payoff, and this payoff will be the value of the game. Denoting a generic searcher strategy s and a generic hider strategy h, where q_i again denotes the probability of the searcher playing strategy s^i, and $p_i = p(h^i)$ the probability of the hider playing h^i:

$$T(h^1, s) = q_1(1+T) + \sum_{i=2}^{L} q_i = T, \tag{16.14}$$

Since:

$$\sum_{i=2}^{L} q_i = 1 - q_1, \tag{16.15}$$

we conclude:

$$q_1 = 1 - \frac{1}{T}, \tag{16.16}$$

Similarly, from the second row we obtain:

$$q_2 = \frac{T - 2 + q_1}{T} = 1 - \frac{1}{T} - \frac{1}{T^2}, \tag{16.17}$$

These two "boundary conditions" on q_1 and q_2 will come in handy shortly. The reader will note that clearly it would be possible to continue in this vein to obtain every q_i as a function of T; we could then find a formula for T by substituting into $\sum_{i=1}^{L} q_i = 1$. One relatively tidy way of doing this is to note that by taking the difference-in-differences between rows, we obtain a homogeneous recurrence relation, which must equal zero in equilibrium:

$$(T(h_{x+1}, s) - T(h_x, s)) - (T(h_x, s) - T(h_{x-1}, s)) = 0, \tag{16.18}$$

$$\Rightarrow q_{x-1} - \frac{1+2T}{T} q_x + q_{x+1} = 0, \tag{16.19}$$

Let:

$$M = \left(\frac{1+2T}{2T} - \frac{1}{2} \left(\left(\frac{1+2T}{T} \right)^2 - 4 \right)^{\frac{1}{2}} \right), \quad (16.20)$$

and:

$$N = \left(\frac{1+2T}{2T} + \frac{1}{2} \left(\left(\frac{1+2T}{T} \right)^2 - 4 \right)^{\frac{1}{2}} \right), \quad (16.21)$$

The solution to our recurrence relation is:

$$q_n = C_1 M^n + C_2 N^n, \quad (16.22)$$

where C_1 and C_2 are constants. To determine their value, we use our initial conditions, equations (16.16) and (16.17). We thus obtain the constants as functions of T:

$$C_1 = \frac{1 + T - T\sqrt{\frac{1+4T}{T^2}} + T^2\sqrt{\frac{1+4T}{T^2}}}{-1 + 2T^2\sqrt{\frac{1+4T}{T^2}} + T\left(\sqrt{\frac{1+4T}{T^2}} - 4\right)}, \quad (16.23)$$

$$C_2 = (-1)\frac{1 + T + T\sqrt{\frac{1+4T}{T^2}} - T^2\sqrt{\frac{1+4T}{T^2}}}{1 + 2T^2\sqrt{\frac{1+4T}{T^2}} + T\left(4 + \sqrt{\frac{1+4T}{T^2}}\right)}, \quad (16.24)$$

Note that $C_1 + C_2 = 1$. So, though we will not write it out in full, we have obtained an expression in terms of T to express the probability of the hider moving or searcher searching in any particular period. Using the notation above, and substituting into $\sum_{i=1}^{L} q_i = 1$, we conclude:

$$C_1 \sum_{i=1}^{L} M^n + C_2 \sum_{i=1}^{L} N^n = 1, \quad (16.25)$$

$$\Rightarrow C_1 \left(\frac{1 - M^{L+1}}{1 - M} \right) + C_2 \left(\frac{1 - N^{L+1}}{1 - N} \right) - C_1 - C_2 = 1, \quad (16.26)$$

Since C_1, C_2, M and N are all functions of L or T, this equation implicitly solves for $\hat{T}$ as a function of L. It can be shown that $C_1 M + C_2 N$ is always positive over the range of values for T we are interested in, and thus q_n is always positive and we do indeed have a valid probability distribution. Since C_1, C_2, M and N are all functions of L or T, we conclude the following:

Theorem 4. *With C_1, C_2, M and N as defined previously, the value, $\hat{T}$, of the silent search-ambush game as a function of L (where $K = 1$) is given implicitly by the equation:*

$$C_1 \left(\frac{1 - M^{L+1}}{1 - M} \right) + C_2 \left(\frac{1 - N^{L+1}}{1 - N} \right) = 2, \quad (16.27)$$

16.5.3 The Game with an Infinite Number of Periods

In the case where L is infinite, the above solution for the value of the single cell simplifies to:

$$\hat{T}_{\infty,1} = 2 , \tag{16.28}$$

If q_n and p_n represents the a priori probability of the hider and searcher electing to move or search, respectively, in the n-th period, we also find that $q_n = p_n = (1/2)^n$. This calculation assumes continuity in the limit of equation (16.27); this can be shown to be reasonable through an alternate (and much simpler) solution technique, namely directly solving the following 2×2 matrix:

$$\begin{array}{cc} & \begin{array}{cc} Search & Ambush \end{array} \\ \begin{array}{c} Move \\ Stay \end{array} & \left(\begin{array}{cc} 1 + T_{\infty,1} & 1 \\ 1 & 1 + T_{\infty,1} \end{array} \right) \end{array}$$

This formulation models the hider and searcher as making a decision at the beginning of each period, rather than each round. This is possible since, as noted, in the infinite one-cell game the hider cannot have any information about the proportion of the space searched, nor is anything gained by knowing what time period they are in (since there is always an infinite number of periods remaining). This is not the case for games with higher values of K or finite L.

16.6 Ambush in Two Cells

Once we move into the more complex set of games where $K > 1$, the situation is complicated by the rapid growth in the number of strategies available to the searcher.[3] For the case of $K = 2$, we restrict ourselves to setting $L = \infty$, rather than attempting to derive an explicit equation for the value of the game under finite L, as we did for the case where $K = 1$. This is not to say that such an equation is impossible; indeed, any game with $K = 2$ and finite L should theoretically be explicitly solveable. The solutions, though, tend to involve large sums of high-order polynomial equations, and, given the awkwardness involved in their manipulation, and the

---

[3] As an aside, note that when L is infinite, we can take every searcher strategy from the case where $K = 1$, which involved searching in a particular period, and associate it with the subset of searcher strategies when $K = 2$ that search for the first time in that same period. That is, we can associate s^1 with $\{s^{1,2}, s^{1,3}, s^{1,4} \ldots\}$, s^2 with $\{s^{2,3}, s^{2,4}, s^{2,5} \ldots\}$, and so on. Thus every searcher strategy when $K = 1$ can be associated with a subset of searcher strategies when $K = 2$ of cardinality $\aleph_0$. The set of searcher strategies when $K = 1$ is of cardinality $\aleph_0$; the set of all searcher strategies when $K = 2$ is thus of cardinality $\aleph_0 \times \aleph_0 = \aleph_0$. An inductive argument demonstrates that for any K, the set of searcher strategies is always of countable cardinality: very briefly, given that the searcher strategy space for $K - 1$ is countable, the addition of a new cell to search in allows each strategy in the previous space to be associated with a countably infinite subset of strategies (note these subsets need not be non-intersecting with one another).

fact that any specific solution would almost certainly be calculated numerically in any case, it seems preferable to examine instead the comparatively tidy infinite case.

In general, the matrix for the two-cell game, $A^{\infty,2}$, for some suitably large but finite value of L, has the following form:

$$
A^{\infty,2} = \begin{array}{c}
 \\
h^1 \\
h^2 \\
h^3 \\
h^4 \\
h^5 \\
h^6 \\
\vdots
\end{array}
\begin{pmatrix}
s^{1,2} & s^{1,3} & s^{1,4} & \cdots & s^{2,3} & s^{2,4} & \cdots \\
1+T & 1+T & 1+T & \cdots & 1 & 1 & \cdots \\
(3+T)/2 & 3/2 & 3/2 & \cdots & 2+T & 2+T & \cdots \\
3/2 & (4+T)/2 & 2 & \cdots & (5+T)/2 & 5/2 & \cdots \\
3/2 & 2 & (5+T)/2 & \cdots & 5/2 & 6+T/2 & \cdots \\
3/2 & 2 & 5/2 & \cdots & 5/2 & 3 & \cdots \\
3/2 & 2 & 5/2 & \cdots & 5/2 & 3 & \cdots \\
\vdots & \vdots & \vdots & \vdots & & \vdots & \vdots
\end{pmatrix}
$$

It can be shown that there is an equilibrium where the searcher gives positive weight only to what we term *wait-then-exhaustive* strategies. These are those such that, once the searcher starts searching, they do not stop until the entire space has been inspected; i.e., if they search for the first time in period n, then they also search in period $n+1$.

Given this behaviour by the searcher, the payoffs faced by the hider are then as follows:

$$
T_{\infty,2}(h^m, s^{n,n+1}) = \begin{cases} m+T & \textbf{if } m=n \\ m+\frac{1}{2}+\frac{T}{2} & \textbf{if } m=n+1 \, , \\ \min\{m, n+\frac{1}{2}\} & \textbf{otherwise} \end{cases} \tag{16.29}
$$

Denote by $\bar{A}^{2,\infty}$ the reduced matrix for infinite L that excludes any strategies other than those that are wait-then-exhaustive. Thus:

$$
\bar{A}^{2,\infty} = \begin{array}{c}
 \\
h^1 \\
h^2 \\
h^3 \\
h^4 \\
h^5 \\
h^6 \\
\vdots
\end{array}
\begin{pmatrix}
s^{1,2} & s^{2,3} & s^{3,4} & s^{4,5} & \cdots \\
1+T & 1 & 1 & 1 & \cdots \\
(3+T)/2 & 2+T & 2 & 2 & \cdots \\
3/2 & (5+T)/2 & 3+T & 3 & \cdots \\
3/2 & 5/2 & (7+T)/2 & 4+T & \cdots \\
3/2 & 5/2 & 7/2 & (9+T)/2 & \cdots \\
3/2 & 5/2 & 7/2 & 9/2 & \cdots \\
\vdots & \vdots & \vdots & \vdots & \ddots
\end{pmatrix}
$$

This game can be solved using a very similar approach to that which was employed for the case where $K = 1$, but, as the resulting equations are substantially more complex and space-consuming, we will omit the details and state only that it is possible to prove the following, where φ is the golden ratio ($1.618\ldots$), and Φ is the inverse of the golden ratio ($0.618\ldots$):

Theorem 5. $\hat{T}_{\infty,2}$, *the value of the silent search-ambush game when $K = 2$, $L = \infty$, is equal to* $\frac{1}{1-\Phi}$, *or equivalently to* φ^2, *that is, approximately* 2.618.

16.6.1 Optimal Searcher and Hider Behaviour

Optimal hider and searcher strategies that support this game value (we do not
address the question of whether the same value can be obtained through alternate
strategies; that is, whether the equilibrium described in the following is unique) are
as follows:

$$h^1 = 1 - \Phi, \tag{16.30}$$

$$h^j = \Phi h^{j-1}, \ \forall j > 1, \tag{16.31}$$

Again, $q(s^{n,n+1})$ represents the probability of the searcher playing the wait-then-
exhaustive strategy that commences in period n, which for this section we will
simply shorten to q_n. Then, to 3 decimal places:

$$q_n = -0.555(-0.521)^n + 0.555(0.594)^n, \tag{16.32}$$

The presence of the $(-0.521)^n$ term in the formula for q_n results in the
probabilities attached to the searcher's strategies oscillating: $q_n > q_{n-1}$ for odd n,
and $q_n < q_{n-1}$ for even n, while we always have $q_{n-2} > q_n$.

These raw probabilities on their own, however, are not particularly informative.
More revealing questions concern the equilibrium beliefs of the hider: to start with,
in each period the hider must have a certain belief, conditional on that period being
reached, regarding the likelihood that the searcher will search. In the first period
of any round, this is easily calculated: the probability of search is equal to the sum
of the probabilities attached to all searcher strategies that search in the first period.
Where $K = 2$, the only such strategy with a positive weight is $s^{1,2}$.

If a round progresses to the second period, however, the hider should discount
their belief in the probability that the searcher is playing $s^{1,2}$, as the fact of their
survival is evidence against that proposition (and if the third period is reached, the
hider should naturally dismiss any possibility of $s^{1,2}$).[4] An application of Bayes'
Theorem allows us to calculate $P_h(\bar{s}|R_z)$, that is, the belief held by the hider (de-
noted P_h) regarding the probability of any particular searcher strategy (denoted $\bar{s}$)
conditional on period z being reached (denoted R_z). The a priori probabilities that
the searcher actually attaches to their strategies $s \in S$ are denoted $q(s)$ (remembering
that in equilibrium the hider can be treated as if they were aware of these probabilities). Reiterating that s_j is either 0 or 1, representing whether ambush or search
occurs in the j-th period, for general K the appropriate formula is the following:

$$P_h(\bar{s}|R_z) = \frac{\left(1 - \frac{\sum_{j=1}^{z-1} \bar{s}_j}{K}\right) q(\bar{s})}{\sum_{s \in S} \left(1 - \frac{\sum_{j=1}^{z-1} s_j}{K}\right) q(s)}, \tag{16.33}$$

---

[4] Under the predator-prey interpretation, the prey is thus employing a variant of the anthropic
principle: that they have not yet been eaten allows them to draw certain conclusions regarding the
likely states of the world (specifically the predator's choice of strategy).

For clarity, note that the term $\sum_{j=1}^{z-1} s_j$ represents the number of times a strategy s searches before period z is reached. For example, at the start of the second period when $K = 2$:

$$P_h(s^{1,2}|R_2) = \frac{\frac{1}{2}q(s^{1,2})}{(1 - q(s^{1,2})) + \frac{1}{2}q(s^{1,2})}, \qquad (16.34)$$

On a brief tangent, once we calculate the updated probability the hider attaches to each strategy, we can calculate the hider's belief regarding the expected number of cells remaining unsearched, conditional on any particular period in a round being reached (essentially, by taking the sum of the cells remaining under each strategy while using the updated hider beliefs in the likelihood of each strategy as weights). This is shown in Fig. 16.1, where in the first period the expectation is, naturally, 2, with a 2-period oscillation converging to a value close to 1.745 in the limit. This is interesting as when first hearing the description of the game it might seem intuitive that the hider must believe the searcher is coming ever-closer as time passes, but the hider's actual beliefs are in fact substantially more complicated.

A related and in some ways more significant calculation is the hider's beliefs concerning the conditional probability that they will be captured in each period assuming they remain stationary. To calculate this, suppose we are looking at period z. Take all the searcher strategies which search in that period. Multiply the conditional probability the hider attaches to each strategy by the probability that, if that strategy genuinely is what the searcher is playing, the hider will be caught (which will be equal to $1/(K - m)$, where m is the number of cells that have been inspected in periods prior to z under each strategy). The sum of these products is the value we are interested in, and is depicted, along with the conditional probability that the searcher is searching, in Fig. 16.2.

This is a rather interesting chart: as with Fig. 16.1, both series oscillate in a cycle of length 2, converging to values of around 0.558 (for the probability of search) and 0.406 (for the probability of capture if stationary).

There is an intuitive interpretation of this behaviour that is worth emphasising. Note that the oscillations are out of phase; that is, the probability of search is high when the probability of capture if stationary is low and vice versa. Since the probability of search is equal to the probability of successfully starting a new round if the hider chooses to move, the logic here is that, to keep the hider indifferent between moving and not moving, the searcher ensures that an increase in the likelihood of being able to start a new round is balanced by a decrease in the chance of capture if stationary. The hider always knows that no cells have been inspected at the start of a round, and thus the searcher is required to attach a relatively high probability to searching in the first period. The hider's beliefs then converge to constant values (which represent a sort of equilibrium within the broader equilibrium of the game), though in practice the chance of any round in the $K = 2$ game lasting more than a handful of periods is remote.

Numerical analysis suggests similar cycles occur for games with higher values of K, where the cycles are in turn of length K. This is suggested by the results shown in Fig. 16.3, for values of K between 1 and 6 and $L = 18$.

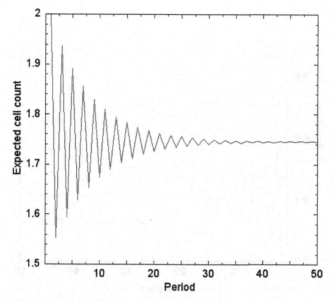

Fig. 16.1 Hider's expectation of number of cells remaining ($K = 2, L = \infty$)

The numerical results will be discussed slightly further in the next section.

16.7 Numerical Results and Conjectures

We start by presenting a conjecture concerning the value of the game in the case where L is infinite.

Conjecture 1 The value of the silent discrete search-ambush game for finite K, $L = \infty$, satisfies:

$$\hat{T}_{\infty,K} \geq \frac{1}{1 - \sigma_K},\qquad(16.35)$$

where σ_K is the unique real solution between 0 and 1 to the polynomial:

$$\sum_{i=1}^{K} x^i - \frac{K}{2} = 0,\qquad(16.36)$$

It should be noted that an earlier version of this conjecture had an equality in place of this inequality; however, further numerical investigation has made that possibility seem less likely. The equality does hold, at least, for the cases of $K = 1$

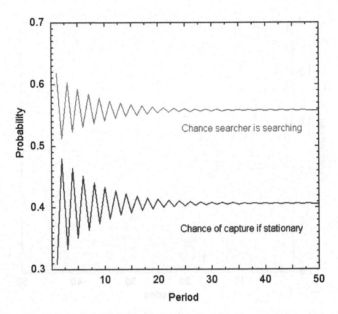

Fig. 16.2 Conditional beliefs of hider ($K = 2, L = \infty$)

and $K = 2$, where σ_K is $\frac{1}{2}$ and Φ, respectively, ensuring consistency with our conclusions in Sects. 16.5 and 16.6. For interest, the polynomial that generates σ_K only ever has at most two real solutions, and always exactly one real solution between 0 and 1.

Table 16.1 gives some detail on specific values of σ_K and the corresponding conjectured game lower bound up to $K = 6$, comparing them to the numerical estimates when $L = 18$.

In the limit, as $K \to \infty$, $\sigma_K \to 1$, and in turn the conjectured lower bound on the game also approaches ∞. This is intuitively sensible: in an infinite search space, the hider should be able to evade capture indefinitely. In addition, for large K the bound on the value of the game appears to converge to a linear function, increasing by around 0.62749 for every unit increase in K.

To convert the discrete search-ambush game into its continuous equivalent, we can assume that each cell, rather than taking one time period to search, takes time $1/K$; this is equivalent to dividing up a continuous search space of area unity that takes one time period to search into K subsections. We re-obtain the continuous search space by taking $K \to \infty$, assuming continuity in the game values in the limit. If the conjecture is true, this suggests the value of the continuous search-ambush game with a silent predator is greater than or equal to 0.62749, while Alpern et al. [3] have previously determined that the value of the game with a noisy predator is 0.666....

We also make a further conjecture regarding hider and searcher behaviour in equilibrium.

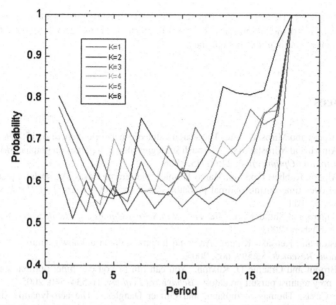

Fig. 16.3 Hider's beliefs in conditional probability that search is occurring (numerical estimates) ($K = 1$ to 6, $L = 18$)

Table 16.1 Conjectured and estimated game values and supporting data

	$K=1$	$K=2$	$K=3$	$K=4$	$K=5$	$K=6$
σ_K	0.5000	0.6180	0.6914	0.7413	0.7773	0.8046
$\frac{1}{1-\sigma_K}$	2.0000	2.6180	3.2406	3.8651	4.4905	5.1165
Estimated $\hat{T}_{18,K}$	2.0000	2.6180	3.2420	3.8715	4.5057	5.1496

> **Conjecture 2** In any silent discrete search-ambush game for finite K where $L = \infty$, $\exists x$ such that for all hider probabilities p_j where $j > x$, $\frac{p_j}{p_{j-1}} = z$ for some constant z which is a function of K. Further, $\exists y$ such that all searcher strategies that involve search in periods $i > y$ are wait-then-exhaustive.

Naturally, substantial work remains to be done regarding investigation of these conjectures. A reasonable starting step – and one that ought to be achievable using mathematics no deeper than that used in obtaining the above results – would be to analytically solve the case where $K = 3$, $L = \infty$. This would either substantially strengthen or undermine the patterns identified above, and in particular determining a value for the $K = 3$ case may enable more precise speculation for the value under general K than Conjecture 1 offers (or even lend weight again to the possibility that Conjecture 1 holds as an equality).

Acknowledgements Thanks to Steve Alpern who, besides inventing the problem under investigation, provided an extremely high level of support throughout. Thanks also to Mathew Abbott and Edward Cavanagh for their able commentary.

References

1. Steve Alpern and Miroslav Asic. The search value of a network. *Networks*, 15:229–238, 1985.
2. Steve Alpern and Miroslav Asic. Ambush strategies in search games on graphs. *SIAM Journal on Control and Optimization*, 24:66–75, 1986.
3. Steve Alpern, Robbert Fokkink, Marco Timmer, and Jerome Casas. Ambush frequency should increase over time during optimal predator search for prey. *Journal of The Royal Society interface*, 8, 2011.
4. Steve Alpern and Shmuel Gal. *The Theory of Search Games and Rendezvous*. Kluwer Academic Publishers, 2003.
5. Vic Baston and Kensaku Kikuta. An ambush game with an unknown number of infiltrators. *Operations Research*, 52:597–605, 2004.
6. Mark Broom and Graeme D. Ruxton. You can run or you can hide: optimal strategies for cryptic prey against pursuit predators. *Behavioral Ecology*, 16:534–540, 2005.
7. Jerome Casas, Thomas Steinmann, and Olivier Dangles. The aerodynamic signature of running spiders. *PLoS ONE*, 3, 2008.
8. Shmuel Gal. *Search Games*. Academic Press, 1980.
9. James W. A. Grant and David L. G. Noakes. Movers and stayers: foraging tactics of young-of-the-year brook charr, salvelinus fontinalis. *Journal of Animal Ecology*, 56:1001–1013, 1987.
10. Raymond B. Huey and Eric R. Pianka. Ecological consequences of foraging mode. *Ecology*, 62:991–999, 1981.
11. Tamiji Inoue and Toshiaki Marsura. Foraging strategy of a mantid, paratenodera angustipennis s.: mechanisms of switching tactics between ambush and active search. *Oecologia*, 56:264–271, 1983.
12. Farmey A. Joseph. Path-planning strategies for ambush avoidance. Master's thesis, 2005.
13. S.P. Lalley. A one-dimensional infiltration game. Technical report, Purdue University, 1987.
14. Cader W. Olive. Behavioral response of a sit-and-wait predator to spatial variation in foraging gain. *Ecology*, 63:912–920, 1982.
15. William Ruckle, Robert Fennell, Paul T. Holmes, and Charles Fennemore.
16. Thomas W. Schoener. Theory of feeding strategies. annual review of ecology and systematics. *Annual Review of Ecology and Systematics*, 2:369–404, 1971.
17. Alan R. Washburn. An introduction to evasion games. Technical report, United States Naval Postgraduate School, 1971.
18. N. Zoroa, M.J. Fernandez-Saez, and P. Zoroa. A foraging problem: sit-and-wait versus active predation. *European Journal of Operational Research*, 208:131–141, 2011.

Chapter 17
A Model of Partnership Formation with Friction and Multiple Criteria

Stephen Kinsella and David M. Ramsey

Abstract We present a game theoretical sequential search problem modelling partnership formation based on two discrete character traits. There are two classes of individual. Each individual observes a sequence of potential partners from the other class. The traits are referred to as attractiveness and character, respectively. All individuals prefer partners of high attractiveness and similar character. Attractiveness can be measured instantly. However, in order to observe the character of an individual, a costly interview (or date) is required. On observing the attractiveness of a prospective partner, an individual must decide whether he/she wishes to proceed to the interview stage. During the interview phase, the prospective pair observe each other's character and then decide whether they wish to form a pair. Mutual acceptance is required for both an interview to occur and a pair to form. An individual stops searching on finding a partner. A set of criteria based on the concept of a trembling hand perfect equilibrium is used to define an equilibrium of this game. It is argued that under such a general formulation there may be multiple equilibria. For this reason, we define a specific formulation of the game, the so called symmetric version, which has a unique symmetric equilibrium. The form of this equilibrium has some similarities to the block separating equilibrium derived for classical models of two-sided mate choice and job search problems, but is essentially different.

S. Kinsella (✉)
Department of Economics, University of Limerick, Limerick, Ireland
e-mail: stephen.kinsella@ul.ie

D.M. Ramsey
Department of Mathematics and Statistics, University of Limerick, Limerick, Ireland
e-mail: david.ramsey@ul.ie

S. Alpern et al. (eds.), *Search Theory: A Game Theoretic Perspective*,
DOI 10.1007/978-1-4614-6825-7__17, © Springer Science+Business Media New York 2013

17.1 Introduction

This chapter presents a game theoretical sequential search problem modelling partnership formation based on two traits. There are two classes of individual, and each individual wishes to form a partnership with one of the opposite class. Each individual observes a sequence of potential partners. Mutual acceptance is required for a partnership to form. On finding a partner, an individual ceases searching. One measure describes 'attractiveness'. Preferences are common according to this measure: i.e. each individual prefers highly attractive partners, and all individuals of a given class agree as to how attractive individuals of the opposite class are. Preferences are homotypic with respect to the second measure, referred to as 'character', i. e. all individuals prefer partners of a similar character.

For convenience, it is assumed that individuals know their own attractiveness and character. Also, the distributions of character and attractiveness are assumed to be discrete with a finite support, and constant over time.

Together, the attractiveness, character and the class of an individual determine their 'type'. Individuals can observe the attractiveness and character of prospective partners perfectly. However, in order to measure character, a costly interview is required. In addition, individuals incur search costs at each stage of the search process.

At each moment an individual is paired with a prospective partner. First, both individuals must decide whether they wish to proceed to the interview stage on the basis of the attractiveness of the prospective partner. The final decision on pair formation is based on both the attractiveness and character of the prospective partners. Mutual acceptance is required for an interview to occur and a pair to form. At equilibrium, each individual uses a strategy appropriate to their type. The set of strategies corresponding to such an equilibrium is called an equilibrium profile.

Such a problem may be interpreted as a mate choice problem in which the classes are male and female, or a job search problem in which the classes are employer and employee. The assumption that attractiveness can be observed very quickly, but an interview (dating) is required to observe someone's character is obviously a simplification. However, in the case of human mate choice many traits that can be thought of as defining attractiveness (physical attractiveness, employment, economic status) are usually measured quickly, whilst observation of traits defining character (political and religious views, tastes and emotions) are generally more difficult to measure.

17.1.1 Related Literature

Our model has a resemblance to 'speed dating' models recently developed and tested by Fisman et al. [16]. The types of preferences we study have been analyzed (in continuous time) by Marimon and Zilibotti [26]. They found that these types of preferences, in this context, were quite tractable and somewhat equivalent to the formulation with ex post idiosyncratic uncertainty. Other recent directed search papers in the same vein are Albrecht et al. [2] and Galenianos and Kircher [19], looking at directed search in job applications.

The benchmark model in this literature is Smith [35], which focuses on assortative matching and block segregation in 'marriage' models (see also Burdett [7], Burdett and Coles [8], Coles and Burdett [10], Chade and Smith [9]). The costly application/invitation-stage of the model, follows Shimer [33] . The novelty of our paper lies in the multi-dimensionality of agents' types.

SOME UNOBSERVABLE TRAITS

In the case of such a model of job search, it is assumed that a job's attractiveness (interpreted as pay, conditions, status) can be observed from an advert. The attractiveness of a job seeker (interpreted as qualifications and experience) are readily seen from his/her application. It is assumed that in order to observe character a costly interview is required. In terms of job search, this would seem reasonable when there is not much information regarding the workers in and the ethos of a firm (e.g. it may be a small or new firm) and the skills required from a worker rather generic. Similarity of character may be interpreted as the ability of the employer (or department) and employee to work together as a team. Several labour market studies have found empirical evidence that employers and employees are happiest with labour market choices they view as similar to themselves in some respects (e.g. labour market type, an organisation's ethos, educational level), see Peterson[30], Beller [6], Albelda [1].

In the literature to date, job search models typically adopt a matching function approach, where the employer and employee search for the perfect 'fit' using a set of costly criteria, see Coles and Burdett [10]. Equilibrium conditions are derived and tested for robustness once the model is built, and policy recommendations follow Burdett [7], Pissarides [31], Drewlanka [14], Shimer and Smith [34]. Jovanovic [24], Hey [21], and MacMinn [25] are the classic studies. Devine and Piore [12] and Shimer and Smith [34] survey the more recent developments.

Another strand in the literature is the search-theoretic literature developed by McCall [27] and extended by Diamond and Maskin [13] and others, where the job search problem is conceived of as a dynamic program which has to be solved in finite time, so labour markets are best described by optimal control problems. The literature here is vast and well studied.

Job search has also been modelled as a mating or network game, with representative contributions being Albelda [1], Beller [6], Peterson et al. [30], Coles and Francesconi [11], Fisman et al. [17], Pissarides [31]. In the biological literature, mate choice is modelled as sequential observation of prospective mates. The model presented here is a development of this strand. For models of mutual mate choice based on common preferences see Johnstone [23], Alpern and Reyniers [4], and Janetos [22].

For a model of mate choice based on homotypic preferences see Alpern and Reyniers [3]. Their models assume that the distribution of the value of available mates changes over time as partnerships form and individuals leave the mating pool. Ramsey [32] considers a similar problem interpreted as a job search problem.

The general approach is presented in Sect. 17.2. Section 17.3 compares the approach used here and classical models of two-sided job search and mate choice problems. It is intended that this section will give an intuitive feel for the approach to solving such problems and the added complexity involved when common and homotypic preferences are combined. Section 17.4 describes a model of partnership formation in which character forms a 'circle' and neither the distributions of the traits nor the search costs depend on the class of a searcher. This model is called symmetric. Section 17.5 gives the set of criteria that we wish an equilibrium to satisfy. These conditions are based on the concept of a trembling hand perfect equilibrium (a refinement of the concept of Nash equilibrium). Section 17.6 describes a general method for calculating the expected utilities of each individual under a given strategy profile. Section 17.7 considers the dating subgame (when individuals decide whether to form a partnership) and the soliciting subgame (when individuals decide whether to go on to the dating subgame). Section 17.8 presents results on the existence and uniqueness of a symmetric equilibrium in the symmetric game. An algorithm for determining this equilibrium is presented, together with an example. Section 17.9 illustrates the problems involved in adopting such an approach to the more general formulation. Section 17.10 gives a brief conclusion and suggests directions for further research.

17.2 General Formulation

We present a model of a sequential decision process leading to the formation of a long term partnership. It is assumed that there are two classes of player and individuals in a partnership have to be of different classes (e.g. in job search problems employees form partnerships with employers, in mate choice problems males form partnerships with females). Individuals view a sequence of prospective partners. It is assumed that costs are incurred during the search process, so in general an individual should not continue searching until he finds his/her best possible partner. The following assumptions are made:

(A) We consider the formation of long term relationships between two classes of player (e.g. marriage, employment). When a partnership is formed, the two individuals involved leave the population of searchers. Henceforth, we will refer to these two classes of players as males and females.

(B) Interactions are bilateral and occur between a male and a female. The length of an interaction is assumed to be small (effectively of zero length) compared to the time between interactions. The pair must decide whether to form a partnership or continue searching. Mutual acceptance is required for a partnership to form. Individuals cannot return to a previously met prospective partner.

(C) When an individual leaves the population, he/she is replaced by a clone, i.e. an individual of the same sex, attractiveness and character enters the population of searchers. Hence, the joint distribution of attractiveness and character is fixed. It might be more realistic to consider a steady-state model in which individuals enter the pool of searchers at a steady rate according to their sex, attractiveness and character. However, due to the novelty of the approach used and some of the issues involved in deriving equilibria, for the present we adopt the simpler clone replacement approach. It is intended that the steady state approach will be adopted in future work.

(D) It is assumed that time is discrete and the search costs of males and females per unit time are c_1 and k_1, respectively. At each moment in time a player encounters a prospective partner. We assume random matching, i.e. the attractiveness and character of the female encountered by a male is chosen at random from the joint distribution of attractiveness and character among females. Using the assumption of clone replacement, it is also easy to adapt the model to assume that encounters occur as a Poisson process. It may be assumed that individuals find prospective partners at rate 1 and males and females pay search costs of c_1 and k_1 per unit time, respectively. Empirical evidence on search costs in real-life job search problems abounds in the literature. For instance, Peterson et al. [30] consider job search costs to be the primary cause of 'sticky' wages and low labour market mobility, when contrasting the European and US labour markets. Devine and Piore [12] also present empirical evidence on search costs, and in the extensive literature incorporating job search costs into macroeconomic models, Andolfatto [5] is representative.

(E) Encounters have two stages. In the first stage, both individuals must decide whether they wish to date based on the attractiveness of the prospective partner. For convenience, it is assumed that these decisions are made simultaneously. Dating only occurs by mutual consent. The costs of dating to males and females are c_2 and k_2, respectively. During a date, each observes the character of the other and decides whether to accept the other as a partner. Again, it is assumed that these choices are made simultaneously. A partnership is formed only by mutual consent. In some scenarios—for example, when character is unimportant—it might pay a female to immediately accept a male without dating. However, to keep the strategy space as simple as possible, it is assumed that individuals must always date before forming a pair. The total utility of an individual is taken to be the utility gained from the partnership minus the sum of the search costs and dating costs incurred. It is assumed that utility is not transferable and individuals maximise their expected total utility from search.

This approach implicitly assumes that there is the same number of males as females. However, the model can be easily adapted to allow the number of males and females to differ. Suppose that there are r times as many males as females. In this case, we may assume that at each moment a proportion $(r-1)/r$ of males meet a prospective partner who would give them an expected utility of $-\infty$. In reality, such males do not meet a prospective partner.

The supergame Γ is defined to be the game in which each player observes a sequence of prospective partners as described above. An encounter between two prospective partners will be referred to as the encounter game. The encounter game is split into two subgames, the soliciting subgame, when players decide whether to date or not and the dating subgame, when they decide whether to form a partnership or not.

17.3 Comparison with the Classical Partnership Formation Game

Two-sided problems are by nature game-theoretic and so we look for a Nash equilibrium solution at which no individual can improve their expected utility by changing their strategy. Note that there may be multiple Nash equilibria. For example, suppose that mate choice is based only on attractiveness and there are only two levels of attractiveness: high and low. Suppose individuals of high attractiveness only accept individuals of low attractiveness as partners. Similarly, individuals of low attractiveness only accept individuals of high attractiveness as partners. It can be seen that this is a Nash equilibrium, since e.g. a male of high attractiveness cannot gain by accepting a female of high attractiveness, since she would not accept him. Also, he could not gain by rejecting a female of low attractiveness, since he would not find a partner. However, one would expect that if a male accepts a female of attractiveness x, then he would accept any female of attractiveness $>x$. McNamara and Collins [28] derive an equilibrium for a game in which choice is based only on attractiveness which satisfies such a condition, referred to as the **optimality criterion**. This criterion states that any individual accepts a prospective partner if and only if the utility from such a partnership is as least as great as the expected utility of the individual given that he/she continues searching. Such an equilibrium can be derived inductively. Consider a female of maximum attractiveness. She will be acceptable to any male. Hence, such females face a one-sided problem and their equilibrium strategy is of the form: accept the first male of attractiveness $\geq x_1$. Call such males first class. It follows that males of attractiveness $\geq x_1$ are acceptable to any female (i.e. face a one-sided problem) and their equilibrium strategy is of the form: accept the first female of attractiveness $\geq y_1$. Call such females first class. It follows that first class males will only pair with first class females. The problem faced by the rest of the population then reduces to a problem in which first class individuals are not present. Define $x_0 = y_0 = \infty$. Arguing iteratively, it can be shown that k classes of males and females can be defined, such that a male of attractiveness x is of class i if $x \in [x_i, x_{i-1})$ and a female of attractiveness y is of class j if $y \in [y_j, y_{j-1})$. Males of class i pair with females of class i. There may be a class of males or females who do not form partnerships.

In the problem considered here, individuals do not always agree on the desirability of a member of the opposite sex as a partner. It would be natural to try and reduce the game considered to a sequence of one-sided choice problems. However, for games within the general framework presented in Sect. 17.2 there are some technical problems associated with such an approach. For example, consider a problem in which attractiveness and character are independent, both have a uniform distribution over the integers $0, 1, 2, \ldots, m$ regardless of sex. It is expected that individuals of maximum attractiveness, m, and close to median character will have a higher expected utility from search (i.e. be choosier) than individuals of attractiveness m and extreme character, either 0 or m (see Alpern and Reyniers [3]). In the problem considered by McNamara and Collins [28] it is relatively easy to order individuals according to how choosy they should be. This ordering is used to derive the unique equilibrium satisfying the optimality criterion. Such an ordering is not so easy in the problem considered here. For example, should a male of attractiveness m and character 0 be more or less choosy than a male of attractiveness $m - 1$ and close to median character? Ramsey [32] shows that multiple equilibria may exist in such a problem, i.e. in general there is no unique sequence of one-sided problems that can be solved to define an equilibrium.

17.4 The Symmetric Model with Character Forming a Circle

Due to the problems outlined in the previous section, we present a model which allows us to adopt a similar (but not identical) approach to the one used by McNamara and Collins [28]. Attractiveness and character are denoted X_a and X_c, respectively. The population is assumed to be large. It will be assumed that

(a) X_a and X_c are independent. The distribution of X_a does not depend on sex. The distribution of X_c in both sexes is uniform on the integers $0, 1, \ldots, m - 1$.
(b) The difference between characters is calculated according to modulo m, i.e. character can be thought of as a circle with 0 and $m - 1$ being neighbouring characters.
(c) Search and dating costs are c_1 and c_2, respectively, independently of sex.
(d) The utility obtained by a type $\mathbf{x} = [x_a, x_c]$ individual from pairing with a prospective partner of type $\mathbf{y} = [y_a, y_c]$ is given by $g(y_a, |x_c - y_c|)$, i.e. the utility function is independent of sex.

Using such an approach, intuitively an individual's mating prospects do not depend on his/her character or sex. Such a game will be referred to as symmetric.

17.5 Equilibrium Conditions

For a game defined within the general framework, introduced in Sect. 17.2, we require an equilibrium profile to satisfy the following generalisation of the optimality criterion for the classical two-sided problem. Namely:

Condition 1: In the dating subgame, an individual accepts a prospective partner if and only if the utility from such a pairing is at least as great as the individual's expected utility from future search (ignoring previous costs).

Condition 2: An individual is only willing to date if their expected utility from the resulting dating subgame minus the costs of dating is as least as great as their expected utility from future search.

Condition 3: The decisions made by an individual do not depend on the moment at which the decision is made.

It should be noted that the expected future utility of an individual from search, and thus the exact form of the dating and soliciting subgames, depends on the strategy profile used in the population as a whole. This dependency will be considered more fully in Sect. 17.7.

The most preferred prospective partners of a type $[x_a, x_c]$ individual are those of maximum attractiveness who have character x_c. Condition 1 states that in the dating subgame an individual will always accept his/her most preferred partner, since an individual's future expected utility from search must be less than the utility from obtaining his most preferred partner. Moreover, if in the dating subgame a male accepts a female who would give him a utility of k, then he must accept any female who would give him a utility of at least k. It follows that the acceptable difference in character is non-decreasing in the attractiveness of a prospective partner.

Condition 3 states that the Nash equilibrium strategy should be stationary. This reflects the following facts:

(a) An individual starting to search at moment i faces the same problem as one starting at moment 1.

(b) Since the search costs are linear, after searching for i moments and not finding a partner, an individual maximises his/her expected utility from search simply by maximising the expected utility from future search (i.e. by ignoring previously incurred costs).

We might also be interested in profiles that satisfy the following condition.

(i) In the soliciting subgame, an individual of attractiveness y_a is willing to date prospective partners of attractiveness above some threshold, denoted $t(y_a)$, such that if $y_1 > y_2$, then $t(y_1) \geq t(y_2)$, i.e. the more attractive an individual, then the choosier he/she is when choosing a dating partner.

However, it seems reasonable that individuals of low attractiveness may not solicit dates from highly attractive prospective partners, as by doing so they might incur dating costs while it is expected that such a date will not lead to pair formation. Hence, we do not require individuals to use threshold rules when deciding whether to solicit a date. For example, Härdling and Kokko [20] argue that in certain circumstances small males should avoid courting attractive females to avoid the possibility of attacks from larger males.

Definition 1. An equilibrium profile, denoted π^*, of Γ is a strategy profile under which the behaviour of all searchers satisfies Conditions 1–3 in each of the possible subgames. The value function of Γ corresponding to π^* is the set of expected utilities of each individual according to type under the strategy profile π^*.

Note that π^* must define the appropriate behaviour in all possible dating subgames, even those that do not occur under π^*.

In the particular case of the symmetric game, we wish to find an equilibrium which is symmetric with respect to character and sex. That is to say:

Condition 4: If an individual of type $[x_a, x_c]$ is willing to date a prospective partner of attractiveness y_a, then any individual of attractiveness x_a is willing to date a prospective partner of attractiveness y_a.

Condition 5: If an individual of type $[x_a, x_c]$ is willing to pair with a prospective partner of type $[y_a, y_c]$ in the dating subgame, then an individual of type $[x_a, x_c + i]$ is willing to pair with a prospective partner of type $[y_a, y_c + i]$ [addition is carried out $\mod(m)$].

An equilibrium which satisfies Conditions 4 and 5 will be referred to as a symmetric equilibrium. Note that under a symmetric strategy an individual's expected utility from search is independent of sex and character, i.e. only depends on attractiveness.

Note that at equilibrium, if an individual of type $[i, k]$ is willing to date an individual of attractiveness j, then he/she must be willing to pair with some prospective partners of attractiveness j (otherwise unnecessary dating costs are incurred). Hence, a type $[j, k]$ prospective partner (the most preferred partner of such attractiveness) must be acceptable in the corresponding dating subgame.

17.6 Deriving the Expected Utilities Under a Given Strategy Profile

Consider the symmetric game described above. We will look for a symmetric equilibrium profile, thus we may assume that the strategy profile used is symmetric (i.e. satisfies Conditions 4 and 5). Given the strategy profile used by a population, we can define which pairs of types of individuals proceed to the dating subgame and which pairs of types of individuals form pairs. From this, it is relatively simple to calculate the expected length of search and the expected number of dates of an individual of a given type.

Let $p(\mathbf{x})$ be the probability that an individual is of type $\mathbf{x}$. Define $M_1(\mathbf{y}; \pi)$ to be the set of types of prospective partners that an individual of type $\mathbf{y}$ will date (under the assumption of mutual acceptance) under the strategy profile π. Define $M_2(\mathbf{y};\pi)$ to be the set of types of prospective partners that eventually pair with an individual of type $\mathbf{y}$. By definition $M_2(\mathbf{y};\pi) \subseteq M_1(\mathbf{y};\pi)$.

The expected length of search of an individual of type $\mathbf{y}$, $L(\mathbf{y};\pi)$, is the reciprocal of the probability of finding a mutually acceptable partner at any given stage. The expected number of dates of such an individual, $D(\mathbf{y};\pi)$, is the expected length of search times the probability of dating at any given stage. Hence,

$$L(\mathbf{y};\pi) = \frac{1}{\sum_{\mathbf{x} \in M_2(\mathbf{y};\pi)} p(\mathbf{x})}; \quad D(\mathbf{y};\pi) = \frac{\sum_{\mathbf{x} \in M_1(\mathbf{y};\pi)} p(\mathbf{x})}{\sum_{\mathbf{x} \in M_2(\mathbf{y};\pi)} p(\mathbf{x})}. \tag{17.1}$$

Note: The number of prospective mates seen and the number of individuals dated by an individual of type $\mathbf{y}$ have geometric distributions with parameters $1/L(\mathbf{y};\pi)$ and $1/D(\mathbf{y};\pi)$, respectively.

The expected utility of a type $\mathbf{y}$ individual from forming a pair under the strategy profile π is the expected utility from pairing given that the type of the prospective partner is in the set $M_2(\mathbf{y};\pi)$. Hence, the individual's expected total utility from search, denoted $R(y_a;\pi)$ since this expected utility depends only on an individual's attractiveness, is given by

$$R(y_a;\pi) = \frac{\sum_{\mathbf{x} \in M_2(\mathbf{y};\pi)} p(\mathbf{x}) g(x_a, |x_c - y_c|)}{\sum_{\mathbf{x} \in M_2(\mathbf{y};\pi)} p(\mathbf{x})} - c_1 L(\mathbf{y};\pi) - c_2 D(\mathbf{y};\pi). \tag{17.2}$$

Note that it is relatively simple to extend these calculations to non-symmetric games.

17.7 The Dating and Soliciting Subgames

Since this game is solved by recursion in the manner developed by Spear [36, 37], we first consider the dating subgame. In this section we consider the general formulation of the game.

17.7.1 The Dating Subgame

Assume that the population are following a symmetric strategy profile π. The male and female both have two possible actions: accept the prospective partner, denoted a, or reject, denoted r. Also, we ignore the costs already incurred by either individual, including the costs of the present date, as they are subtracted from all the payoffs in the matrix, and hence do not affect the equilibria in this subgame.

Suppose the male is of type $\mathbf{x}$ and the female is of type $\mathbf{y}$. The payoff matrix is given by

$$
\begin{array}{cc}
& \text{Female: } a \qquad\qquad\qquad\quad \text{Female: } r \\
\begin{array}{c} \text{Male: } a \\ \text{Male: } r \end{array}
&
\left(
\begin{array}{cc}
[g(y_a,|x_c - y_c|), g(x_a,|x_c - y_c|)] & [R(x_a;\pi), R(y_a;\pi)] \\
[R(x_a;\pi), R(y_a;\pi)] & [R(x_a;\pi), R(y_a;\pi)]
\end{array}
\right).
\end{array}
$$

Note that the payoff matrix depends on the strategy profile used via the expected utilities of the individuals involved in a subgame. These expected utilities were derived in Sect. 17.6.

From Condition 1, at equilibrium an individual accepts a prospective partner if and only if the utility gained from such a partnership is at least as great as the expected utility from future search. Hence, the appropriate Nash equilibrium of this subgame is for the male to accept the female if and only if $g(y_a,|x_c - y_c|) \geq R(x_a;\pi)$ and the female to accept the male if and only if $g(x_a,|x_c - y_c|) \geq R(y_a;\pi)$.

For convenience, we assume that when $g(x_a,|x_c - y_c|) = R(y_a;\pi)$, a female always accepts the male (in this case she is indifferent between rejecting and accepting him). Similarly, if $g(y_a,|x_c - y_c|) = R(x_a;\pi)$, it is assumed that a male always accepts a female. The implications of this assumption are considered at the end of Sect. 17.8.

Note that if a male rejects a female, then the female is indifferent between accepting or rejecting the male. Under a rule satisfying Condition 1, a female will make an optimal response whatever action the male takes. Such an equilibrium is a trembling hand perfect equilibrium (i.e. robust to the other player making a mistake). Hence, there is a unique trembling hand perfect equilibrium satisfying Condition 1. Let $\mathbf{v}(\mathbf{x},\mathbf{y};\pi) = [v_M(\mathbf{x},\mathbf{y};\pi), v_F(\mathbf{x},\mathbf{y};\pi)]$ denote the value of the dating subgame corresponding to this equilibrium, where $v_M(\mathbf{x},\mathbf{y};\pi)$ and $v_F(\mathbf{x},\mathbf{y};\pi)$ are the values of the subgame to the male and female, respectively.

We now consider the soliciting subgame.

17.7.2 The Soliciting Subgame

Once the dating subgame has been solved, we may solve the soliciting subgame and hence the game $G(\mathbf{x},\mathbf{y};\pi)$, played when a male of type $\mathbf{x}$ meets a female of type $\mathbf{y}$. As before, we assume that the population is following a symmetric strategy profile π.

Both players have two actions: a – accept (solicit a date) and r – do not solicit a date. Since the utility an individual expects from a date is independent of his/her character, the payoff matrix can be expressed as follows:

$$
\begin{array}{cc}
& \text{Female: } a \qquad\qquad\qquad\qquad \text{Female: } r \\
\begin{array}{c} \text{Male: } a \\ \text{Male: } r \end{array}
&
\left(
\begin{array}{cc}
[\bar{v}_M(x_a,y_a;\pi) - c_2, \bar{v}_F(x_a,y_a;\pi) - c_2] & [R(x_a;\pi), R(y_a;\pi)] \\
[R(x_a;\pi), R(y_a;\pi)] & [R(x_a;\pi), R(y_a;\pi)]
\end{array}
\right).
\end{array}
$$

Here $[\bar{v}_M(x_a, y_a; \pi), \bar{v}_F(x_a, y_a; \pi)]$ denotes the expected values of the dating subgame to the male and female, respectively, given the strategy profile used by the population, the measures of attractiveness of the prospective partners and the fact that a date followed.

From Condition 2, the female should solicit a date if and only if her expected utility from such a date is at least as great as the expected utility from future search, i.e.

$$\bar{v}_F(x_a, y_a; \pi) - c_2 \geq R(y_a; \pi).$$

Similarly, the male should solicit a date if

$$\bar{v}_M(x_a, y_a; \pi) - c_2 \geq R(x_a; \pi).$$

Note that Condition 2 requires the players to use a trembling hand perfect equilibrium in any soliciting subgame.

In the next section we present an algorithm to find a symmetric equilibrium of the symmetric game, as an algorithmic game in the tradition of Velupillai [38], Nisam et al. [29].

17.8 A Symmetric Equilibrium of the Symmetric Game

Theorems 1–3 describe the form of a symmetric equilibrium of the symmetric game. These results do not fully characterize such an equilibrium. However, they do justify the logic behind the algorithm presented in Sect. 17.8.1. The constructive form of this algorithm allows us to state the key result of this section, Theorem 5, on the existence and uniqueness of a symmetric equilibrium in the symmetric game.

Theorem 1. *At a symmetric equilibrium π^* of the symmetric game, the utility of an individual is non-decreasing in attractiveness.*

Proof. Assume that for some $i > j$, $R(i; \pi^*) < R(j; \pi^*)$. Consider the dating subgame. From Condition 1, a type $[j, k]$ ($[i, k]$) male accepts a female of type $[i_0, j_0]$ only if the male's reward is greater than $R(j; \pi^*)$ ($R(i; \pi^*)$, respectively). Hence, males of type $[i, k]$ accept any female that a male of type $[j, k]$ would accept. Similarly, if a type $[j, k]$ male is acceptable to a female of type $[i_0, j_0]$, then such a female would accept a type $[i, k]$ male (who gives a greater reward from mating). Hence, $M_2([j, k]; \pi^*) \subseteq M_2([i, k]; \pi^*)$. It follows that a female who is willing to date a male of attractiveness j will also be willing to a date one of attractiveness i (who is expected to be a better partner and at least as likely to be mutually acceptable). Hence, a searcher of type $[i, k]$ can obtain the same expected utility as a searcher of type $[j, k]$ as follows:

(a) In the soliciting subgame, solicit dates with any prospective partner who would date an individual of attractiveness j.

(b) In the dating subgame, accept any prospective partner who would pair with an individual of type $[j,k]$.

Hence, the expected utility of a searcher at such an equilibrium must be non-decreasing in his/her attractiveness.

Corollary to Theorem 1. Suppose $i \geq j$. In the dating subgame, if an individual of attractiveness i accepts one of attractiveness j, then acceptance is mutual. This follows from the fact that the individual of attractiveness j obtains a greater utility from the pairing than the individual of attractiveness i, but has a lower expected utility from future search.

Theorem 2. *At a symmetric equilibrium π^* of the symmetric game, searchers of maximum attractiveness, i_{max}, are willing to date prospective partners of attractiveness above a certain threshold.*

Proof. From Theorem 1, if a searcher of attractiveness i_{max} accepts a prospective partner of type $[i,k]$ in the dating subgame, then acceptance is mutual. Hence, the expected utility of such a searcher from the dating subgame is non-decreasing in the attractiveness of the prospective partner (since character is independent of attractiveness). A searcher should be willing to date a prospective partner if the expected utility from such a date is at least as great as the expected utility from future search. If this condition is satisfied for some level of attractiveness i, then it will be satisfied for all higher attractiveness levels. Note that a searcher of attractiveness i_{max} is willing to date a prospective partner of attractiveness i_{max}, since such dates give the highest possible expected utility.

Theorem 3. *At a symmetric equilibrium π^* of the symmetric game, a searcher of attractiveness i solicits dates with prospective partners of attractiveness in $[k_1(i), k_2(i)]$, where $k_2(i)$ is the maximum attractiveness of a prospective partner willing to date the searcher. In addition, $k_1(i)$ and $k_2(i)$ are non-decreasing in i and $k_1(i) \leq i \leq k_2(i)$.*

Proof. The proof of this theorem is by recursion. Theorem 2 states that for $i = i_{max}$ the equilibrium strategies are of the appropriate form. Assume that Theorem 3 is valid for searchers of attractiveness $\geq i+1$, where $i < i_{max}$.

First, suppose no prospective partner of attractiveness $> i$ will date a searcher of attractiveness i. By ignoring meetings with prospective partners of attractiveness $> i$, the game faced by searchers of attractiveness i can be reduced to a game in which they are the most attractive. From Theorem 2, it follows that searchers of attractiveness i are willing to date prospective partners of attractiveness in $[k_1(i), k_2(i)]$, where $k_2(i) = i < k_2(i+1)$ and $k_1(i) \leq i < k_1(i+1)$.

Now assume that $k_2(i) > i$. It follows that $k_1(i+1) \leq i$. Firstly, we show that searchers are willing to date prospective partners of attractiveness j, where $i \leq j \leq k_2(i)$. If such a prospective partner finds a searcher of attractiveness i acceptable in the dating subgame, then acceptance is mutual. It follows that the expected utility of the searcher from such a date is greater than the expected utility of the prospective partner. Hence, from Theorem 1, a searcher of attractiveness i should be willing to date a prospective partner of attractiveness j.

Secondly, the proof that searchers of attractiveness i should be willing to date prospective partners of attractiveness j, where $j \leq i$, if and only if j is above some threshold is analogous to the proof of Theorem 2. Also, if a prospective partner of type $[k_1(i+1), l]$ is mutually acceptable to a searcher of type $[i+1, k]$ in the dating subgame, then from Theorem 1 a prospective partner of type $[k_1(i+1), l]$ accepts a searcher of type $[k_1(i+1), k]$ (and hence any searcher of the same character and greater attractiveness) in the dating subgame. Thus, by accepting exactly the same types of prospective partners of attractiveness $k_1(i+1)$ in the dating subgame as searchers of type $[i+1, k]$ do, a searcher of type $[i, k]$ will ensure the same expected utility from a date with a prospective partner of attractiveness $k_1(i+1)$ as a searcher of type $[i+1, k]$ obtains. This expected utility must be at least $R(i+1; \pi^*)$. It follows from Theorem 1 that $k_1(i) \leq k_1(i+1)$.

Hence, the general form of the equilibrium of the symmetric game is intuitive. Individuals date those who are of a similar level of attractiveness. It should also be noted that at such an equilibrium a type $[i, j]$ male will pair with a type $[i, j]$ female.

Due to the assumption that there are no costs associated with soliciting a date when dating does not follow, at such an equilibrium a searcher of attractiveness i is indifferent between soliciting and not soliciting a date with a prospective partner who is not willing to date. In this case we should check the condition based on the concept of a trembling hand perfect equilibrium. This states that a searcher should solicit a date if the expected utility from dating after a 'mistaken' acceptance is greater than the expected utility from future search. Suppose a prospective partner of attractiveness j would not pair with any searcher of attractiveness i in the dating subgame. A searcher of attractiveness i should not solicit a date with a prospective partner of attractiveness j, in order to avoid the dating costs when there is no prospect of pairing.

It is possible that a prospective partner would wish to pair with a searcher of lower attractiveness in the dating subgame but, due to the costs of dating and the risks of obtaining a prospective partner of inappropriate character, would not solicit a date. This will be considered in the example given in Sect. 17.8.2.

Note: At the equilibrium of the classical problem considered by McNamara and Collins [28], the population is partitioned into classes, such that class i males only form pairs with class i females. For the game considered here, such a partition only exists in very specific cases, e.g.:

1. When the search and dating costs are low enough, type$[i, k]$ males only pair with type $[i, k]$ females.
2. When the costs of dating are high relative to the importance of character, mate choice is based entirely on attractiveness.

The difference between the equilibria of these two games is illustrated by the example in Sect. 17.8.2.

After deriving some of the properties of an equilibrium, we now describe a procedure for deriving the equilibrium itself. Individuals of maximum attractiveness face a one-sided search problem. They should be willing to date a prospective partner if and only if the expected utility obtained from such a date minus the dating costs is at least as great as the expected utility from future search. Similarly, in the dating game a searcher should accept a prospective partner if and only if the utility gained from such a partnership is at least as great as the expected utility from future search. By following such a strategy, such individuals will maximise their expected utility from search, see Whittle [39]. Individuals of a lower level of attractiveness face a similar problem given the strategies followed by those of a higher level of attractiveness.

Since the solution of the corresponding set of inequalities is difficult to present in an explicit form, in Sect. 17.8.1 we describe an algorithm which derives a symmetric equilibrium. The constructive nature of this algorithm leads to the key theorem of the paper on the uniqueness and existence of a symmetric equilibrium in this game.

17.8.1 The Algorithm

Define $r = \lfloor \frac{m}{2} \rfloor$. Since the equilibrium is assumed to be symmetric with respect to character and sex, it suffices to consider males of character r. The advantage of considering such individuals is that the difference between character j and character r is simply the standard absolute difference between the two characters.

The game can be solved as follows

1. Assume that males of maximum attractiveness are only willing to date females of maximum attractiveness. Consider strategy profiles π_t, $t = 0, 1, 2, \ldots, \lfloor m/2 \rfloor$, where under strategy profile π_t males of type $[i_{max}, r]$ pair with females whose characters do not differ by more than t (i.e. as t increases males accept successively less preferred females). We calculate $R(i_{max}; \pi_0), R(i_{max}; \pi_1), \ldots$ in turn until $R(i_{max}; \pi_t) > g(i_{max}, t+1)$ (i.e. the expected utility from search is greater than the utility from mating with any female of maximum attractiveness who is not acceptable) or $t = m/2$. This gives us the optimal rule of the form considered, see Whittle [39], which is a lower bound on $R(i_{max}; \pi^*)$.

2. If this lower bound is less than the utility obtained by a type $[i_{max}, r]$ male from pairing with a type $[i_{max} - 1, r]$ female minus the dating costs (i.e. the maximum possible reward from soliciting a date from a female of attractiveness $i_{max} - 1$), it may be optimal for males of maximum attractiveness to solicit dates with females of the second highest level of attractiveness. We can order females of the top two levels of attractiveness with regard to the preferences of a type $[i_{max}, r]$ male. By considering strategies under which type $[i_{max}, r]$ males are prepared to pair with successively less preferred females as in Point 1, we can derive the optimal strategy of males given they date females of the two highest levels of attractiveness.

3. If required, in a similar way we can derive the optimal strategies of type $[i_{max}, r]$ males given that they date females of the u highest levels of attractiveness, where u is at least three and not more than the number of attractiveness levels. Hence, we can derive a strategy maximizing the expected utility of a type $[i_{max}, r]$ male. This strategy defines what attractiveness levels induce solicitation of a date from a male of maximum attractiveness and what types of females should be paired with after such dates (i.e. the pattern of dates and partnerships exhibited by individuals of maximum attractiveness at equilibrium).

4. The strategy defined in Points 1–3 above should be extended to ensure trembling hand equilibria in all the possible derived dating subgames involving males of maximum attractiveness. The set of acceptable females in dating subgames can be easily found using Condition 1: i.e. a male of maximum attractiveness should accept a prospective partner in the dating subgame if and only if the utility he obtains from such a partnership is greater than his expected utility from search. Note that the behaviour of males in dating subgames that do not occur under the equilibrium profile does not affect their expected utility from search at equilibrium.

Hence, the problem faced by a male of maximum attractiveness can be solved by solving a sequence of one-sided problems. The strategies used by other individuals of maximum attractiveness can be found using the symmetry of the profile with respect to character and sex. Note that it is assumed that if a male is indifferent between two strategies, then he uses the strategy which maximizes the number of attractiveness levels inducing willingness to date, together with the number of types of female that he will eventually pair with.

Suppose we have found the equilibrium strategies of individuals of attractiveness $> i$. The problem faced by a male of type $[i, r]$ reduces to a one-sided problem in which the set of females of higher attractiveness who are willing to date and pair with such males has been derived. The optimal response of such a male can be calculated in a way analogous to the one described in Points 1–4 above, with the following adaptations (which take into account the form of the equilibrium):

(a) The initial strategy of a type $[i, r]$ male is as follows: (A) solicit a date with (i) any female of a greater attractiveness who would solicit a date from him, (ii) any female of the same attractiveness and (iii) females of lower attractiveness who would be solicited by males of attractiveness $i + 1$, (B) in the dating game always pair with a female of the same type and any female who gives at least the same utility as the expected utility of an individual of attractiveness $i + 1$.

(b) The set of prospective partners accepted in the dating subgame and solicited in the soliciting subgame is extended in an analogous way to the case of individuals of maximum attractiveness. Firstly, we find the optimal strategy which involves dating individuals of the attractiveness levels derived in (a). This is done by including successively less preferred mates into the set of those mated with in the dating subgame. If required, we then find the optimal rules obtained when an individual solicits dates with successively less attractive prospective partners.

(c) After determining the set of females that a male of type $[i, r]$ solicits a date with and the females that he would pair with in the dating subgame, the behaviour of the male in the dating subgames that would not occur under the partial strategy profile derived as above is determined using the requirement of a trembling hand equilibrium in the dating subgame.

The full strategy profile can be then defined using the assumption of the symmetry of the profile with respect to sex and character. It should be noted that although this algorithm has some similarities to the one presented by McNamara and Collins [28], it is clearly different. Their algorithm is purely a one-dimensional search, which derives the sets of attractiveness levels that define a partition of the attractiveness levels for each sex. The value of the symmetric game described here can be described by a one-dimensional function. However, in order to derive the equilibrium, a two-dimensional search over levels of both attractiveness and character is required.

The following theorem shows that the profile derived in this way is a symmetric equilibrium profile.

Theorem 4. *The strategy profile derived using the algorithm given above defines a symmetric equilibrium profile.*

Proof. From the definition of the algorithm, the form of strategy profile derived satisfies Theorems 2–2. Also, the behaviour of individuals in dating subgames that do not occur under such a profile explicitly satisfies the equilibrium conditions. This behaviour has no effect on a searcher's expected utility from search.

First, consider males of maximum attractiveness. From Whittle [39], if no female of a given attractiveness level gives a utility as great as the optimal expected reward from search, then it cannot be optimal to date such a female. The algorithm considers soliciting dates with prospective partners of successively decreasing attractiveness, until the most preferred partner of a given attractiveness, say i, gives a lower utility than the greatest expected utility from search found so far. From Whittle's condition, a male of maximum attractiveness should not solicit dates from females of attractiveness $\leq i$. Also, in the dating subgame a male should not pair with a female who gives a utility less than the optimal expected utility from search. Hence, the algorithm considers all the best strategies based on dating females of attractiveness $\geq j$, for all $j > i$ and picks the strategy which maximises the expected utility from search. Hence, this maximises the expected utility of males of maximum attractiveness from search.

Now suppose that this algorithm derives the equilibrium strategy of males of attractiveness $\geq i$. From Theorems 1 and 3 a male of type $[i-1, r]$ should solicit dates from any female of attractiveness $\geq i$ who solicits dates with him and pair with any female of attractiveness $\geq i$ who would pair with him in the dating game. The strategy of a male of type $[i-1, r]$ derived by the algorithm extends the sets of those females solicited and those paired with starting from a strategy which satisfies this condition. Given the strategies of the females of attractiveness $\geq i$, a male of attractiveness $i-1$ faces a one-sided search problem and the optimal strategy in this problem is derived in an analogous way as for individuals of maximum attractiveness (using the fact that a male of type $[i-1,r]$ should pair with any female that a male of type $[i,r]$ would pair with). Hence, the algorithm derives the equilibrium strategy of individuals of attractiveness $i-1$ given the strategies used by individuals of attractiveness $\geq i$. It follows by induction, the symmetry of the strategy profile derived and the form of the equilibrium strategy from Theorems 1–3 that the algorithm gives a symmetric equilibrium profile.

Due to the form of the equilibrium and the finite number of types, the resulting strategy profile is well defined and unique. The theorem below follows directly from the construction of the symmetric equilibrium.

Theorem 5. *Assume that if any individual is indifferent between two strategies, then he/she uses the strategy which maximises the number of attractiveness levels inducing the solicitation of a date, together with the number of types of prospective partners that he/she will eventually pair with. There exists exactly one symmetric equilibrium of the symmetric game.*

One might ask whether an asymmetric equilibrium exists. Consider a finite-horizon game where each individual can observe up to n prospective partners. Suppose that in addition to Conditions 1–3, we require that an equilibrium profile in Γ is the limit of an equilibrium search profile in the finite-horizon game when $n \to \infty$. When $n = 1$, at the unique equilibrium profile each individual accepts any prospective mate (i.e. the equilibrium is symmetric). When n steps remain, an individual (a) should solicit dates from prospective partners of attractiveness i if the expected utility from such a date is greater than the future expected utility from search (i.e. when $n-1$ steps remain) and (b) pair with prospective partners in the dating subgame if the expected utility from pairing is greater than the future expected utility from search. Given the equilibrium profile in the $(n-1)$-step game is symmetric, all these expected utilities are independent of sex and character. Hence, the unique equilibrium profile in the n-step game is symmetric. It follows that we can strengthen Theorem 5 to the following theorem:

Theorem 6. *If, in addition to the assumptions of Theorem 5, it is assumed that the solution to the infinite-horizon game must be the limit of the solution to the appropriately defined finite-horizon game, then there is a unique equilibrium profile of the symmetric game Γ, which itself is symmetric.*

One might consider what equilibria are possible when there is equality between the future expected reward of a searcher of type $[i,j]$ and the reward obtained by mating with a prospective partner of type $[i_0, j_0]$. This will be of importance when $i > i_0$. If searchers of type $[i,j]$ do not accept prospective partners of type $[i_0, j_0]$, then searchers of type $[i_0, j_0]$ will have a lower expected reward from search than at the equilibrium considered above and thus become less choosy than at the original equilibrium. This may well have knock on effects on the equilibrium strategy of individuals of attractiveness below i_0. Suppose $i_0 > i_1 > i_2$. Those of attractiveness i_1 may become more choosy at such an equilibrium (as those of slightly higher attractiveness are more likely to accept them), which in turn might lead to those of attractiveness i_2 becoming less choosy at such an equilibrium (as those of slightly higher attractiveness are less likely to accept them), and so on.

However, suppose the distribution of the types of individuals is fixed and consider the space of possible cost vectors $[c_1, c_2] \in R^+ \times R^+$. For nearly all cost vectors (i.e. apart from on a set of measure 0), it is expected that there will be a unique equilibrium of the game Γ.

17.8.2 Example

Suppose that the support of both X_a and X_c is $\{0, 1, 2, 3, 4, 5, 6\}$ and the distributions of attractiveness and character are uniform. The search costs, c_1, and the interview costs, c_2 are equal to $\frac{1}{7}$. The utility obtained from a partnership is defined to be the attractiveness of the partner minus the distance (modulo 7) between the characters of the pair.

Since the expected payoff of a male does not depend on whether he is willing to date females who are unwilling to date with him, in order to derive the expected pay-offs of individuals under any strategy profile it suffices to consider strategy profiles of the following form: a searcher of attractiveness i is willing to date prospective partners of attractiveness $\geq a_i$ and in the dating subgame will pair with a prospective partner who gives a reward of at least b_i, $i = 0, 1, 2, 3, 4, 5, 6$. Denote such a strategy by $\{(a_6, b_6), (a_5, b_5), \ldots, (a_0, b_0)\}$. We derive the equilibrium strategies of individuals in the order of most attractive to least attractive. Hence, for example, by $\{(a_6, b_6), (a_5, b_5), \bullet, \bullet, \bullet, \bullet\}$ we denote the set of strategy profiles such that individuals of attractiveness levels 5 and 6 use the strategies defined by (a_5, b_5) and (a_6, b_6), respectively, and the strategies of the remaining individuals are undefined, **but satisfy Conditions 1 and 3**, i.e. a trembling hand equilibrium is always played in the dating subgame and the strategy profile is stationary. Other similar sets of strategy profiles will be denoted in an analogous manner.

First we consider males of maximum attractiveness. Suppose they only solicit dates with females of attractiveness 6. The ordered preferences of a $[6, 3]$ male are as follows: first (group one) – $[6, 3]$, second equal (group two) – $[6, 2], [6, 4]$, fourth equal (group 3) $[6, 1], [6, 5]$ and sixth equal (group 4) – $[6, 0], [6, 6]$. Group 1, 2, 3 and 4 females give a utility from pairing of 6, 5, 4 and 3, respectively. The sets of strategy profiles in which type three males mate with (a) only those from group 1, (b) those from groups 1 and 2, (c) those of groups 1, 2 and 3 and (c) those from all four groups are $\{(6, 6), \bullet, \bullet, \bullet, \bullet, \bullet\}$, $\{(6, 5), \bullet, \bullet, \bullet, \bullet, \bullet\}$, $\{(6, 4), \bullet, \bullet, \bullet, \bullet, \bullet\}$ and $\{(6, 3), \bullet, \bullet, \bullet, \bullet, \bullet\}$, respectively. We successively include females into the set of acceptable partners starting from the most preferred until no female of attractiveness 6 outside this set gives a greater utility than the current expected utility of a type $[6, 3]$ male.

$$R(6; \{(6,6), \bullet, \bullet, \bullet, \bullet, \bullet\}) = 6 - 49 \times \frac{1}{7} - 7 \times \frac{1}{7} = -2$$

$$R(6; \{(6,5), \bullet, \bullet, \bullet, \bullet, \bullet\}) = \frac{16}{3} - \frac{49}{3} \times \frac{1}{7} - \frac{7}{3} \times \frac{1}{7} = \frac{8}{3}$$

$$R(6; \{(6,4), \bullet, \bullet, \bullet, \bullet, \bullet\}) = \frac{24}{5} - \frac{49}{5} \times \frac{1}{7} - \frac{7}{5} \times \frac{1}{7} = \frac{16}{5}.$$

The expected utility of a type $[6,3]$ male under $\{(6,4), \bullet, \bullet, \bullet, \bullet, \bullet\}$ is greater than 3. It follows that $[6,0]$ and $[6,6]$ females should not be accepted in the dating subgame.

Since our lower bound (3.2) on the expected utility of a type $[6,3]$ male is less than the utility obtained from pairing with a female of the same character and the second highest attractiveness minus the costs of dating, $\frac{34}{7}$, we now consider strategy profiles in which type $[6,3]$ males solicit dates with females of attractiveness 5 and 6. We only have to consider:

1. Strategy profiles in which females of type $[5,3]$ are acceptable in the dating subgame. If this were not the case, then a type $[6,3]$ employer would be incurring unnecessary dating costs.
2. Prospective partners who give a utility higher than the current lower bound on the expected utility of a type $[6,3]$ male from search.

The ordered preferences of a type $[6,3]$ male among the set of females of attractiveness at least 5 who satisfy criterion 2 above is given by: group 1 is $\{[6,3]\}$, group 2 is $\{[6,2],[5,3],[6,4]\}$ and group 3 $\{[6,1],[6,5],[5,2],[5,4]\}$. We only need to consider strategy profiles of the following two types: (a) type $[6,3]$ males pair with females from groups 1 and 2 above, i.e. profiles from the set $\{(5,5), \bullet, \bullet, \bullet, \bullet, \bullet\}$, (b) type $[6,3]$ males pair with females from all three groups, i.e. profiles from the set $\{(5,4), \bullet, \bullet, \bullet, \bullet, \bullet\}$. We have

$$R(6; \{(5,5), \bullet, \bullet, \bullet, \bullet, \bullet\}) = \frac{21}{4} - \frac{49}{4} \times \frac{1}{7} - \frac{14}{4} \times \frac{1}{7} = 3$$

$$R(6; \{(5,4), \bullet, \bullet, \bullet, \bullet, \bullet\}) = \frac{37}{8} - \frac{49}{8} \times \frac{1}{7} - \frac{14}{8} \times \frac{1}{7} = \frac{7}{2}.$$

We now consider strategy profiles in which type $[6,3]$ males solicit dates with females of attractiveness at least 4. Since the present lower bound on $R(6; \pi^*)$ is 3.5, we only need to consider strategy profiles in which type $[6,3]$ males pair with the same types of females as in $\{(5,4), \bullet, \bullet, \bullet, \bullet, \bullet\}$ with the addition of type $[4,3]$ females, i.e. strategy profiles from the set $\{(4,4), \bullet, \bullet, \bullet, \bullet, \bullet\}$. We have

$$R(6; \{(4,4), \bullet, \bullet, \bullet, \bullet, \bullet\}) = \frac{41}{9} - \frac{49}{9} \times \frac{1}{7} - \frac{21}{9} \times \frac{1}{7} = \frac{31}{9} < \frac{7}{2}.$$

It follows that type $[6,3]$ males should not solicit dates with females of attractiveness 4. Hence, at a symmetric equilibrium, type $[6,3]$ males solicit dates with females of attractiveness 5 and 6 and pair with females of type in M_6, where

$$M_6 = \{[6,1],[6,2],[6,3],[6,4],[6,5],[5,2],[5,3],[5,4]\}.$$

It should be noted that a type $[6,3]$ male should pair with a type $[4,3]$ female given that they are dating. However, due to the costs of dating, the low probability of finding an acceptable partner of attractiveness 4 and the relatively small gain obtained from such a partnership compared to the expected utility from future search, a type $[6,3]$ male should not solicit a date with such a female. Define $M_i + [s,t]$ to be the set of $[k+s, j+t]$ where $[k, j] \in M_i$. From the symmetry of the equilibrium with respect to character and sex, an individual of type $[6, 3+t]$ is willing to date prospective partners of attractiveness 5 and 6 and pair with those in $M_6 + [0, t]$.

Note that searchers are not matched in a block-separated way as in McNamara and Collins [28]. For example, a type $[6,3]$ male will pair with a type $[6,1]$ female, who would mate with a type $[6,0]$ male. However, a type $[6,3]$ male will not pair with a type $[6,0]$ female.

Now we consider males of type $[5,3]$ and assume that individuals of maximum attractiveness follow the strategies derived above and males of attractiveness 5 solicit dates with females of attractiveness 5 and 6. Note that from the form of the equilibrium, a male should always solicit dates with females of the same attractiveness, together with females of higher attractiveness who solicit dates with him. From the symmetry of the game with respect to sex and character, since males of type $[6,3]$ pair with females of type $[5,2], [5,3]$ and $[5,4]$, it follows that males of type $[5,3]$ will pair with females of type $[6,2], [6,3]$ and $[6,4]$. They must also pair with females of type $[5,2], [5,3]$ and $[5,4]$, as such females give a type $[5,3]$ male a utility of $4 \geq R(6; \pi^*) \geq R(5; \pi^*)$. The expected utility of a type $[5,3]$ male under such a strategy profile, i.e. from the set $\{(5,4), (5,4), \bullet, \bullet, \bullet, \bullet\}$, is

$$R(5; \{(5,4), (5,4), \bullet, \bullet, \bullet, \bullet\}) = \frac{29}{6} - \frac{49}{6} \times \frac{1}{7} - \frac{14}{6} \times \frac{1}{7} = \frac{10}{3}.$$

This is greater than the expected utility from accepting the next most preferred types ($[5,1]$ and $[5,5]$). Hence, we can now consider strategy profiles in which males of type $[5,3]$ solicit dates with females of attractiveness at least 4. The only case we need to consider is extending the set of acceptable females to include those of type $[4,3]$, i.e. the set of strategy profiles $\{(5,4), (4,4), \bullet, \bullet, \bullet, \bullet\}$. We have

$$R(5; \{(5,4), (4,4), \bullet, \bullet, \bullet, \bullet\}) = \frac{33}{7} - \frac{49}{7} \times \frac{1}{7} - \frac{21}{7} \times \frac{1}{7} = \frac{23}{7} < \frac{10}{3}.$$

It follows that males of type $[5,3]$ should solicit dates with females of attractiveness 5 and 6 and pair with females of a type in $\{[5,2], [5,3], [5,4], [6,2], [6,3], [6,4]\}$. In these cases acceptance is mutual. It should also be noted that males of type $[5,3]$ should accept females of type $[6,1]$ or $[6,5]$ in the dating subgame. However, in these cases acceptance is not mutual. Females of type $[4,3]$ would be accepted in the dating subgame by a type $[5,3]$ male, but such males would not solicit a date with such a female.

Thus males and females of the top two levels of attractiveness do not date individuals of any lower level of attractiveness. The problem faced by males of attractiveness 4 thus reduces to a problem analogous to the one faced by those of attractiveness 6 (they are the most attractive of the remaining males). It follows that $M_4 = M_6 - [2,0]$. Arguing iteratively, $M_i = M_{i+2} - [2,0]$ for $i = 1,2,3,4$. Males of attractiveness 2 or 4 solicit dates with females of the same attractiveness or of attractiveness one level lower. Males of attractiveness 1 or 3 solicit dates with females of the same attractiveness or of attractiveness one level higher. We have $R(4; \pi^*) = 3/2$ and $R(3; \pi^*) = \frac{4}{3}$, so in the dating subgame males of attractiveness 3 and 4 accept any females giving them a utility of at least 2 (in this game the utility from a pairing is by definition an integer). Also, $R(2; \pi^*) = -1/2$ and $R(1; \pi^*) = -\frac{2}{3}$, so in the dating subgame males of attractiveness 1 and 2 accept any female giving them a utility of at least 0.

It follows that the value of the game to a player of given attractiveness from 1 to 6 must be given by the expected reward of such a player under any strategy profile from the set $\{(5,4),(5,4),(3,2),(3,2),(1,0),(1,0),\bullet\}$.

Since females of attractiveness 5 and 6 do not solicit dates with males of attractiveness 4, such males are indifferent between soliciting and not soliciting dates with such females. We should check the relevant equilibrium condition based on the concept of a trembling hand equilibrium, i.e. if a female of attractiveness 6 did 'by mistake' accept a date with a male of attractiveness 4, should the male solicit a date? In the dating subgame, only a female of type $[6,3]$ would accept a male of type $[4,3]$. Hence, the expected utility of a type $[4,3]$ male from dating a female of attractiveness 6 is

$$\bar{v}_M(4,6; \pi^*) = \frac{1}{7} \times 6 + \frac{6}{7} \times 1.5 - \frac{1}{7} = 2 > R(4; \pi^*).$$

It follows that males of attractiveness 4 should solicit dates with females of attractiveness 6. Arguing similarly, such males should solicit dates with females of attractiveness 5 and males of attractiveness 2 should solicit dates with females of attractiveness 3 or 4.

No females of attractiveness 5 or 6 would pair with a male of attractiveness 3 in the dating subgame. It follows that males of attractiveness 3 should not solicit dates with females of attractiveness 5 or 6. Arguing similarly, a male of attractiveness 1 should not solicit dates with females of attractiveness above 2.

It remains to determine the strategy used by a type $[0,3]$ male. Since no female of greater attractiveness will date such a male, we only have to consider strategy profiles where a male pairs with successively less preferred partners of attractiveness 0. Note that $R(0; \pi^*) \le R(1; \pi^*) = -2/3$, thus in the dating game individuals of attractiveness 0 must pair with any prospective partner who gives them a utility of 0 (i.e. with those of the same type). The expected rewards from the game under a strategy profile where individuals of attractiveness at least 1 use the strategies derived above and males of type $[0,3]$ pair with females of types in (a) $\{[0,3]\}$, (b) $\{[0,2],[0,3],[0,4]\}$ and (c) $\{[0,1],[0,2],[0,3],[0,4],[0,5]\}$ are equal to the expected rewards from the game under the respective strategy profiles

$$\{(5,4),(5,4),(3,2),(3,2),(1,0),(1,0),(0,0)\},$$
$$\{(5,4),(5,4),(3,2),(3,2),(1,0),(1,0),(0,-1)\},$$
$$\{(5,4),(5,4),(3,2),(3,2),(1,0),(1,0),(0,-2)\}$$

We now have that

$$R(0;\{(5,4),(5,4),(3,2),(3,2),(1,0),(1,0),(0,0)\}) = -49 \times \frac{1}{7} - 7 \times \frac{1}{7} = -8$$

$$R(0;\{(5,4),(5,4),(3,2),(3,2),(1,0),(1,0),(0,-1)\}) = -\frac{2}{3} - \frac{49}{3} \times \frac{1}{7} - \frac{7}{3} \times \frac{1}{7} = -\frac{10}{3}$$

$$R(0;\{(5,4),(5,4),(3,2),(3,2),(1,0),(1,0),(0,-2)\}) = -\frac{6}{5} - \frac{49}{5} \times \frac{1}{7} - \frac{7}{5} \times \frac{1}{7} = -\frac{14}{5}.$$

The expected utility of a type $[0,3]$ male under

$$\{(5,4),(5,4),(3,2),(3,2),(1,0),(1,0),(0,-2)\}$$

is greater than the utility obtained from pairing with females of type $[0,0]$ and $[0,6]$. Hence, at a symmetric equilibrium males of type $[0,3]$ should not accept such females.

It remains to consider the set of females that a male of attractiveness 0 should solicit a date with according to the conditions based on the concept of a trembling hand perfect equilibrium. At equilibrium, no female of attractiveness greater than 2 would ever pair with a male of attractiveness 0 in the dating subgame. Hence, males of attractiveness 0 should never solicit dates with females of attractiveness above 2. Females of type $[2,3]$ and $[1,3]$ would pair with a type $[0,3]$ male in the dating subgame. Arguing as in the case of males of attractiveness 4 soliciting dates with females of attractiveness 5 and 6, males of attractiveness 0 should solicit dates with females of attractiveness 1 and 2.

Table 17.1 gives a synopsis of the equilibrium strategy profile. Each individual should accept a prospective partner in the dating subgame if the utility from such a matching is at least as great as the utility from search. For ease of presentation, the set of such partners is not presented.

17.9 Generalizing the Model

Suppose that character is placed along a line instead of around a circle, i.e. the difference between two characters is calculated according to the standard absolute difference. Considering the game presented in Sect. 17.8.2 (with unspecified search and dating costs), there is still a large degree of symmetry with respect to sex and character (e.g. the character levels j and $6 - j$ can be interchanged without essentially changing the game).

Attractiveness	Solicits dates with prospective partners of attractiveness	Expected Utility
6	{ 5,6 }	7/2
5	{ 5,6 }	10/3
4	{ 3,4,5,6 }	3/2
3	{ 3,4 }	4/3
2	{ 1,2,3,4 }	−1/2
1	{ 1,2 }	−2/3
0	{ 0,1,2 }	−14/5

Table 17.1 Brief description of the symmetric equilibrium for the example considered

We wish to derive an equilibrium which reflects this inherent symmetry. Suppose a type $[i, j]$ male solicits a date with a female of attractiveness k and pairs with a female of type $[k, l]$. Firstly, a type $[i, j]$ female should be willing to date a male of attractiveness k and pair with a male of type $[k, l]$. Secondly, a type $[i, 6 - j]$ male should solicit a date with a female of attractiveness k and pair with a female of type $[k, 6 - l]$.

It is expected that males of type $[6, 3]$ have the highest expected utility from search and so we can treat the problem they face as a one-sided problem. However, it is unclear whether in a specific problem individuals of type $[6, 2]$ or those of type $[5, 3]$ should have the higher expected utility from search at such an equilibrium. Hence, it is unclear how the algorithm should proceed.

In order to solve more general problems, the algorithm presented in Sect. 17.8 must be further developed. However, it seems that the general approach of solving a sequence of appropriately defined one-sided problems could be useful in deriving a strategy profile which is very similar to an equilibrium strategy profile (see Ramsey [32] for a similar approach). Also, the form of the general problem and the usefulness of such an approach indicate that if there are multiple equilibria, then the behaviour observed at such equilibria should be qualitatively similar.

17.10 Conclusion and Directions for Further Research

This chapter has presented a model of partnership formation where both common and homotypic preferences are taken into account. The preferences of all searchers are common with respect to the attractiveness of prospective partners and homotypic with respect to character. Attractiveness can be assessed immediately, but in order to assess character a costly date (or interview) is required.

We have considered a particular type of such problems in which the distribution of attractiveness and character, as well as search and interview costs, were independent of the class (sex) of a player. Character was assumed to form a circle, such that the 'extreme' levels of character are neighbours. For convenience, the supports of attractiveness and character were assumed to be finite sets of integers. The distribution of character is uniform.

The form of a symmetric equilibrium profile which satisfies various criteria based on the concept of a trembling hand perfect equilibrium was derived and an algorithm to find such a profile described. These criteria are a generalization of the optimality criterion used by McNamara and Collins [28] to define the unique equilibrium in the classical two-sided job search problem. It is shown that such an equilibrium exists and is unique (assuming that if an individual is indifferent between accepting or rejecting a prospective partner at any stage, then he/she accepts). Although the equilibrium derived here does have some similarities to the equilibrium derived by McNamara and Collins [28], it is essentially different, since it is not a block separating equilibrium.

The use of this combination of preferences would seem to be logical in relation to job search and mate choice. Although there is no perfect correlation in individuals' assessment of the attractiveness of members of the other class, there is normally a very high level of agreement, particularly among males in mate choice problems. These 'mixed' preferences seem to be both reasonably tractable within the framework of searching for a partner within a relatively large population and allow a general enough framework to model the preferences of individuals reasonably well (although it would seem that modelling character as a one-dimensional variable is rather simplistic). By using a larger number of types, we could approximate continuous distributions of attractiveness and character.

For simplicity, it was assumed that individuals know their own attractiveness and character, whereas in practice they may have to learn about these measures over time (see Fawcett and Bleay [15]).

Also, it was assumed that individuals are able to measure attractiveness and character perfectly, although at some cost. It would be interesting to consider different ways in which information is gained during the search process. For example, some information about the character of a prospective partner may be readily available. Hence, an improved model would allow some information to be gained on both the attractiveness and character of a prospective partner at each stage of an interaction.

In terms of the evolution of such procedures, it is assumed that the basic framework is given, i.e. the model assumes that the various search and dating (interview) costs are given. Hence, this model cannot explain why such a system has evolved, only the evolution of decisions within this framework.

Individuals may lower their search costs by joining some internet or social group. Such methods can also lead to biasing the conditional distribution of the character of a prospective partner in a searcher's favour. It is possible that dating (interview) costs are dependent on the types of the two individuals involved. For example, two individuals of highly different characters might incur low dating costs, as they realize very quickly that they are not well matched.

Also, the ability to incur dating costs may well transfer information regarding the attractiveness and/or character of an individual. In this case, it may be more costly to successfully date highly attractive prospective partners, since they would only accept partners who can pay high dating costs (i.e. are attractive).

In addition, it would be useful to investigate how the utility functions, together with the relative costs of searching and interviewing, affect the importance of attractiveness and character in the decision process. It should be noted that using attractiveness as an initial filter in the decision process will lead to attractiveness becoming relatively *more* important than character, especially if the costs of dating are relatively high.

It would also be useful to adapt the algorithm to problems in which the distribution of character is not uniform and/or the set of character levels do not form a circle. In this case, it is expected that individuals of extreme character will usually be less choosy than those of a central character for a given level of attractiveness. Two major problems result from this. Firstly, the form of an equilibrium will be more complex than the form of the symmetric equilibrium given here. Any algorithm to derive an equilibrium in this case will certainly be more complex than the algorithm outlined in this chapter, which uses the fact that the problem can be reduced to a sequence of one-sided problems. The unique equilibrium derived here would be useful as a point of reference.

Finally, it would be interesting to consider games in which the distributions of traits and/or search costs depended on class. In this case, it would be natural to assume that the equilibrium is asymmetric with respect to class. In the spirit of the derivation of equilibrium points in the classical matching problem (see Gale and Shapley [18]), it would be interesting to see whether equilibria analogous to male-choice and female-choice equilibria exist. For the types of model considered here, to find a male-choice equilibrium we would try to maximize the expected utility of males while adapting female choice to male choice.

References

1. Albelda R. P.: Occupational segregation by race and gender: 1958–1981. Industrial and Labour Relations Review. **39**: 404–411 (1981)
2. Albrecht J., Gautier P. A. and Vroman S.: Equilibrium directed search with multiple applications. Rev. Econ. Studies. **73**: 869–891 (2006)
3. Alpern S. and Reyniers D.: Strategic mating with homotypic preferences. J. Theor. Biol. **198**, 71–88 (1999).
4. Alpern S. and Reyniers D.: Strategic mating with common preferences. J. Theor. Biol. **237**, 337–354 (2005).
5. Andolfatto D.: Business cycles and labor-market search. Am. Econ. Rev. **86**: 112–132 (1996)
6. Beller A. H.: Occupational segregation by sex: Determinants and changes. J. Hum. Res. **17**: 371–392 (1982)
7. Burdett K.: A theory of employee job search and quit rates. Am. Econ. Rev. **84**: 1261–1277 (1978).
8. Burdett K. and Coles M. G.: Long-term partnership formation: marriage and employment. The Economic Journal, **109**, 307–334 (1999).
9. Chade H. and Smith L.: Simultaneous search. Econometrica. **74**: 1293–1307 (2006).
10. Coles M. G. and Burdett K. Marriage and class. Quart. J. Econ. **112**: 141–168 (1997).
11. Coles M. G. and Francesconi M.: On the emergence of toyboys: Equilibrium matching with ageing and uncertain careers. IZA Discussion Papers 2612, Institute for the study of labor (2007).

12. Devine T. J. and Piore M. J.: Empirical Labor Economics: The Search Approach. Open University Press, New York (1991).
13. Diamond P. and Maskin E.: An equilibrium analysis of search and breach of contract 1: Steady states. Bell J. Econ. **10**: 282–316 (1979).
14. Drewlanka S.: A generalized model of commitment. Math. Soc. Sci. **52**: 233–251 (2006)
15. Fawcett T. W. and Bleay C.: Previous experiences shape adaptive mate preferences. Behav. Ecol. **20**, 68–78 (2009)
16. Fisman R., Iyengar S. S., Kamenica E. and Simonson I.: Gender differences in mate selection: Evidence from a speed dating experiment. Quart. J. Econ. **121**: 673–679 (2006).
17. Fisman R., Iyengar S. S., Kamenica E. and Simonson I.: Racial preferences in dating. Rev. Econ. Stud. **75**: 117–132 (2008).
18. Gale D. and Shapley L. S.: College admissions and the stability of marriage. Am. Math. Monthly, **69**, 9–15 (1962).
19. Galenianos M. and Kircher P.: Directed search with multiple job applications. J. Econ. Theory. **144**: 445–471 (2009).
20. Härdling R. and Kokko H.: The evolution of prudent choice. Evol. Ecol. Res. **7**, 697–715 (2005).
21. Hey J. D.: Search for rules of search. J. Econ. Behav. Organ. **3**: 65–81 (1982).
22. Janetos A. C.: Strategies of female mate choice: a theoretical analysis. Behav. Ecol. Sociobiol. **7**, 107–112 (1980).
23. Johnstone R. A.: The tactics of mutual mate choice and competitive search. Behav. Ecol. Sociobiol. **40**, 51–59 (1997).
24. Jovanovic B.: Job matching and the theory of turnover. J. Pol. Econ. **87**: 972–990 (1979).
25. MacMinn R. D.: Search and market equilibrium. J. Pol. Econ. **88**: 308–327 (1980).
26. Marimon R. and Zilibotti F.: Unemployment vs. mismatch of talents: Reconsidering unemployment benefits. Econ. J. **109**: 255–291 (1999).
27. McCall J. J.: Information and job search. Quart. J. Econ. **84**: 113–126 (1970).
28. McNamara J. M. and Collins E. J.: The job search problem as an employer-candidate game. J. Appl. Prob. **28**, 815–827 (1990).
29. Nisam N., Roughgarden T., Tardos E. and Vazirani V. V.: Algorithmic Game Theory. Cambridge University Press, New York (2007).
30. Peterson T., Saporta I. and Seidel M. -D. L.: Offering a job: Meritocracy and social networks. Am. J. Sociol. **106**: 763–816 (2000).
31. Pissarides C.: Search unemployment with on-the-job search. Rev. Econ. Stud. **61**: 457–475 (1994).
32. Ramsey D. M.: A large population job search game with discrete time. Eur. J. Oper. Res. **188**, 586–602 (2008).
33. Shimer R.: The assignment of workers to jobs in an economy with coordination frictions. J. Pol. Econ. **113**: 996–1025 (2005).
34. Shimer R. and Smith L.: Assortative matching and search. Econometrica. **68**: 343–369 (2000).
35. Smith L.: The marriage model with search frictions. J. Pol. Econ. **114**: 1124–1144 (2006).
36. Spear S. E.: Learning rational expectations under computability constraints. Econometrica. **57**: 889–910 (1989).
37. Spear S. E.: Growth, externalities and sunspots. J. Econ. Theory. **54**: 215–223 (1991).
38. Velupillai K. V.: Expository notes of computability and complexity in (arithmetical) games. J. Econ. Dynam. Cont. **21**: 955–979 (1997).
39. Whittle P.: Optimization Over Time, Vol. 1. Wiley, New York (1982).

Chapter 18
Applications of Search in Biology: Some Open Problems

Jon Pitchford

Abstract The theory of search and rendezvous can be applied to answer real world problems which are both interesting and of practical importance. Here I provide a personal account of where existing theories may need modification in order to tackle the uncomfortable complexities of biology, and argue that in many cases these modifications are tractable. Finally, three open problems in the application of search theory to biological systems are presented: how should a fish swim; how do plant roots exploit patchy nutrients; and why do animals form groups?

18.1 Introduction

Biology is an exciting place to do research. While Euclid's mathematical proofs are immutable, current technological advances force biologists to rewrite their textbooks every few years. The basic elements do not change: life involves things making entire-but-imperfect copies of themselves by encoding information chemically in DNA or RNA. Evolutionary forces within and between species and environment shape long-term changes. However, our ability to quantify and systematise details of the underlying processes is expanding at a tremendous rate. The emerging fields of post-genomic biology are shifting the emphasis away from simply (!) reading sequences of DNA code, and towards complex systems of feedbacks across a range of time scales and embedded within heterogeneous and dynamic environments.

The mathematics of search and rendezvous is elegant and compelling in its own right, and is further enhanced by accurate computations and careful data analysis. If these methods can be adapted to intersect with questions emerging from biology, then the prospects for important biological breakthroughs driven by mathematical

J. Pitchford (✉)
York Centre for Complex Systems Analysis and Departments of Mathematics and Biology, University of York, York, UK
e-mail: jon.pitchford@york.ac.uk

S. Alpern et al. (eds.), *Search Theory: A Game Theoretic Perspective*,
DOI 10.1007/978-1-4614-6825-7_18, © Springer Science+Business Media New York 2013

reasoning are bright. Importantly, any such insight would emerge from rigorous theory, where the assumptions are transparent and the model ingredients precisely defined (even if they do not perfectly mimic biological reality). This is a useful counter to the fashion for ever-larger supercomputer simulations of biological systems, which can be visually compelling but which are prone to statistical misinterpretation, hidden parameter assumptions, and errors in computational implementation. There is room, and need, for both flavours of research.

In this brief article I first try to identify where the existing theories of search and rendezvous may need rethinking when challenged by the uncomfortably dirty realities of real-world biology. These are not criticisms. Indeed, many of the issues raised are readily tackled within the context of existing theory. Where modifications are needed, there is every chance that these are both tractable and intellectually satisfying for those with large enough brains.

Finally, three open problems are described. In stark similarity to the "blind" and "stupid" foragers considered below, these are constrained by the author's local knowledge of the mathematical and biological research environment, and biased heavily by those with whom he is fortunate to collaborate. Solving these problems will confer uncertain intellectual fitness benefits within the complex and stochastic research landscape, but is likely to provide avenues into still richer problems emerging from the flood of technology-driven data.

18.2 Biological Complications Relating to Search and Rendezvous

Individuals Move in Interesting Ways

Movement in biology was traditionally modelled using biased random walks; the animal (or cell, or chemical) takes randomly oriented steps of a constant size, and its location is described via a diffusive process with a Gaussian probability density. Variations of this paradigm have been applied with much success [8]. However, where random walks fail, this is likely to be for interesting reasons.

Firstly, organisms can sense their local environment directly, they can make changes to this local environment, and they may gain more global knowledge via visual or chemical cues. Global information will always, however, be uncertain relative to local knowledge [6]. Such elaborations can be built into search theories; strategies employing 'tokens' (see Kiniwa et al., Chap. 8, and Chalopin et al., Chap. 12, for games that involve tokens), and the ideas of reinforced random walks, have been usefully explored. Behaviour also depends on state; a hungry solitary lion moves differently to a well-fed member of a pride. Again, existing theories can be adapted (Broom, Chap. 15) or [1, 15].

Simple random walks are challenged by recent literature synthesising the data and theory behind Lévy walks. Individuals across a broad range of species are observed to make small local random movements interspersed with rare long-distance jumps. When the probability distribution for the step size follows a power law the resulting process is called a Lévy walk. Theory borrowed from theoretical physics has been used to infer optimality of such super-diffusive movement strategies for sparse patchily distributed prey. Careful analyses of theory, computation, and inference from data have cast doubt on any such global assertion of optimality [16, 17]. Nevertheless, search theories should be freed from the constraints of the diffusive paradigm.

Environments Are Interesting

Biological environments are seldom, if ever, homogeneous, isotropic, and static. Complex and dynamic structure is visible at scales ranging from the sub-cellular to the ecological. The consequences of this complexity are intimately related to those concerning modelling movement as simple random walks versus sub- or super-diffusive processes outlined above, and there are mathematical analogies in their methods of solution. Ecology involves, by definition, interactions between life and environment, and so very often the feedbacks between searcher and search arena will dictate behaviour.

One exciting direction is to escape Euclidean space in favour of environments described by graphs; a collection of locations connected by weighted edges. Interactions between biology and complex environments can be captured succinctly and elegantly in this way (see, for example, Durham et al. [11, 12]) and the expanding theory of search on networks (Chap. 2 of this volume) has great potential to be applied. Linking such individual-based studies to larger-scale properties at the population level (sensu [18]) is an important and tractable challenge.

Some Princesses Are Monsters

It is tempting to transfer theories directly between disciplines, for example by taking the classic Princess and Monster game [14] and identifying the Princess as "prey" and the Monster as "predator". However, even a princess must eat, and only the rarest and most savage monster can safely consider itself invulnerable to attack. In other words, the actors in biological systems may be motivated by factors not considered by the modeller, either by choice or through ignorance.

Conversely, observing a particular behaviour in nature and attributing this to optimisation of some externally imagined metric could be misleading – ambiguities between pattern and process are common in the literature. It may be possible to show compelling statistical differences between the movement of male and female butterflies, and to correlate these with environmental factors, but to infer the mechanisms driving the observations is a much greater challenge. Further practical challenges of predicting the conservation consequences of changes in habitat structure can be answered only speculatively at this stage (Preston et al., [25]). State-based models offer some natural avenues of progress; the key challenge is more likely to be the framing of the biological question and interpreting the data, rather than the mathematical technicalities.

Some Individuals Are Not Individuals

An ant colony might contain thousands of foraging non-reproductive clones, serving the interests of a single fertile female. In effect the only individual of interest is the queen, and ecological success of foraging workers needs to be measured in this context. This is an extreme case, but sociality is common in animal studies and is perhaps mirrored at smaller scales by quorum sensing in bacteria. This is a large and active research area (see, for example, [27], which contains excellent ecological background spanning the remit of this article) and notions of inclusive fitness in social systems are well developed, though not without controversy [20].

Taking plants as a less obvious example, each root tip could be thought of as an individual forager (Problem 2, below). Neighbouring roots are not independent, however, and the metric of interest is the cumulative foraging success of each plant's dynamic population of roots. The scope for practical problems and interesting mathematics is enormous.

Sometimes Being Optimal Is Not Good Enough

Much of the theory of search and rendezvous concerns finding mathematical solutions which minimise the expected value of some property. This makes sense for long-lived and "valuable" humans, but is probably too anthropocentric a view for biology in general. Evolution by natural selection involves, by necessity, probabilistic forces. "Survival of the fittest" is a statement about extremes, not averages. Put simply, if an individual is almost certain to die soon anyway, then it has no interest in maximising its long term average fitness – it only needs to get very lucky, very quickly. This has been long understood in human societies [21], and so-called "risk sensitive foraging" is not a new theory in ecology [26].

This is not necessarily a difficult problem to overcome mathematically; one simply optimises the property of interest in the surviving tail of the distribution rather than globally [9]. However, its consequences are possibly very large (see Problem 1, below). New technologies are revealing that natural systems can evolve dramatically so as to exploit stochasticity [2]. In this author's biased opinion, of the complications listed here, this last is likely to require the smallest modifications to existing theory, but has the potential to provide the most revealing new insights.

18.3 Some Open Problems

These problems benefit from being both practically relevant and mathematically tractable. The author, with co-workers, has made attempts to solve each of them but there is wide scope for fresh thinking and new approaches.

Problem 1: How Fast Should a Fish Swim? (Complications 1, 2, 3, and 5)

Fish are important; we eat them, and they play crucial roles in wider ecosystem function. In a sustainable fishery one would hope that each fish removed from the adult population (the "stock") is replaced by a juvenile entering that population (a "recruit"). Unfortunately, the stock-recruitment relationship is notoriously unpredictable [5, 19]. One of the main reasons for this is the peculiar reproductive strategy used by many pelagic species, where each female will produce millions of small and seemingly useless eggs over her lifetime. Only a tiny proportion of these survive the egg and the larval stages to reach adulthood and contribute to future generations. Most die of starvation or are devoured by anything with a larger mouth – including their own brothers and sisters.

So, how fast should a fish larva swim? The problem is ostensibly a simple balance between energy costs and foraging benefits. For example, swimming twice as fast might double the predator-prey encounter rate, but may also incur a four-fold increase in the cost of swimming (assuming Stokes drag, appropriate for small foragers in water). Simple optimisation would reveal a quadratic fitness landscape with a fixed optimal swimming speed.

Reality, however, is more interesting. For a small animal in a large ocean, prey are not homogeneously distributed and turbulence cannot be ignored. Effectively, local random stirring brings the predator into contact with prey regardless of active swimming – perhaps the best strategy is to sit and wait?

The temporal mean-field problem was tackled by Pitchford et al. [22, 23]. They use a simple multi-scale approach using (appropriately enough) Poisson processes: the forager, the patches of prey, and the prey individuals are regarded as independent spheres, with their relative speeds and encounter rates governed by physiology and physics. A predator finds, forages within, and leaves a patch at Poisson rates which combine swimming speeds with turbulent speeds at the appropriate length scales (visual range for forager-prey encounters, patch size for forager-patch encounters). Pitchford et al. derive analytical expressions for the optimal swimming speeds, and extend the study to ask when a predator should alter its behaviour according to whether it thinks it is in a patch.

These models can be improved. When there is a large number of predator and prey, then intraspecific competition and population dynamics may play a role, with consequences for the elusive stock-recruitment relationship. Swimming may be more saltatory or Levy-like than a simple constant swim [4]. The environment is also more complex than simple independent spheres; the interaction between individual swimming and larger-scale fluid flow can induce structures and patterns requiring modified theory [11]. Also, swimming faster makes the larva more visible to its predators, a further (but quantifiable) blurring of princesses and monsters.

Evolution adds a further, and potentially large, twist. Typically only <1 % of hatched larvae survive the early juvenile stage; the average fish is thoroughly dead. "Optimality" therefore must be firmly rooted in the luckiest tails of the probability distributions [24]. The mathematical frameworks exist to allow this to be quantified [9]; all that is needed is a careful definition of the ecological and evolutionary problem, and intelligence and ambition in framing the research within the context of sustainable fisheries management.

Problem 2: What Is the Difference Between a Plant and a Fish? (Complications 2, 4, and 5)

Plants are important; we eat them, and through photosynthesis they are fundamental to terrestrial life as we know it. The diversity of plant species, and the variety of elaborate chemicals they produce, is staggering [13]; if plants need simply to absorb nutrients and eat sunshine, then why is there not some dominant superspecies? Thinking more practically, agricultural use of artificial fertilisers is thought to be necessary for sustained food production, but if applied inefficiently this is financially expensive and incurs severe detrimental downstream ecological costs. A better knowledge of how plants search for, and exploit, nutrient patches is therefore of both practical and intellectual value. There are obvious similarities with fish: a "predator" (root tip) with only local knowledge moves through a complex and possibly dynamic environment seeking "prey" (patches of nutrient). However, the differences cannot be ignored; a plant has many roots and can preferentially proliferate root growth towards regions of higher nutrient concentration, but unlike most animals it cannot completely relocate itself in response to threat from competitors, consumers, or the environment.

A simple model by Croft et al. [10] imagines a plant growing in one-dimensional soil, whose growth is enhanced by finding discrete patches of nutrient, and which can "choose" to proliferate roots preferentially to the left or right depending on the location of the most recently acquired patch. For an isolated plant in a uniformly random environment all proliferation strategies are equal. In a patchy environment it becomes strongly favourable to proliferate in the direction of the most recently acquired patch. None of this is surprising. However, when a population of identical plants competes for resources, things change. Even in a uniformly random environment there is an evolutionary pressure to proliferate towards the most recent patch.

This is because, although the environment contains no information per se, the fact that a patch has been acquired provides implicit information that a competitor has not already visited this location. In a patchy world the evolutionary pressure for directed proliferation becomes stronger.

The results in Croft et al. are mainly simulation-driven, and therefore rather narrow in scope and limited by the available computational power. Initial analytical understanding of the foraging problems faced by roots may be possible by analogy with the fish foraging results above. Perhaps more exciting is the opportunity to expand the remit to encompass more realism: The soil environment is complex. Models based on three spatial dimensions allow competitors to overlap. Alternatively, network-based models of the soil environment may better capture the fractal-like structure in which roots grow. Roots can also form symbioses with soil fungi (mycorrhizas), allowing the plant to forage cheaply across a larger area at the cost of providing carbon to the fungal partner. Finally, while agriculture may rely on monocultures, the plants themselves have typically evolved within complex competitive communities. How can knowledge of the strategies evolved in the wild be exploited so as to more efficiently exploit managed crop systems? These issues are described more thoroughly in [10], but a firm mathematical grasp remains elusive and the practical problems are unsolved.

Problem 3: Safety in Numbers, or Presenting a Bigger Target? *(Complications 1, 2, 3, 4)*

The notion of "safety in numbers" imagines princesses gathering in groups of size n so that, even if the monster finds the group, each individual only has a probability 1/n of meeting a grisly demise. This verbal reasoning is used to explain many natural instances of prey aggregation, from minnows to gazelles. There are additional factors such as the increased vigilance afforded by many sets of eyes, and complications such as individuals only visiting the periphery of the group when physiology dictates [3].

These problems have received attention from ecologists and strong theory exists, but perhaps in directions somewhat orthogonal to the mathematics presented in this volume. The simple question of when it pays for two princesses to collaborate in hiding is naturally game-theoretic, and will generalise naturally to larger groups, complex environments, or group-induced changes in speed of movement or detectability.

For small wet princesses, the turbulent encounter theories of Problem 1 allow quantification of the processes involved. A group of fish larvae or zooplankton is larger than an individual and therefore subject to faster random turbulent advection; encounter rates with predators therefore increase. When grouping confers another advantage, for example navigational precision via a "many wrongs" principle [7], then further trade-offs emerge; depending on the levels of turbulence a group of larvae reach safety of a coral reef more quickly, but present a larger target whilst doing so. In this case, chemical changes downstream may provide further quantifiable cues for a searching predator.

The problems sketched above are all essentially self-contained; precise questions can be formulated and solved in an abstract world where, even when complications exist, they are known and quantifiable. They should, however, be viewed in the context of the practical challenges motivating them. There is scope for search and rendezvous theory to make a direct impact on issues such as habitat conservation, sustainable management, or biodiversity and ecosystem-level function. Similar sets of problems and opportunities exist at the cellular and microbiological scales, with rapid technology-driven changes in volume and specificity of data. The challenges will be both fascinating and important. The key ingredient is the determined, and possibly slow and painful, communication across the disciplinary boundaries.

References

1. Alpern, S., Fokkink, R., Timmer, M. and Casas, J., Ambush frequency should increase over time during optimal predator search for prey, Journal of the Royal Society Interface 7–8(64), 2011, 1665–72.
2. Beaumont H.J., Gallie J., Kost C., Ferguson G.C., and Rainey PB. Experimental evolution of bet hedging. Nature 462(7269), 2009, 90–3.
3. Brierley, A. S. and Cox, M. J. Shapes of krill swarms and fish schools emerge as aggregation members avoid predators and access oxygen. Current Biology 20:19, 2010, 1758–1762.
4. Burrow, J.F., Baxter P.D., and Pitchford, J. W. Levy processes, saltatory foraging, and superdiffusion. Mathematical Modelling of Natural Phenomena 3 (3), 2008, 115–130.
5. Burrow, J. F., Horwood, J., and Pitchford, J. W. Variable variability: difficulties in estimation and consequences for fisheries management. Fish and Fisheries (in press; published online March 2012).
6. Codling, E. A., Hill, N. A., Pitchford, J. W. and Simpson, S. D. Random walk models for the movement and recruitment of reef fish larvae. Marine Ecology Progress Series 279, 2004, 215–224.
7. Codling, E. A., Pitchford, J. W. and Simpson, S. D. Group navigation and the "many-wrongs principle" in models of animal movement. Ecology 88 (7), 2007, 1864–1870
8. Codling, E. A., Plank, M. J. and Benhamou, S. Random walk models in biology. Journal of the Royal Society Interface 5 (25), 2008, 813–834.
9. Currey, J. D., Pitchford, J. W., and Baxter, P. D. Variability of the mechanical properties of bone, and its evolutionary consequences. Journal of The Royal Society Interface 4 (12), 2007, 127–135.
10. Croft, S. A., Pitchford, J. W., and Hodge, A. Optimal root proliferation strategies: the roles of nutrient heterogeneity, competition and mycorrhizal networks. Plant and Soil 351 (1–2), 2012, 191–206.
11. Durham, W. M., Kessler, J. O., and Stocker, R. Disruption of vertical motility by shear triggers formation of thin phytoplankton layers. Science 323, 2009, 1067–1070.
12. Durham, W. M., Tranzer, O., Leombruni, A., Stocker, R. Division by fluid incision: biofilm patch development in porous media. Physics of Fluids (in press).
13. Firn, R. Nature's Chemicals: The Natural Products that Shaped Our World. Oxford University Press, 2010.
14. Gal, S. Search games with mobile and immobile hider, SIAM J. Control Optim. 17 (1), 1979, 99–122.

15. Houston, A. I. and McNamara, J. M. Models of Adaptive Behaviour: An Approach Based on State (Cambridge Studies in Mathematical Biology), Cambridge University Press, 1999.

16. James, A., Pitchford, J. W., and Plank, M.J. Efficient or Inaccurate? Analytical and Numerical Modelling of Random Search Strategies. Bulletin of Mathematical Biology 72, 2010, 896–913.

17. James, A., Plank, M.J., and Edwards, A.M.: Assessing Lévy walks as models of animal foraging. Journal of the Royal Society Interface, 8(62), 2011,1233–1247.

18. Kopelman, R. Fractal reaction kinetics. Science 241(4873), 1988, 1620–1626.

19. Minto, C., Myers, R. A. and Blanchard, W. Survival variability and population density in fish populations. Nature 452, 2008, 344–347.

20. Nowak, M. A., Tarnita, C. E. and Wilson, E. O. The evolution of eusociality. Nature 466, 2010, 1057–1062.

21. Orwell, G. The Road To Wigan Pier. Penguin Modern Classics, 1937.

22. Pitchford, J. W. and Brindley, J. Prey patchiness, predator survival and fish recruitment. Bulletin of Mathematical Biology 63 (3), 2001, 527–546.

23. Pitchford, J. W., James A., and Brindley, J. Optimal foraging in patchy turbulent environments. Marine Ecology Progress Series 256, 2003, 99–110

24. Preston, M.D., Pitchford, J. W., and Wood, A.J. Evolutionary Optimality in Stochastic Search Problems. Journal of the Royal Society Interface, 7 (50), 2010, 1301–1310.

25. Preston, M. D., Armsworth, P., Forister, M. and Pitchford J. W. Movement and reproductive dynamics of the Melissa Blue butterfly. In preparation.

26. Real, L. and Caraco, T. Risk and foraging in stochastic environments. Annual Review of Ecology and Systematics, 1986, 371–390.

27. Stephens, D. W., Brown, J. S., and Ydenberg, R. C. , editors. Foraging: Behavior and Ecology. University of Chicago Press, 2007.

Printed in the United States
By Bookmasters